WALLY AMOS
Founder/ Chairman
The Uncle Noname Cookie Compar

Communication is important, not only in business, but in our everyday experiences. Lack of communication creates assumptions and false beliefs which create separation and disharmony. Clear, honest communication creates an environment that allows everyone to profit.

JOSEPH S. MONTGOMERY
Founder and President
Cannondale
(Shown with son Scott Montgomery,
Cannondale Vice President of Marketing)

Strong communication skills are essential in business. The people you interact with—whether it's colleagues, customers, bankers, the media—will judge you not only on the content of your ideas, but on the quality with which they are presented.

HOLLY STIEL
Service Consultant
Holly Speaks

Learning the skills of a good communicator will be the deciding factor in your success or failure in your business as well as your personal relationships. To be understood requires that you take the time to listen and to understand to whom you are communicating. Only then, can you choose words and body language that will best work in each situation.

BILL GATES
Chief Executive Officer
Microsoft Corportation

We are poised on the start of the next wave of computing. There is a lot of uncertainty about what this is going to look like. The idea of interactivity goes way beyond anything we've done to date. The chance here is to improve the way people work, the way they learn, the way they play and, literally, the way that society communicates. I'm as excited today about the opportunities for everyone in this business as I have ever been.

SUSAN TAYLOR
Editor-in-Chief
Essence Magazine

The ability to communicate effectively is essential in today's demanding business climate. Whether it's writing a business memo or articulating an idea during a presentation—your success will be determined greatly by your ability to inspire others by expressing yourself clearly, concisely, and with warmth and passion.

JOHN HETTERICK
President and CEO
Rollerblade, Inc.

A critical part of my job is gathering and processing information from every area of the company in order to make the best and most timely decisions. In fact, communicating is the most important thing I do.

APPLIED ENGLISH

LANGUAGE SKILLS
FOR BUSINESS
AND EVERYDAY USE

ROBERT E. BARRY
Emeritus, Mount San Antonio College, Walnut, California

LORETTA SCHOLTEN
Editorial Consultant, Saint Paul, Minnesota

DONNA J. COCHRANE, ED.D.
Associate Professor, Bloomsburg University, Bloomsburg, Pennsylvania

BARBARA COX, Ph.D.
Editorial Consultant, Laguna Niguel, California

PRENTICE HALL
ENGLEWOOD CLIFFS, NEW JERSEY 07632

Library of Congress Cataloging-in-Publication Data

Applied English: language skills for business and everyday use /
    Robert E. Barry . . . [et al.].
        p.   cm.
    Includes index.
    ISBN 0-13-606047-1 (student ed.).—ISBN 0-13-606450-7 (Instruct. ed.)
    1. English language—Textbooks for foreign speakers.   2. English
language—Business English—Problems, exercises, etc.   I. Barry,
Robert E., 1921–
    PE1128.A58   1995
808'.066651—dc20   94-37186
                                                         CIP

Senior Editor: *Elizabeth Sugg*
Design Supervisor: *Marianne Frasco*
Cover Design: *Mary Jo DeFranco*
Interior Design: *Sheree Goodman*
Managing Editor: *Mary Carnis*
Photo Researcher: *Lydia Silvas*
Supplements Editor: *Judy Casillo*
Director of Production and Manufacturing: *Bruce Johnson*
Manufacturing Manager: *Ed O'Dougherty*

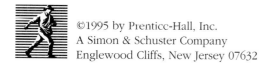
©1995 by Prentice-Hall, Inc.
A Simon & Schuster Company
Englewood Cliffs, New Jersey 07632

Printed in the United States of America
10 9 8 7 6 5 4 3 2 1

ISBN 0-13-606047-1

Prentice-Hall International (UK) Limited, *London*
Prentice-Hall of Australia, Pty. Limited, *Sydney*
Prentice-Hall Canada, Inc., *Toronto*
Prentice-Hall Hispanoamericana, S. A., *Mexico*
Prentice-Hall of India Private Limited, *New Delhi*
Prentice-Hall of Japan, Inc., *Tokyo*
Simon & Schuster Asia Pte. Ltd., *Singapore*
Editora Prentice Hall do Brasil, Ltda., *Rio de Janeiro*

Contents in Brief

Section 4: Words That Modify and Connect   259

Section 5: Sentence Mechanics; The Writing Process   329

CONTENTS

PREFACE

What skills will you need to adapt to today's and tomorrow's changing work environment? You may not be able to predict the exact technology you will use, but you should be aware that there will be an increased emphasis on communications skills—the ability to communicate thoughts, ideas, information, and messages in writing. No matter what occupation you pursue, you will find that a need exists for people with effective communication skills. The computer has freed workers from many routine tasks, making it imperative that future workers know how to plan, research, and communicate ideas.

What makes *Applied English* an effective text? How is it different from other texts? *Applied English* guides you through the instruction, reinforces concepts, and provides appropriate practice exercises. You will learn how to use reference materials as vital tools for improving your writing. The importance of good dictionary skills is introduced in Chapter 1 and reinforced throughout the book. The use of the dictionary assures that you will become an independent learner and be able to solve many writing problems on your own.

To become a better writer, you must understand basic writing principles and grammar rules. *Applied English* is organized with this in mind. The text introduces the process of recognizing major sentence parts through the use of an easy-to-use flowchart. The flowchart method gives you another tool to use when learning and applying grammar principles. The emphasis on the recognition and analysis of the major parts of a sentence is a unique feature of this text.

Textbook Features

Levels of Learning. *Applied English* is divided into five sections. Within each section are chapters relating to the section topic. The chapters are organized for easy use because each chapter is divided into learning units with clearly focused learning goals. Each section, chapter, and learning unit is designed to relate, reinforce, and build on previously introduced material. The first two learning units of each chapter usually contain the basic level material, while the third and sometimes fourth learning units usually contain advanced level material on the topic.

Rules and Examples. The grammar rules and language skills taught in *Applied English* are reinforced with numerous examples. When examples point out common grammatical errors or poor writing habits, you will see this symbol ⊘ .

Writing Tips.   Throughout the text concise suggestions are given to help you apply the grammar rules to your writing.

Study Tips.   These easy-to-remember methods will help you recall grammar and punctuation rules.

Chapter Summary: A Quick Reference Guide.   This feature appears at the end of each chapter. It lists in chart form a summary and examples of each key point. This reference guide can serve as a quick review or reference as you progress through the text.

You Try It.   These exercises appear at the end of each learning unit and provide the opportunity to master the material before continuing on. The answers are provided so you can get immediate feedback and reinforcement.

Quality Editing.   At the end of each chapter are short documents for practice of proofreading and editing. Practice includes attention to detail, recognition of poor sentence structure, and identification of errors in grammar, punctuation, and spelling. The difficulty level of these exercises increases as you progress through the course. Information learned in previous chapters is integrated into the exercises.

Applied Language.   These problem-solving applications provide many real-world examples. You will practice writing for specific situations. For example, an exercise might require writing an answer to a customer complaint letter.

Worksheets.   Every chapter ends with a variety of exercises keyed to each learning unit. The popular workbook format makes it easy to remove and hand in the worksheet exercises for checking or grading.

Supplements

A variety of supplements are provided to enhance your understanding of the material.

Blue Pencil Software.   It would be more efficient if you could compose and edit at the computer rather than editing in longhand. With this in mind, the *Blue Pencil Software* lets you do editing at the computer.

Multimedia Study Guide.   You will have an opportunity to review and learn with this exciting package of multiple choice, true-false, and fill-in review questions with immediate feedback for reinforcement.

The entire instructional package for *Applied English* will enable you to achieve success in written communication through the study of grammar, punctuation, and language usage. Written communication skills are logically developed and reinforced throughout *Applied English* within a framework which encourages you to apply what you learn to editing and writing different kinds of documents. You will be prepared for the twenty-first century with the ability to utilize good communication skills in your chosen occupation.

Acknowledgments

The authors extend their appreciation to the instructors who reviewed this text in the course of its development: Jane Berg, Chabot College, California; Jackie Arteno, Blackfeet Community College, Montana; Christine Foster, Grand Rapids Community College, Michigan; Ginger Guzman, J. Sargeant Reynolds Community College, Virginia; Dani Jones-Phelps, National Education Center—Allentown Business School, Pennsylvania; Stanley Kozikowski, Bryant College, Massachusetts; Diane LaSala, Johnson & Wales University, Rhode Island; Paul Martin, Aims Community College, Colorado; Bettye Matier, Walla Walla Community College, Washington; Carol McGonagil, Pierce College, Washington; Stuart Meisel, National School of Technology, Florida; Anita Musto, Utah Valley State College, Utah; Jane Rada, Western Wisconsin Technical College, Wisconsin; Judith Rice, Chippewa Valley Technical College, Wisconsin; Terry Strauss-Thacker, Coastline Community College, California; Janet Hough Taylor, Spokane Community College, Washington; Kathleen Troy, Mohawk College, Ontario, Canada; Carol Williams, Pima Community College, Arizona.

Special acknowledgment to Jeffrey Slater, North Shore Community College, Massachusetts, and Ray H. Garrison, School of Accountancy, Brigham Young University, Utah, for their contributions to *Applied English*.

Donna J. Cochrane

ABOUT THE AUTHORS

Robert E. Barry is a graduate of Salem State College in Salem, Massachusetts. He earned his Master of Arts degree at California State University in Los Angeles, California. Mr. Barry spent several years in banking and in selling before he entered the field of education. During his 30 years at Mount San Antonio College in Walnut, California, he taught 12 different courses in the business area, including teaching Business English for 15 years. He is best known, however, for his expertise and research in the field of communications. Throughout his life he has taken a keen interest in the use and misuse of the English Language.

Loretta Scholten earned a Bachelor of Arts degree and a Secondary Teaching Certificate from Calvin College, Grand Rapids, Michigan. For the past 30 years, Ms. Scholten has been an editor and consultant in the fields of accounting and business mathematics at the college level. In working with textbook authors, she has concentrated on developing effective learning and teaching tools by presenting subjects in a way that helps students become critical thinkers and effective workers. To this textbook, Ms. Scholten brings a student-focused learning procedure based on the principle that learning is possible only through the association of ideas.

Donna J. Cochrane is an associate professor at Bloomsburg University of Pennsylvania, Department of Business Education and Office Administration, where she teaches courses in office systems. Prior to teaching at Bloomsburg University, she taught business subjects at the secondary level in New York State and Pennsylvania. Her doctorate in vocational education is from Temple University and her B.S. and M.S. in business education are from the State University of New York at Albany.

Dr. Cochrane is active in the Eastern Business Education Association, the National Business Education Association, and the Office Systems Research Association.

Barbara Cox earned her Ph.D. in education and psychology from Stanford University. In addition to 12 years of conducting research and developing language curricula in academic settings, she has worked as an editor with several major publishers.

Dr. Cox is an author of 70 books in education and has written numerous training manuals and research reports. She conducts workshops for business and academic writing through her firm, Cox Marketing Services, and is also involved in developing interactive multimedia educational programs.

ASSIGNMENT SHEET

Use this sheet to track your progress by checking off the assignments as you complete them or filling in the date of completion.

	You Try It	Quality Editing	Applied Language	Worksheet
CHAPTER 1 Learning Unit 1-1 Learning Unit 1-2				
CHAPTER 2 Learning Unit 2-1 Learning Unit 2-2 Learning Unit 2-3				
CHAPTER 3 Learning Unit 3-1 Learning Unit 3-2 Learning Unit 3-3 Learning Unit 3-4				
CHAPTER 4 Learning Unit 4-1 Learning Unit 4-2				
CHAPTER 5 Learning Unit 5-1 Learning Unit 5-2 Learning Unit 5-3				
CHAPTER 6 Learning Unit 6-1 Learning Unit 6-2 Learning Unit 6-3				
CHAPTER 7 Learning Unit 7-1 Learning Unit 7-2 Learning Unit 7-3 Learning Unit 7-4				
CHAPTER 8 Learning Unit 8-1 Learning Unit 8-2 Learning Unit 8-3				
CHAPTER 9 Learning Unit 9-1 Learning Unit 9-2 Learning Unit 9-3				

	You Try It	Quality Editing	Applied Language	Worksheet
CHAPTER 10 Learning Unit 10-1 Learning Unit 10-2 Learning Unit 10-3				
CHAPTER 11 Learning Unit 11-1 Learning Unit 11-2 Learning Unit 11-3 Learning Unit 11-4				
CHAPTER 12 Learning Unit 12-1 Learning Unit 12-2 Learning Unit 12-3				
CHAPTER 13 Learning Unit 13-1 Learning Unit 13-2 Learning Unit 13-3				
CHAPTER 14 Learning Unit 14-1 Learning Unit 14-2 Learning Unit 14-3				
CHAPTER 15 Learning Unit 15-1 Learning Unit 15-2 Learning Unit 15-3				
CHAPTER 16 Learning Unit 16-1 Learning Unit 16-2 Learning Unit 16-3				
CHAPTER 17 Learning Unit 17-1 Learning Unit 17-2 Learning Unit 17-3				
CHAPTER 18 Learning Unit 18-1 Learning Unit 18-2 Learning Unit 18-3 Learning Unit 18-4				
CHAPTER 19 Learning Unit 19-1 Learning Unit 19-2 Learning Unit 19-3				
CHAPTER 20 Learning Unit 20-1 Learning Unit 20-2 Learning Unit 20-3				

INTRODUCTION TO WRITING—FROM WORDS TO SENTENCES

CHAPTER 1

WRITING PRINCIPLES AND REFERENCE TOOLS

The first and most important step in writing is to know who you're writing to—who they are and what they want.
—David G. Lyon, *The XYZ'S of Business Writing*

LEARNING UNIT GOALS

LEARNING UNIT 1-1: OVERVIEW OF WRITING PRINCIPLES

- Understand seven writing principles. (pp. 3–7)
- Use these seven principles in your writing. (pp. 3–7)

LEARNING UNIT 1-2: WRITING REFERENCE TOOLS

- Select a suitable dictionary. (p. 9)
- Explain the general dictionary usage guidelines. (pp. 9–12)
- Understand the purpose of a thesaurus and special dictionaries. (p. 13)
- Use these reference tools in your writing. (pp. 9–13)

As a unique individual, you have special interests, ambitions, desires, and goals that influence your choice of an occupation. For example, if you enjoy working with computers, you probably will select an occupation involving computers. Your future success in the occupation of your choice is directly related to your desire to succeed and your ability to communicate.

How do people communicate? They communicate by writing, speaking, and listening. While communication—the process of exchanging information through messages—is vital to all human endeavors, it is particularly important in business. Effective communication—that is, effective writing, speaking, and listening skills—affects every aspect of business. As Lee Iacocca, the former chief executive officer of Chrysler Corporation, has said, "The ability to communicate is everything."

In addition to the communication skills of writing, speaking, and listening, we should add one more skill—effective reading. You must read effectively to acquire the information you need to communicate. Although these four skills are interrelated, the focus of this text is on the skill of writing. Whether you are conducting personal business or occupational business, the ability to write effectively will often result in financial reward. A well-written personal complaint letter can have positive results, such as receiving a cash refund. Impressive job-related writing usually establishes you as a person with a future in business.

The only way to improve your writing is *to write*. The overview of writing principles in Learning Unit 1-1 will give you some basic guidelines. As you proceed through this text, you will learn how the correct use of grammar can fine-tune your writing. The appendix contains procedures for writing letters and memos. An organized writing procedure for larger documents is explained in Chapter 20.

LEARNING UNIT 1-1
OVERVIEW OF WRITING PRINCIPLES

Business writing is writing with a purpose. Usually, business writers write to inform or persuade. Throughout this text, you will be given exercises in business writing. By using the seven principles that follow, you can immediately improve the effectiveness of all your writing.

PRINCIPLE 1. Organize your thoughts before you write.

Before you begin to write a business document, you must first know the purpose of the document and the intended audience. These two factors will affect the organization you choose for your communication.

Organized writing begins with organized thoughts. Only when your thoughts are organized can you present your message in a direct and efficient way. If readers believe the information is important, they will tend to read a poorly written report before they read a poorly organized report.

How should business writing be organized? Most business writing is organized from the *general to the specific*. To do this, writers begin with a general subject statement. Then, they proceed with the specific information such as history, examples, or reasons. For a dramatic effect, business writing can be organized from the *specific to the general*. In this form of organization, writers give the history, examples, or reasons first and give the general subject statement as the conclusion.

The details of a subject statement can be developed in several ways. The following five methods are often used: (1) *comparison or contrast—*

describe the similarities or differences; (2) *time*—present the history, schedule, or sequence of events; (3) *space*—describe locations, directions, layout, floor plans; (4) *cause-effect*—explain why something happened followed by the result; and (5) *problem-solution*—present facts or examples to indicate a problem and then give the solution to the problem.

Once you know the purpose, the audience, and the method of organization, what should you do next? You must develop a plan or outline. Many people do not like to make formal outlines with topics and subtopics. They feel confined and complain that they do not know how a piece of writing will develop. If this is your problem, you can solve it by listing your ideas on paper as you think of them. When you are finished, study these ideas. Decide how to group them so they will fit together and flow. Imagine you are the reader and decide what you would want to know first, second, third, and so on. Grouping ideas is so important that Robert Louis Stevenson said, "If a man can group his ideas, he is a good writer."

PRINCIPLE 2. Prefer the active voice.

Successful writers prefer the active voice to the passive voice. The active voice is more direct and gives strength to the writing. When you write a sentence in the **active voice**, the subject performs the action named by the verb. For example, the sentence *Kim installed a ceiling fan* is in the active voice. Kim performs the action. To write this sentence in the **passive voice**, you would say, *A ceiling fan was installed by Kim.* Now *fan* becomes the subject and receives the action named by the verb.

The words *who* (or *what*) *did what* will help you write in the active voice. Whenever possible, use the following order to describe the parts of a sentence: WHO (or WHAT) . . . DID . . . WHAT. This is the simple and direct subject-verb-object construction—the active voice.

The passive voice uses a *what was done by whom* construction. It *reverses* the subject-verb-object sequence of the active voice. In the passive construction, the word *by* often appears. This can help you identify passive-voice writing.

Active Voice: A broken cable stopped the printer. (A *broken cable* [what] is the subject; *stopped* [did] is the verb; and *the printer* [what] is the object.)

Passive Voice: The printer was stopped by a broken cable. (The object *the printer* in the active-voice sentence becomes the subject [what] and receives the action of the verb *was stopped* [was done]; *a broken cable*, the subject in the active-voice sentence, becomes the object [by whom]. In the passive voice, the verb needs a helper.)

Active Voice: Ace Hardware sold five bicycles. (*Ace Hardware* [who] is the subject; *sold* [did] is the verb; and *five bicycles* [what] is the object.)

Passive Voice: Five bicycles were sold by Ace Hardware. (The object *five bicycles* in the active-voice sentence becomes the subject [what] and receives the action of the verb *were sold* [was done]; *Ace Hardware*, the subject in the active-voice sentence, becomes the object [by whom].)

To write in the active voice, you must start the action by placing a person or thing—the doer—at the beginning of the sentence. When *you* are the doer, write in the first person, which means starting a sentence with the word *I*. When you do not know the doer or do not want to name the doer, you will have to use the passive voice.

Using the active voice is the quickest way to improve the conciseness, readability, and precision of your writing. In business writing, you should arrange most of your sentences so the reader immediately knows *who did what.*

PRINCIPLE 3. Use a simple writing style that avoids unbusinesslike expressions.

Usually, experienced writers reveal their most powerful thoughts in simple, familiar language. You may think that when you write about technical subjects that appeal to a special audience, you should use complex words. Although some subjects may require the use of technical language, you should always prefer to express your thoughts in simple language.

Prefer short sentences, but use some sentences of various lengths to avoid choppy writing. Construct all sentences so that ideas build on each other and flow.

Use informal or conversational expressions sparingly. (These terms are sometimes identified as *colloquial* in the dictionary.) Some of these expressions are not precise enough to assure understanding, while others suggest an unbusinesslike tone. Avoid the italicized words in the following sentences:

The other *guys* have a better product.

We are *dying* to hear the news.

Our dispatcher is a *dud.*

She is *crazy* about your idea.

The new marketing representative made an *awful* impression.

Her presentation was really *far out.*

All six of us felt that the report was *horrible.*

Often, you will know if a word or term is inappropriate. If you want to be completely certain, however, you can consult a dictionary. Usage labels such as *slang* or *colloq.*, which is an abbreviation for the word *colloquial*, identify words that should be avoided. Although the G. & C. Merriam Company no longer uses such labels, most dictionary publishers continue to provide them. For example, *Webster's New World Dictionary* labels the noun "slew" as colloquial. Therefore, you should avoid an expression such as a "slew of people." Of course, you may still use an occasional informal or conversational expression to emphasize a point or to create a desired effect.

PRINCIPLE 4. Omit unnecessary words.

Perhaps nothing is more annoying to a reader than wordiness. A reader is less likely to read long, rambling memos, letters, and reports that do not express a writer's thoughts clearly and concisely. People respond more positively to *Please try our product for one month* than *Will you be so kind as to try our product for the duration of one month.* When you write, *make every word count.* Remember that wordy writing *hides your message.*

Watch for the following wordiness in your writing (sometimes called trite phrases) and try to substitute the concise words at the right. (Chapter 20 gives additional examples of trite phrases.)

Wordy	Concise
along the lines of	like
at a later date	later
at this point in time	now
for the duration of a month	although (though)
for the reason that	because
in spite of the fact that	although (though)
on the basis of	by
on behalf of	for
will you be so kind as to	please

This does not mean that you should *never* use the phrases in the left column. Sometimes, they are useful for emphasis or to give rhythm to a sentence. Usually, however, you will find that concise words give more strength to your writing.

WRITING TIP

Write to *express*, not *impress*.

Wordiness is also caused when two words that have overlapping meanings are used together. Writers use some words together so often that they do not realize they are using two or three words when one word is sufficient. For example, have you automatically used some of the overlapping words in the left column that follows instead of the single word in the right column?

Instead of	Use
absolutely complete	complete
advance planning	planning
assembled together	together
consensus of opinion	consensus
enclosed herewith	enclosed
exactly identical	identical
join together	join
same identical	identical

Once you become aware of wordy expressions, you will notice other expressions that can be reduced to a single word. Remember that concise writing is strong writing.

PRINCIPLE 5. Use positive words and specific words.

Whenever possible, use positive words instead of negative words. If you must give a negative message, choose your words carefully and include positive statements.

> AVOID: Managers should tell employees not to wear jeans.
>
> USE: Managers should explain the accepted dress code to employees.
>
> AVOID: Do not forget to lock the file.
>
> USE: Be sure to lock the file.

Use specific, or concrete, words. Business usually deals with specifics. When a business suffers a $4 million loss, avoid saying the business suffered a *substantial* loss. The word *substantial* means different things to different people. If possible, say the exact amount that the business lost. Business writing should leave no unanswered questions in the mind of the reader. (Chapters 6 and 20 also discuss concrete words.)

PRINCIPLE 6. Use grammatically correct constructions.

Grammatical correctness is necessary for effective writing. Although most people do not enjoy the study of grammar, its mastery is important. People who express themselves well grammatically are more likely to be respected and chosen for jobs and promotions.

Probably one of the reasons people avoid grammar is because they remember grammar as a list of unrelated rules. This is true to some extent. The game of grammar does follow some rules that may seem unrelated. If you think of these rules as *accepted guidelines*, you may be mentally a little more receptive to the study of grammar.

The study of grammar teaches the logical relationship of words in a sentence. This knowledge will help you word a sentence so the reader can clearly follow your thinking. You must spend a little time, however, learning to understand grammar guidelines if you want this knowledge to become a habit that produces effective business writing.

PRINCIPLE 7. Use a dictionary.

As you study this text, you will realize the value of a reliable dictionary. You should find yourself using a dictionary more frequently than you have in the past.

Because the English language is constantly undergoing subtle changes, to be of maximum value a dictionary should be less than ten years old. New words are being added, and the meanings of old words are changing. This increases the need for a current dictionary. Learning Unit 1-2 gives more information about dictionary usage.

STUDY TIP

The dictionary is a virtual treasure chest of useful information.

YOU TRY IT

Each of the following sentences violates one of the seven writing principles discussed in Learning Unit 1-1. In the space provided at the left, write the number of the principle that is violated. Then, in the space provided below, rewrite the sentence so that it no longer violates the principle.

EXAMPLE:

__2__   The airline tickets were picked up by Ms. Stevens.
         Ms. Stevens picked up the airline tickets.

_____ **1.** The ticket agent gived her the receipt.

_____ **2.** Before you buy the ticket, check the schedule. But before you do that, get approval for the purchase.

_____ **3.** In spite of the fact that the manager was gone for the duration of a month, the job was completed.

_____ **4.** Make decisions *a priori.*

_____ **5.** The schedules were all messed up.

_____ **6.** The engines were checked by the mechanics.

_____ **7.** Do not neglect to notify everyone on the list.

Answers: 1. 6—The ticket agent gave her the receipt. 2. 1—Get approval for the purchase and check the schedule before you buy the ticket. 3. 4—Although the manager was gone for a month, the job was completed. 4. 7—Okay as is. Optional: Do not make decisions before examining the facts. 5. 3—The schedules were disorganized. 6. 2—The mechanics checked the engines. 7. 5—Be sure to notify everyone on the list.

LEARNING UNIT 1-2
WRITING REFERENCE TOOLS

Words are the mainsprings of language. Choosing the right word, however, can be difficult. This learning unit discusses the writer's most valuable reference tools—a general dictionary, a thesaurus, and special dictionaries. Later, when you study capitalization and punctuation, you also will learn about another tool called a style manual.

GENERAL DICTIONARY USAGE

A dictionary suitable for business and college use contains a lengthy explanation of how to use the dictionary. Here are some guidelines for using a general dictionary.

Dictionary Choice

The best dictionaries for business are abridged, meaning they do *not* contain every word in the English language. Unabridged dictionaries, as the name implies, *do* contain every word and are cumbersome to use. Pocket dictionaries are popular because they are small and easily portable. However, you should have at least one desk or college dictionary available containing 150,000 or more entries. Any of the following college dictionaries can be used:

> *The Random House Webster's College Dictionary* (latest)
>
> *The American Heritage Dictionary of the English Language, Second College Edition*
>
> *Webster's New World Dictionary, Third College Edition*
>
> *Merriam Webster's Collegiate Dictionary, Tenth Edition*

Guide Words

Two guide words appear at the top of each dictionary page. The word on the left is the first entry word on the page. The word on the right is the last entry word. All words in the dictionary, of course, are arranged alphabetically.

Figure 1-1 shows a sample page from a college dictionary. The guide words *slay* and *slick* indicate that the page begins with a definition of *slay* and ends with a definition of *slick*. By looking at the guide words, a person can quickly decide whether a particular word is on that page.

Syllabication

Dots usually divide boldface entry words into syllables. Syllabication helps determine whether a word is written as one word. For example, in Figure 1-1 the entry *sleep • walk • ing* tells us that we should write the word without a space or a hyphen (*sleepwalking*). The space between the words *slice bar* shows that we should write the term as two words.

Syllabication also shows whether we can break a word at the end of a line. The hairline in words such as *sleep|er* means that the word should not be broken before the *er*.

slay (slā) *vt.* **slew** or for 2 **slayed**, **slain**, **slay'ing** ‖ ME *slean* < OE < **slahan*, akin to Ger *schlagen*, Du *slagen* < IE base **slak-*, to hit > MIr *slacc*, sword ‖ **1** to kill or destroy in a violent way **2** [Slang] to impress, delight, amuse, etc. with overwhelming force **3** [Obs.] to strike or hit —***SYN.*** KILL[1] —**slay'er** *n.*

SLBM submarine-launched ballistic missile

sld **1** sailed **2** sealed

sleave (slēv) *n.* ‖ < OE *-slæfan*, to separate; akin to *slifan*: see SLIVER ‖ **1** [Obs.] *a)* a fine silk thread separated from a large thread *b)* untwisted silk that tends to mat or tangle; floss **2** [Rare] any tangle, as of ravelings —*vt.* **sleaved**, **sleav'ing** [Obs.] to separate or pull apart (twisted or tangled threads)

sleaze (slēz) *n.* ‖ back-form. < fol. ‖ [Slang] **1** the quality or condition of being sleazy; sleaziness **2** anything cheap, vulgar, shoddy, etc. **3** a shady, coarse, or immoral person: also **sleaze'bag'** (-bag') or **sleaze'ball'** (-bôl')

slea·zy (slē'zē) *adj.* **-zi·er**, **-zi·est** ‖ < *slesia*, var. of SILESIA ‖ **1** flimsy or thin in texture or substance; lacking firmness *[a sleazy rayon fabric]* **2** shoddy, shabby, cheap, morally low, etc. Also [Slang] **slea'zo** (-zō) —**slea'zi·ly** *adv.* —**slea'zi·ness** *n.*

sled (sled) *n.* ‖ ME *sledde* < MLowG or MDu, akin to Ger *schlitten*: for IE base see SLIDE ‖ any of several types of vehicle mounted on runners for use on snow, ice, etc.: small sleds are used in the sport of coasting, large ones (also called *sledges*), for carrying loads —☆*vt.* **sled'ded**, **sled'ding** to carry on a sled —☆*vi.* to ride or coast on a sled —**sled'der** *n.*

☆**sled·ding** (-iŋ) *n.* **1** a riding or carrying on a sled **2** the condition of the ground with reference to the use of sleds: often used figuratively *[the work was hard sledding]*

sledge[1] (slej) *n.*, *vt.*, *vi.* **sledged**, **sledg'ing** ‖ ME *slegge* < OE *slecge* < base of *slean*, to strike, SLAY ‖ SLEDGEHAMMER

sledge[2] (slej) *n.* ‖ MDu *sleedse*, akin to *sledde*, SLED ‖ a sled or sleigh for carrying loads over ice, snow, etc. —*vi.*, *vt.* **sledged**, **sledg'ing** to go or take by sledge

sledge·ham·mer (-ham'ər) *n.* ‖ see SLEDGE[1] ‖ a long, heavy hammer, usually held with both hands —*vt.*, *vi.* to strike with or as with a sledgehammer —*adj.* crushingly powerful

sleek (slēk) *adj.* ‖ var. of SLICK, with Early ModE vowel lengthening ‖ **1** smooth and shiny; glossy, as a highly polished surface, well-kept hair or fur, etc. **2** of well-fed or well-groomed appearance *[fat, sleek pigeons]* **3** polished in speech and behavior, esp. in a specious way; unctuous **4** highly fashionable, or stylish; elegant —*vt.* to make sleek; smooth: also **sleek'en** —**sleek'ly** *adv.* —**sleek'ness** *n.*

sleek·it (slēk'it) *adj.* ‖ Scot var. of pp. of SLEEK ‖ [Scot.] **1** sleek, or smooth and shiny **2** sly, crafty, or sneaky

sleep (slēp) *n.* ‖ ME *slep* < OE *slæp*, akin to Ger *schlaf*, sleep, *schlaff*, loose, lax < IE **slab* < base **(s)leb-*, **(s)lab-*, loose, slack > LIP, LIMP[1], L *labor*, to slip, sink ‖ **1** *a)* a natural, regularly recurring condition of rest for the body and mind, during which the eyes are usually closed and there is little or no conscious thought or voluntary movement, but there is intermittent dreaming *b)* a spell of sleeping **2** any state of inactivity thought of as like sleep, as death, unconsciousness, hibernation, etc. **3** *Bot.* NYCTITROPISM —*vi.* **slept**, **sleep'ing** **1** to be in the state of sleep; slumber **2** to be in a state of inactivity like sleep, as that of death, quiescence, hibernation, inattention, etc. **3** [Colloq.] to have sexual intercourse (*with*) **4** [Colloq.] to postpone a decision (*on*) to allow time for deliberation *[let me sleep on it]* **5** *Bot.* to assume a nyctitropic position at night, as petals or leaves —*vt.* **1** to slumber in (a specified kind of sleep) *[to sleep the sleep of the just]* **2** to provide sleeping accommodations for *[a boat that sleeps four]* —**last sleep** death —**sleep around** [Slang] to have promiscuous sexual relations —**sleep away** **1** to spend in sleeping; sleep during **2** to get rid of by sleeping —**sleep in** **1** to sleep at the place where one is employed as a household servant **2** to sleep later in the morning than one usually does —**sleep off** to rid oneself of by sleeping —**sleep out** **1** to spend in sleeping; sleep throughout **2** to sleep outdoors —**sleep over** [Colloq.] to spend the night at another's home

sleep·er (slē'pər) *n.* ‖ ME *slepere* < OE *slæpere* ‖ **1** a person or animal that sleeps, esp. as specified *[a sound sleeper]* **2** a timber or beam laid horizontally, as on the ground, to support something above it **3** [Chiefly Brit., etc.] a tie supporting a railroad track ☆**4** SLEEPING CAR ☆**5** a previously disregarded person or thing that unexpectedly achieves success, assumes importance, etc. ☆**6** *a)* [*usually pl.*] a kind of pajamas for infants and young children, that enclose the feet *b)* BUNTING[1] (sense 3) ☆**7** *Bowling* a pin concealed by one in front of it, in bowling for a spare

sleep·i·ly (-pə lē) *adv.* in a sleepy or drowsy manner

sleep·i·ness (-pē nis) *n.* a sleepy quality or state

☆**sleeping bag** a large, warmly lined, zippered bag, often waterproof, in which a person can sleep, esp. outdoors

☆**sleeping car** a railroad car equipped with berths, compartments, etc. for passengers to sleep in

sleeping partner *Brit., etc. term for* SILENT PARTNER

☆**sleeping pill** a pill or capsule containing a drug, esp. a barbiturate, that induces sleep

sleeping sickness **1** an infectious disease, esp. common in tropical Africa, caused by either of two trypanosomes (*Trypanosoma gambiense* or *T. rhodesiense*) that are transmitted by the bite of a tsetse fly: it is characterized by fever, weakness, tremors, and lethargy, usually ending in prolonged coma and death **2** inflammation of the brain, caused by a virus and characterized by apathy, drowsiness, and lethargy

sleep·less (slēp'lis) *adj.* **1** unable to sleep; wakeful; restless **2** marked by absence of sleep *[a sleepless night]* **3** constantly moving, active, or alert —**sleep'less·ly** *adv.* —**sleep'less·ness** *n.*

1261 **slay / slick**

sleep·walk·ing (-wôk'iŋ) *n.* the act or practice of walking while asleep; somnambulism —**sleep'walk'** *vi.* —**sleep'walk'er** *n.*

sleep·wear (-wer') *n.* NIGHTCLOTHES

sleep·y (slē'pē) *adj.* **sleep'i·er**, **sleep'i·est** **1** ready or inclined to sleep; needing sleep; drowsy **2** characterized by an absence of activity; dull; idle; lethargic *[a sleepy little town]* **3** causing drowsiness; inducing sleep **4** of or exhibiting drowsiness

SYN.—**sleepy** applies to a person who is nearly overcome by a desire to sleep and, figuratively, suggests either the power to induce sleepiness or a resemblance to this state *[a sleepy town, song, etc.]*; **drowsy** stresses the sluggishness or lethargic heaviness accompanying sleepiness *[the drowsy sentry fought off sleep through the watch]*; **somnolent** is a formal equivalent of either of the preceding *[the somnolent voice of the speaker]*; **slumberous**, a poetic equivalent, in addition sometimes suggests latent powers in repose *[a slumberous city]*

sleep·y·head (-hed') *n.* a sleepy person

sleet (slēt) *n.* ‖ ME *slete* < OE **sliete*, akin to Ger *schlosse*, hail < IE base **(s)leu-*, loose, lax > SLUR, SLUG[1] ‖ **1** partly frozen rain, or rain that freezes as it falls **2** transparent or translucent precipitation in the form of pellets of ice that are smaller than 5 mm (.2 in.) **3** the icy coating formed when rain freezes on trees, streets, etc. —*vi.* to shower in the form of sleet —**sleet'y** *adj.*

sleeve (slēv) *n.* ‖ ME *sleve* < OE *sliefe*, akin to Du *sloof*, apron: for IE base see SLIP[3] ‖ **1** that part of a garment that covers an arm or part of an arm **2** a tube or tubelike part fitting over or around another part **3** a thin paper or plastic cover for protecting a phonograph record, usually within a JACKET (*n.* 2*b*) **4** a drogue towed by an airplane for target practice —*vt.* **sleeved**, **sleev'ing** to provide or fit with a sleeve or sleeves —**up one's sleeve** hidden or secret but ready at hand

sleeved (slēvd) *adj.* fitted with sleeves: often in hyphenated compounds *[short-sleeved]*

sleeve·less (slēv'lis) *adj.* having no sleeve or sleeves *[a sleeveless sweater]*

sleeve·let (-lit) *n.* a covering fitted over the lower part of a garment sleeve, as to protect it from soiling

☆**sleigh** (slā) *n.* ‖ Du *slee*, contr. of *slede*, a SLED ‖ a light vehicle on runners, usually horse-drawn, for carrying persons over snow and ice —*vi.* to ride in or drive a sleigh

☆**sleigh bells** a number of small, spherical bells fixed to the harness straps of an animal drawing a sleigh

sleight (slīt) *n.* ‖ ME < ON *slœgth* < *slœgr*, crafty, clever: see SLY ‖ **1** cunning or craft used in deceiving **2** skill or dexterity

sleight of hand **1** skill with the hands, esp. in confusing or deceiving onlookers, as in doing magic tricks; legerdemain **2** a trick or tricks thus performed

slen·der (slen'dər) *adj.* ‖ ME *slendre, sclendre* < ? ‖ **1** small in width as compared with the length or height; long and thin **2** having a slim, trim figure *[a slender girl]* **3** small or limited in amount, size, extent, etc.; meager *[slender earnings]* **4** of little force or validity; having slight foundation; feeble *[slender hope]* —**slen'der·ly** *adv.* —**slen'der·ness** *n.*

slen·der·ize (-īz') *vt.* **-ized'**, **-iz'ing** to make or cause to seem slender —*vi.* to become slender

slept (slept) *vi.*, *vt.* *pt. & pp. of* SLEEP

Sles·vig (sles'vikh) *Dan. name of* SCHLESWIG

sleuth (slo͞oth) *n.* ‖ ME, a trail, spoor < ON *slóth*, akin to *slothra*, to drag (oneself) ahead: for IE base see SLUG[1] ‖ **1** [Rare] a dog, as a bloodhound, that can follow a trail by scent: also **sleuth'hound'** (-hound') ☆**2** [Colloq.] a detective —☆*vi.* to act as a detective

☆**slew**[1] (slo͞o) *n. alt. sp. of* SLOUGH[2] (sense 4)

slew[2] (slo͞o) *n.*, *vt.*, *vi. alt. sp. of* SLUE[1]

☆**slew**[3] (slo͞o) *n.* ‖ Ir *sluagh*, a host ‖ [Colloq.] a large number, group, or amount; a lot

slew[4] (slo͞o) *vt. pt. of* SLAY

slice (slīs) *n.* ‖ ME < OFr *esclice* < *esclicier*, to slice < Frank *slizzan*, akin to SLIT ‖ **1** a relatively thin, broad piece cut from an object having some bulk or volume *[a slice of apple]* **2** a part, portion, or share *[a slice of one's earnings]* **3** any of various implements with a flat, broad blade, as a spatula **4** *a)* the path of a hit ball that curves away to the right from a right-handed player or to the left from a left-handed player *b)* a ball that follows such a path —*vt.* **sliced**, **slic'ing** **1** cut into slices **2** *a)* to cut off as in a slice or slices (often with *off, from, away*, etc.) *b)* to cut across or through like a knife **3** to separate into parts or shares *[sliced up the profits]* **4** to use a SLICE (*n.* 3) or slice bar to work at, spread, remove, etc. **5** to hit (a ball) in a SLICE (*n.* 4*a*) —*vi.* **1** to cut (*through*) like a knife *[a plow slicing through the earth]* **2** to hit a ball in a SLICE (*n.* 4*a*) —**slic'er** *n.*

slice bar an iron bar with a broad, thin end, used in a coal furnace to loosen coals, clear out ashes, etc.

slick (slik) *vt.* ‖ ME *slikien* < OE *slician*, to make smooth, akin to ON *slikr*, smooth < IE **(s)leig-*, slimy, to smooth, glide < base **(s)lei-*: see SLIDE ‖ **1** to make sleek, glossy, or smooth **2** [Colloq.] to make smart, neat, or tidy: usually with *up* —*adj.* ‖ ME *slike* < the v. ‖ **1** sleek; glossy; smooth **2** slippery; oily, as a surface **3** accomplished; adept; clever; ingenious **4** [Colloq.] clever in deception or trickery; deceptively plausible; smooth *[a slick alibi]* **5** [Colloq.] having or showing skill in composition or technique but little depth or literary significance *[a slick style of writing]* **6** [Old Slang] excellent, fine, enjoyable, attractive, etc. —*n.* ☆**1** *a)* a smooth area on the surface of

at, āte, cär; ten, ēve; is, īce; gō, hôrn, look, to͞ol; oil, out; up, fur; ə *for unstressed vowels, as* a *in* ago, u *in* focus; ' *as in* Latin (lat''n); chin; she; zh *as in* azure (azh'ər); thin, *the*; ŋ *as in* ring (riŋ)
In etymologies: * = unattested; < = derived from; > = from which
☆ = Americanism **See inside front and back covers**

Figure 1-1 Page from *Webster's New World Dictionary, Third College Edition*

Pronunciation

Immediately after an entry word you will usually find its phonetic spelling in parentheses. The diacritical marks tell you how to pronounce the word. In most dictionaries, an accent mark appears *after* any accented syllable. In *Merriam Webster's Collegiate Dictionary, Tenth Edition*, however, an accent mark appears *before* an accented syllable. Many dictionaries have a pronunciation key at the bottom of every odd-numbered page. (Such a key appears in Figure 1-1.) Recent editions of *The American Heritage Dictionary* have pronunciation notes on all pages.

Etymology

The etymology of a word, which is its history or derivation, may appear in brackets near the beginning of an entry (see the first entry in Figure 1-1). Some dictionaries put the etymology near the end of an entry, as is done in *The American Heritage Dictionary.* The front section of a dictionary explains the abbreviations used in etymologies, such as "ME" for "Middle English."

Many of our words have been passed from one language to another, and then finally to English. As you read the information that appears between brackets, remember that the symbol < means derived from. The language mentioned last is probably the language in which the word originated, possibly in a form no longer recognized.

Order of Entries

If a word has more than one meaning, dictionaries usually present the definitions in historical order, giving the oldest definition, which most people have forgotten, first. A few dictionaries, however, present definitions in the order of frequency of use. That is, the dictionary gives the most common meaning of a word first and the least common meaning last. You should know how your dictionary presents its definition.

Parts of Speech

Before dictionaries present the definitions for a particular word, they give its part of speech with an abbreviation such as *adj., adv., conj., interj., n., pron., vt.,* or *vi.* Many words in our language function as more than one part of speech. The entry for *contract*, for example, includes definitions for both the noun and verbs (*vt.* and *vi.*). The division of verbs into transitive (*vt.*) and intransitive (*vi.*) is important when you analyze and construct sentences.

Usage Labels

As stated earlier, a dictionary entry may include a label that restricts the use of a word. The label *obs.*, which is an abbreviation for *obsolete*, tells us that the definition given is now out of date. The labels *colloq.* or *informal* mean that the word as defined may not be suitable for *standard English* (English well established and widely recognized as acceptable). Make sure that you understand the meanings of labels such as *rare, archaic, U.S., Brit., illit., dial., substand., slang,* or *bot.* Words considered as standard English have no usage labels.

Two-Word Expressions

A dictionary provides entries for many common expressions containing two or more words. For example, you may find entries for *pipe cutter, half note,* and *social security.* If your dictionary does not list a particular expression, do not write it as a single word. The lack of an entry for *peach tree* suggests that this expression is two words. In some instances, however, words that are still listed as two words in a general dictionary may have become one word in practice.

Prefixes (Word Beginnings) and Suffixes (Word Endings)

College dictionaries include prefixes, such as *ante, bi,* and *circum,* and suffixes, such as *-able,* and *-ous* in their main listing. If your dictionary does not list a word such as *acrophobia,* you may have to look up the two parts of the word. *Acro* (the prefix or beginning) means top, summit, or height. Your dictionary will tell you that *phobia,* the root word, means an abnormal or illogical fear. Therefore, acrophobia means fear of heights. Most dictionaries show *phobia* as a word and as a prefix.

Synonyms

Synonyms are words that have similar meanings. If the meanings are not identical, your dictionary may provide a paragraph that points out any slight differences. This paragraph always appears after the entry for one of the synonyms discussed. In Figure 1-1, note the synonyms (SYN.) following the entry for the word *sleepy.* In addition to *sleepy,* the dictionary discusses the words *drowsy, somnolent,* and *slumberous.* After the entry for each of these words, you usually find this notation: —SYN. see *sleepy.*

Inflected Forms

As used in grammar, *inflection* means a change in the form of words to indicate a change in meaning. Nouns, verbs, adjectives, and adverbs can have *inflected forms.* The noun change from *box* to *boxes* is a *regular* inflected form that is not shown in the dictionary. The noun change from *mouse* to *mice* is an *irregular* inflected form that is shown in the dictionary.

The dictionary also gives irregular verb tenses and irregular adjective and adverb comparatives. For now, you need only remember that most dictionaries give irregular inflected forms but not regular inflected forms.

Signs and Symbols

Most dictionaries contain a special appendix listing signs and symbols used in astronomy, chemistry, mathematics, and other scientific disciplines. In *The American Heritage Dictionary,* you will find signs and symbols included in the main listing.

Miscellaneous

The main listings of most dictionaries include abbreviations, biographical names, and geographical names. In *Merriam Webster's Collegiate Dictionary, Tenth Edition* and in *The American Heritage Dictionary,* however, all biographical names and geographical names are given in special sections that follow the main listing. *The American Heritage Dictionary* also provides a special section for abbreviations.

THESAURUS

A thesaurus is a book of words with their synonyms and antonyms listed by categories. Earlier in this learning unit, we defined a **synonym** as a word that has a similar meaning to another word. The words *curious* and *inquisitive* are synonyms. An **antonym** is a word that has an opposite meaning of another word. The words *good* and *bad* or *black* and *white* are antonyms. The St. Martin's *Roget's Thesaurus of English Words and Phrases* (New York: St. Martin's Press, 1965) is a popular thesaurus.

SPECIAL DICTIONARIES

Many disciplines have special dictionaries. These dictionaries list only the words used in a particular field. For example, a well-known computer dictionary is *Que's Computer User's Dictionary* by Bryan Pfaffenberger (Carmel, IN: Que Corporation, 1990). When a special dictionary is available for your subject, use it as your first word source.

YOU TRY IT

The sentences below contain difficult word pairs. Use your dictionary to check the meaning of each of the words. Then, circle the word that best fits the meaning of the sentence.

1. The president made an allusion/illusion to the year's profitability.

2. The plane preceded/proceeded to enter the runway for takeoff.

3. The pilot implied/inferred to the passengers that the flight might be rough.

4. The cargo supervisor complimented/complemented the loading crew on their efficiency.

5. The crowd was denied excess/access to the boarding area.

6. From what the reporter said, I inferred/implied that the problem was due to the weather.

7. The computer refused to except/accept my credit card number.

8. A basic understanding of navigation is a prerequisite/perquisite for in-flight training.

Answers: 1. allusion, 2. proceeded, 3. implied, 4. complimented, 5. access, 6. inferred, 7. accept, 8. prerequisite.

ABOUT THE CHAPTER SUMMARY

At the end of each chapter, you will find a chapter summary that is also a quick reference guide. This summary gives the page number on which each chapter topic begins, the key points of each topic, and usually one or two examples. After you have studied the chapter, look at the topics listed in the chapter summary. How many of the key points can you explain?

If you have difficulty recalling and explaining the key points of a topic, go back to the text and review the topic. When you know the key points of each topic, and understand the topics, you are ready for the Quality Editing, Applied Language, and Worksheet activities.

CHAPTER SUMMARY: A QUICK REFERENCE GUIDE

PAGE	TOPIC	KEY POINTS	EXAMPLES
3	**Writing principles**	1. **Organize your thoughts before you write.** Know your purpose and audience; then organize your thoughts. Begin with a general subject statement; follow with specific information. Develop details by comparison or contrast, time, space, cause-effect, and problem-solution. List your ideas; group your ideas.	
		2. **Prefer the active voice.** In the active voice, the subject performs the action: *who* (or *what*) *did what.* In the passive voice, the subject receives the action: *what was done by whom.*	Active: Diane wrote the last letter. Passive: The last letter was written by Diane.
		3. **Use a simple writing style that avoids unbusinesslike expressions.**	Avoid: He was *sick* about the telephone bill.
		4. **Omit unnecessary words.**	Instead of *prior to*, say *before*.
		5. **Use positive words and specific words.**	Instead of *do not run*, say *walk*.
		6. **Use grammatically correct constructions.**	
		7. **Use the dictionary.**	
9	**General dictionary**	1. **Dictionary choice.** Have at least one desk or college dictionary with 150,000 or more words.	
		2. **Guide words.** Guide word on the left is the first entry word on the page. Guide word on the right is the last entry word on the page.	
		3. **Syllabication.** Dots usually divide entry words into syllables. Syllabication tells when to write a word as one word.	
		4. **Pronunciation.** The phonetic spelling of the word with its diacritical marks tells how to pronounce the word.	
		5. **Etymology.** A word's history or derivation may appear in square brackets either near the beginning or end of an entry.	

PAGE	TOPIC	KEY POINTS	EXAMPLES
		6. **Order of entries.** If a word has more than one meaning, definitions are usually given in historical order. Some dictionaries, however, list the most common meanings first.	
		7. **Parts of speech.** These are given before the definition of the word. Many words function as more than one part of speech.	
		8. **Usage labels.** These tell how a word should be used.	*obs., colloq.* (or *informal*), *slang*
		9. **Two-word expressions.** If a particular expression is not in dictionary, treat it as two words.	*Fish stick, fish fry*
		10. **Prefixes (word beginnings) and suffixes (word endings).** For infrequently used words, look up the prefix or suffix and the root word.	*Ante, bi, circum; able, ous*
		11. **Synonyms.** Words with the same or closely related meanings are synonyms.	*Haste\ SYN. hurry, speed, expedition, dispatch.*
		12. **Inflected forms.** Words with inflected forms change grammatically. Most dictionaries include only unusual inflected forms.	*walk, walked; mouse, mice*
		13. **Signs and symbols.** These occur usually in a special appendix listing signs and symbols of astronomy, chemistry, mathematics, and so on.	*> is greater than (6>5)*
		14. **Miscellaneous.** Most dictionaries list abbreviations, biographical names, and geographical names.	
13	**Thesaurus**	A book of words with their synonyms and antonyms listed by categories.	*Roget's Thesaurus of English Words and Phrases*
13	**Special dictionaries**	These are dictionaries of special disciplines.	*Que's Computer User's Dictionary*

ABOUT THE QUALITY EDITING EXERCISE

Always edit your writing. When editing, look for spelling errors, grammar errors, and errors in meaning or writing style. The writing sample given in the Quality Editing section could have errors discussed in any previous chapters and sometimes common errors, such as misspelled words, not yet discussed.

Remember that often you can edit a sentence in more than one way and several different approaches could be acceptable. See your instructor to check your work.

In your editing use professional proofreader marks. Some common proofreading marks follow. A more complete list is on the inside back cover of this book. Below the table is a paragraph that has been edited.

Table 1-1 Proofreading Marks

Symbol	What It Means	Example
⚥	Delete or omit	plays a role
∧	Insert	several groups large
#	Insert space	high school
⊙	Insert period	He hit Carl
cap ≡	Change lowercase letter to capital	this is a new product.
[	Move left	[Dear Sir:
∧	Insert comma	blue, black, and gray
sp	Spell it out	came in 5 days
⌒	Close up space	pay ment
∿	Transpose	recieve
¶	New paragraph	¶ The group
stet	Keep crossed out characters with dots underneath	You will get your answer. stet
⌿	Change capital to lowercase letter	Check the Lock on the door.
]	Move right	Sincerely,]
∨	Insert apostrophe	dont
ital	Set in italic	Find the key ital

At the end of July Frank Maris decided open a shoerepair shop. He went to his accountant to get inforamtion on the correct procedure The accountant explained that his shoe shop would be considered as a separate unit from his personal business finances. He went on to say that all busines transactions can be organized by something called the basic ac counting equation. Frank had never Heard of the basic accounting equation. He listened carefully as the accounting explained it.

QUALITY EDITING

Use the proofreading marks to make corrections in the following notice. Look for obvious spelling or typographical errors and for violations of the writing principles given in Learning Unit 1-1.

Notice to Ticket Agents

STS Airlines pride itself on its custoemr service. As ticket agents and check-in attendants, You have more business con tact with our passengers than anyone else in the company. During the month of may, we will hold a series of three workshops. plan ahead to attend each workshop. Will you be so kind as to sign up for the time that fits your shift? The first workshop willbe "How to keep your cool when the flight is late and the passengers are angry" The second workshop will be "How to mange stress." The last and final workshop will be "On-the-spot problem solving." In spite of the fact that you are all very busy, you will find the workshops super-duper.

APPLIED LANGUAGE

1. Pretend that you are going to write a short letter to STS Airlines (an imaginary company), inquiring about a job. Choose a job that would be relevant to the airline industry, such as ticket agent, baggage handler, or flight attendant. Make a list of three things that you will put in the letter about yourself. The items could be special qualities or skills that you have, experience you have with the travel industry, educational background, or other information that would be interesting to a person at STS reading your letter.

2. Find three examples of wordiness in print. Check the editorial section of your daily newspaper, stories in a tabloid (usually sold at grocery store checkout stands), club newsletters, and other sources. Edit the examples to reduce wordiness.

3. Describe the contents of an entry in a thesaurus. When do you think you would be most likely to use a thesaurus?

WORKSHEET 1

Name _____

Date _____

WRITING PRINCIPLES AND REFERENCE TOOLS

Part A. Informal Words

Each of the sentences below contains an expression that is inappropriate for business writing. Underline the word or phrase and then suggest one to replace it.

1. The instructions for passengers to follow in case of an emergency landing were all mixed up.

2. The gate agent had no alibi for being late.

3. After working on the report until late at night, the committee was bushed.

4. If you think that I am going to pay this invoice, you're nuts.

5. The airline's television commercials really bug me.

6. The manager was hot under the collar about the missing equipment.

7. Our office manager is sold on the new computer system.

8. The baggage handling at that airport is a disaster.

9. Airport management leased space to some crummy junk-food joints.

10. Joyce says that working in customer service is a piece of cake.

Part B. Wordiness

Each of the following sentences includes an expression that is too wordy. Underline the wordy expression. Then, suggest one to replace it. If the expression is unnecessary and can be deleted, write the word *delete* next to the sentence.

1. The supervisor spoke to the cargo loading staff in reference to procedures for dangerous goods.

2. Every airline with the exception of STS pays the airport exorbitant landing fees.

3. Can you get a ticket to New York for the price of $398?

4. Until such time as airplanes provide larger seats, I shall take the train.

5. The airline has six flights to Salt Lake City at the present time.

6. Due to the fact that the airline has suffered losses for the past three years, it may go out of business.

Part C. The Active Voice

The following sentences have been written in the passive voice. In the space provided, rewrite each one in the active voice.

1. The employees were told by the company leadership that an improved benefits package would soon be available to them.

_____

_____

2. The plane was serviced by the mechanics.

_____

_____

3. The passengers were told by the gate agent that the flight had been overbooked by the computer.

_____

_____

4. The passenger's credit card number was entered into the computer by the ticket agent.

_____

_____

5. The frequent flyer mileage statement from the airline was received by the loyal customer.

_____

_____

6. The report on the results of the last travel promotion was prepared by the marketing department.

_____

_____

7. The account clerk was given instructions by his supervisor to send the invoices immediately.

_____

_____

8. The computer program was edited by the programming consultant so that it could be accessed directly by customers.

_____

_____

9. I was surprised by his candor concerning the company's financial situation.

_____

_____

10. Coupons for 20 percent off any flight taken in May were sent to previous users by the marketing department.

_____

_____

Part D. Overlapping Meanings

Each of the phrases below contains words with overlapping meanings. On the line provided, reduce the phrase with a single word.

EXAMPLE: his own   _____**his**_____

1. final outcome   _____
2. consensus of opinion   _____
3. end result   _____
4. exactly identical   _____
5. serious crisis   _____
6. my own   _____
7. completely different   _____
8. enclosed herewith   _____
9. important essentials   _____
10. advance planning   _____

Part E. Reference Tools

Write the missing word in the space provided. Refer to the information provided in the chapter if necessary.

1. A book that lists words and their synonyms and antonyms by categories

    is a(n) _____.

2. The label *colloquial* has the same meaning as _____.

3. _____ is the abbreviation for obsolete.

4. A dictionary that does *not* contain every word in the language is _____.

5. A person can determine how a word should be pronounced by looking

    at its _____ spelling.

6. The main listing of words contains prefixes and word endings, which are also

    called _____.

7. Between syllables of an entry word, you would expect to see a(n) _____.

8. The etymology of a word is its _____.

9. Words that have similar meanings are known as _____.

10. The history or derivation of a word is generally enclosed in _____.

CHAPTER 2

SPELLING GUIDELINES

The secret to becoming a good speller is simple: pay attention.
—Val Dumond, *Grammar for Grownups*

LEARNING UNIT GOALS

LEARNING UNIT 2-1: SPELLING WITH PREFIXES AND SUFFIXES

- Spell correctly words with prefixes. (p. 25)
- Recognize when to double final consonants before adding suffixes. (p. 25)
- Understand how to add suffixes to words ending with *e* and *y*. (pp. 27–28)
- Use prefixes and suffixes correctly in your writing. (pp. 24–28)

LEARNING UNIT 2-2: MISCELLANEOUS SPELLING GUIDELINES

- Spell correctly words with *e* and *i*. (pp. 28–29)
- Recognize when to use memory devices and the dictionary. (pp. 29–30)
- Use words with *e* and *i* correctly, memory devices to remember difficult words, and the dictionary when you write. (pp. 28–30)

LEARNING UNIT 2-3: A SPELLING LIST

- Spell 200 frequently misspelled words correctly when you write. (pp. 31–32)

Do you recall your grade-school spelling bees? Did your classmates want you on their team because you were a good speller? Or were you, like most of us, a mediocre or poor speller and not a favorite team choice? Do you think it is important to be a good speller?

Spelling errors can quickly discredit a business document. Readers tend to notice only the misspelled words and have difficulty concentrating on the message in the document. This chapter will help you spell correctly.

Today, most computers and electronic typewriters contain spelling checkers that can correct common spelling errors. These spelling checkers, however, do not include many proper names. Also, they do not recognize incorrect word usage. Although a word may be spelled correctly, it may be the wrong word for the sentence.

Since you may not always have spelling checkers available, you should know certain basic spelling guidelines and how to spell common words. Also, a knowledge of spelling guidelines can help you look up words in the dictionary.

Spelling is difficult because the English language borrows words from many languages, including Latin, Greek, French, German, and Welsh. The result is that the words in the English language lack uniformity and can cause problems for even the most conscientious business writers.

Many people believe that you must have a retentive memory to spell correctly. Although a retentive memory is helpful, learning certain spelling guidelines and developing a perceptive power of observation will enable you to avoid most spelling errors. Learn to recognize when the spelling of a word does not "look right." Then, check the spelling in your dictionary.

This chapter presents eight spelling guidelines. They appear in the first two learning units. Learning Unit 2-1 concentrates on spelling guidelines for words with prefixes and suffixes. Learning Unit 2-2 gives three miscellaneous spelling guidelines. Learning Unit 2-3 contains a list of 200 words that are considered difficult to spell. After you complete the chapter, review the chapter summary. If you are unclear about any of the guidelines, study them again.

WRITING TIP

If you have a computer word processing software program with a spelling checker, remember to use it. Many people also find spelling dictionaries helpful. They list the spelling and division of words without giving any definitions. A popular spelling dictionary is *Webster's New World Misspeller's Dictionary* published by Prentice Hall.

LEARNING UNIT 2-1
SPELLING WITH PREFIXES AND SUFFIXES

The spelling of English words would be simple if every word were spelled phonetically (according to sound) or if a strict rule governed each word. Since neither of these is true, we have only spelling guidelines. Unfortunately, most of the guidelines have exceptions that must also be remembered.

WRITING TIP

Often, we misspell some words because we mispronounce them. Take time to pronounce a word correctly. When you pronounce a word correctly, you increase your chances of spelling the word correctly.

Say	Not	Say	Not
ath-letes	*ath-a-letes*	**pre-scribe**	*per-scribe*
di-sas-trous	*di-sas-terous*	**priv-i-lege**	*priv-lege*
griev-ous	*gre-vious*	**real-tors**	*real-a-tors*
hin-drance	*hin-der-ance*	**rep-re-sent**	*rep-er-sent*

Before studying the eight spelling guidelines, you should understand the meaning of prefix, root, and suffix. A **prefix** is a syllable, syllables, or word added *before* a root to change its original meaning. A **root** is the base of a word; it may or may not be a word by itself. A **suffix** is one or more syllables added *after* the root to change its original meaning. By adding prefixes and suffixes to a root, you get new words with new meanings.

SPELLING WITH PREFIXES

GUIDELINE 1.   Learn to recognize the two parts of any word containing a prefix and a root. Adding a prefix usually does not change the spelling of the root word.

Several of the most commonly misspelled words in our language have a prefix, such as *un*, and a familiar root, such as *natural*. When you add the prefix to the root, you have the word *unnatural*. (Note that you spell the word *unnatural* with two *n's*.) Since most prefixes do not change the spelling of the root word, knowing a few common prefixes will help you avoid misspelling such words. Business writers often have difficulty with the prefixes *dis*, *mis*, *im*, and *un*.

Prefix	+	Root	=	New Word	Prefix	+	Root	=	New Word
dis		appear		disappear	mis		spell		misspell
il		legible		illegible	over		ripe		overripe
il		logical		illogical	re		commend		recommend
im		perfect		imperfect	un		necessary		unnecessary
inter		racial		interracial	under		rate		underrate
mis		shape		misshape	un		noticed		unnoticed

SPELLING WITH SUFFIXES

Guidelines 2 and 3 explain when to double the final consonant before adding a suffix. Guideline 3 may save you many trips to the dictionary because it is seldom violated. Guideline 4 explains how to add a suffix if a word ends in *e* preceded by a consonant. In Guideline 5, you learn how to add a suffix if a word ends with *y* preceded by a consonant.

GUIDELINE 2.   If a one-syllable word ends in a consonant-vowel-consonant, double the final consonant before adding a suffix that begins with a vowel.

Root	+	Suffix	=	New Word
bat		ed		batted
bet		or		bettor
big		est		biggest
brag		ing		bragging
drug		ist		druggist
hum		ing		humming
plan		ing		planning
star		ed		starred
stop		ing		stopping
tag		ed		tagged
ton		age		tonnage
wrap		er		wrapper

In such words as *quit*, the *u* has the sound of *w*, a consonant. So you can apply Guideline 2. Note the spellings of the words *quitter* and *quizzes*.

Exceptions.   Guideline 2 does not apply if the final consonant is *x* (as in *boxes, mixed*) or if it is silent. For example, in the words *row* and *tow*, the final consonant is silent. Therefore, you do not double the *w* in the words *rowing* and *towing*.

GUIDELINE 3.   If a two-syllable word ends in a consonant-vowel-consonant and is accented on the second syllable, double the final consonant before adding a suffix that begins with a vowel.

Root	+	Suffix	=	New Word
admit		ance		admittance
allot		ed		allotted
begin		er		beginner
compel		ed		compelled
defer		ed		deferred
expel		ing		expelling
occur		ence		occurrence
refer		ed		referred
repel		ing		repelling
transmit		er		transmitter

Exceptions.   In some words, such as reference (*REF er ence*), the accent changes to the first syllable when adding the suffix. In this case, you do not double the final consonant of the root word. Guideline 3 does not apply to words such as *bestow*, in which the final consonant is not pronounced. Note the spellings of the words *bestowed* and *bestowing*. The word *programmer* is usually spelled with two *m*'s even though it is accented on the first syllable. However, the alternative spelling of *programer* is also acceptable.

GUIDELINE 4. If a word ends in *e* preceded by a consonant, drop the *e* before adding a suffix that begins with a vowel.

Root	+	Suffix	=	New Word
choose		ing		choosing
excite		able		excitable
excuse		able		excusable
desire		ous		desirous
freeze		ing		freezing
hope		ing		hoping
like		able		likable
obese		ity		obesity
please		ant		pleasant
scarce		ity		scarcity
value		able		valuable

Guideline 4 implies that you keep the final *e* when adding a suffix beginning with a consonant.

Root	+	Suffix	=	New Word
appease		ment		appeasement
move		ment		movement
remote		ness		remoteness
strange		ly		strangely
tire		less		tireless

Exceptions. Since English has few, if any, spelling guidelines that are completely dependable, Guideline 4 has several noteworthy exceptions, such as the following:

acknowledgment	judgment	truly
argument	ninth	wholly

Other exceptions include words that end with the soft sound of *ge* or *ce*. Usually, the *e* is kept when adding *able* or *ous*, as in the following words:

advantageous	courageous	noticeable
changeable	manageable	peaceable

GUIDELINE 5. If a word ends with *y* preceded by a consonant, generally change the *y* to an *i* before adding a suffix other than one beginning with *i*.

An understanding of Guideline 5 will help you spell many nouns, verbs, and adjectives correctly. Note the following correct spellings:

Root	+	Suffix	=	New Word	Root	+	Suffix	=	New Word
carry		es		carries	merry		ly		merrily
company		es		companies	plenty		ful		plentiful
happy		ness		happiness	tasty		er		tastier
hurry		es		hurries	vary		es		varies
jury		es		juries	worry		er		worrier
lazy		est		laziest	worry		some		worrisome

BUT:

Root	+	Suffix	=	New Word
baby		ish		babyish
carry		ing		carrying
cry		ing		crying
dry		ness		dryness
modify		ing		modifying
pity		ous		piteous
plenty		ous		plenteous
shy		ly		shyly
worry		ing		worrying
wry		ly		wryly

YOU TRY IT

For each root and suffix combination below, write the resulting new word in the space provided.

	Root	Suffix	New Word
1.	sit	ing	_____
2.	force	ing	_____
3.	hungry	ly	_____
4.	believe	able	_____
5.	dip	er	_____
6.	remit	ance	_____
7.	penny	less	_____
8.	bid	able	_____
9.	confer	ing	_____
10.	shape	ing	_____

Answers: 1. sitting 2. forcing 3. hungrily 4. believable 5. dipper 6. remittance 7. penniless 8. biddable 9. conferring 10. shaping

LEARNING UNIT 2-2
MISCELLANEOUS SPELLING GUIDELINES

Learning Unit 2-2 gives three miscellaneous spelling guidelines. Guideline 6 discusses how to spell words after a soft *c*. Guideline 7 explains how to create a **mnemonic**, which is a memory device used to recall information, in this case, the spelling of difficult words. Guideline 8 is probably the most important guideline: Use a dictionary whenever you question the spelling of a word.

Table 2-1 *ie* and *ei* Words

i before e		ei after c	ei when sounded like a	Exceptions i before e	ei after c
achieve	mischievous	ceiling	beige	caffeine	ancient
audience	niece	conceit	eight	counterfeit	conscience
belief	patient	conceive	freight	either	efficient
believe	relief	deceit	heinous	feisty	financier
brief	shriek	deceive	heir	foreigner	science
cashier	spiel	perceive	neighbor	seizure	sufficient
chief	thief	receipt	reign	sleight	
field	tier	receive	skein	sovereign	
friend	view		sleigh		
grief	wield		vein		
grieve	yield				

GUIDELINE 6. **After a soft *c*, write the *e* before the *i* when these two letters appear in sequence.**

> Use *i* before *e* except after *c*
> Or when sounded as *a*
> As in *neighbor* and *weigh*;
> But *their*, *weird*, and *either*,
> *Foreign*, *seize*, *neither*,
> *Leisure*, *forfeit*, and *height*,
> Are exceptions spelled right.[1]

GUIDELINE 7. **Create a memory device (a mnemonic) for each word that is difficult to spell.**

The average troublesome word usually has only one challenging letter or combination of letters. Consider the word *asphalt*. The most common misspelling of this word is *asfalt*. Writers who habitually write *f* instead of *ph* should create a *mnemonic* (a technique intended to assist memory) that will make the correct spelling difficult to forget. They may decide to associate the word *asphalt* with the word *pavement*. If they concentrate for a moment on the expression *asphalt pavement* and note the use of the letter *p* in each of those words, they will probably never again spell *asphalt* with an *f*. The simplicity of such memory devices makes them effective. Often, a mnemonic that might seem completely nonsensical is the easiest to remember. Here are a few mnemonics that have proved helpful for many writers.

---

[1]Mellie John, Paulene M. Yates, Edward N. De Laney, *Basic Language IV Messages and Meanings* (New York: Harper & Row, 1973).

	To spell	**Use this mnemonic**
	pursue *not* persue	*Pursue* the *purse* snatcher. (Note the *pur* in each word.)
	parallel *not* paralell	*All* the lines are par*all*el. (The word *all* appears in the word *parallel*.)
	personnel *not* personal	*Personnel* is usually a noun; *personal* is an adjective. (Note the two *n's* in both *noun* and *personnel*.)
	privilege *not* priviledge	It will be your privi*lege* to stand on one *leg*. (The word *leg*, of course, appears in the word *privilege*.)
	principle or principal	The princi*pal* is your *pal*. (A *principal* is a person who has controlling authority or is in a leading position. A *principle* is a fundamental law, doctrine, or assumption—a rule or code of conduct.)
	separate *not* seperate	*Separate* means to take something *apart*. If you think *part*, you will remember to spell *separate* with *ar*.

GUIDELINE 8.   Keep a good dictionary handy and use it whenever you question the spelling of a word.

Many words in our language have more than one spelling. Dictionaries show preferred forms in a variety of ways. In *Merriam Webster's Collegiate Dictionary, Tenth Edition* spellings that are equally acceptable are joined by the word *or* and printed in boldface, or heavy, type. If the first spelling given is more acceptable, the word *also* precedes the secondary spelling in both *Merriam Webster's Collegiate Dictionary, Tenth Edition* and *The American Heritage Dictionary.* Your first choice in spelling a word should be the preferred spelling. Using the secondary spelling, however, is not a spelling error.

In *Webster's New World Dictionary, Third College Edition,* definitions are given with the form known or judged to be the one most frequently used. Other spellings are likely to be cross-referred to that entry. A variant spelling that is used less frequently than the main-entry spelling may appear at the end of the entry block. If you read the guide to usage at the beginning of your dictionary, you will know where to find the preferred spelling of a word.

WRITING TIP

Often writers repeatedly misspell the same words. To avoid this, keep a list of words you misspell on 3 × 5 index cards. When you must look up a word in the dictionary, write it correctly several times and add it to your list.

YOU TRY IT

A. Review the lists of words spelled with *ei* and *ie* shown in Table 2-1. Select four words from each of the five categories (*i* before *e*; *ei* after *c*; *ei* when sounded like *a*; and the exceptions to *i* before *e* and *ei* after *c*) that you think are the most difficult to remember. Write these words in the spaces provided below.

1. *i* before *e*: **a.** _____ **b.** _____ **c.** _____ **d.** _____

2. *ei* after *c*: **a.** _____ **b.** _____ **c.** _____ **d.** _____

3. *ei* like *a*: **a.** _____ **b.** _____ **c.** _____ **d.** _____

4. Exception
i before *e*: **a.** _____ **b.** _____ **c.** _____ **d.** _____

5. Exception
ei after *c*: **a.** _____ **b.** _____ **c.** _____ **d.** _____

B. Try to think of additional mnemonic devices for remembering how to spell four of the eight exceptions you selected for *i* before *e* and *ei* after *c*. For example, *caffeine was in the coffee that was served in the cafe*. (In the words *caffeine, coffee*, and *cafe*, the *f* is followed by *e*.) Write your four mnemonics in the spaces provided below. Be creative!

1. _____

2. _____

3. _____

4. _____

LEARNING UNIT 2-3
A SPELLING LIST

The following list of 200 words are commonly misspelled by business writers. Study the list and learn the words that give you difficulty. If necessary, try to develop a mnemonic to help you remember the words you repeatedly misspell.

Table 2-2 200 Frequently Misspelled Words

absence	apparently	believable	choose	consensus	disappoint	families
absorption	approaches	believing	chose	convenience	disastrous	February
accede	appropriate	benefited	clientele	convenient	disbursement	financially
accessible	approximately	bought	collateral	convertible	discrepancy	foreign
accommodate	arguing	brought	column	cooperation	discriminate	fortieth
accumulate	arrangement	bulletin	commission	corroborate	dissatisfied	forty
achievement	assistant	calendar	commitment	council	duplicate	fourth
acknowledgment	athletics	canceled	committee	counsel	efficient	freight
acquire	attendance	capital	companies	deficit	eligible	fulfill
acquisition	attendants	capitol	comparative	definitely	embarrassing	further
advantageous	attorneys	cashier	comparison	description	environment	government
alleged	auxiliary	catalog	competitor	desirable	exaggerate	guarantee
all right	bankruptcy	category	concede	despair	excellence	haphazard
analysis	basically	census	conscience	development	existence	illegible
analyze	beginning	changeable	consciousness	disappearance	familiar	immediately

Table 2-2 200 Frequently Misspelled Words *(Continued)*

importance	library	nickel	personnel	psychology	safety	supersede
incidentally	license	noticeable	persuade	pursuing	salable	technique
independent	likely	occurred	physical	quantity	schedule	thorough
indispensable	loose	occurrence	political	receiving	secretaries	thought
influential	lose	omission	possession	recipient	seize	through
interference	maintenance	omitted	precede	recognize	separation	transferred
intermediary	mathematics	pamphlet	predictable	recommend	sergeant	truly
interrupt	mediocre	parallel	preferred	reference	significant	until
irresistible	misspelling	pastime	principal	repetition	straight	usage
irresponsible	mortgage	patience	principle	responsible	strait	vacuum
judgment	necessary	peaceable	privilege	rhythmical	succeed	waiver
labeled	negligence	permanent	procedure	ridiculous	success	Wednesday
laboratory	negotiation	permitted	proceed	sacrifice	sufficient	weird
leisurely	neither	persistent	professor			

YOU TRY IT

Read the list of commonly misspelled words carefully. Select the ten that give you the most trouble. Write these below. Then, try to develop a separate mnemonic device for each word.

1. _____

2. _____

3. _____

4. _____

5. _____

6. _____

7. _____

8. _____

9. _____

10. _____

For example, a mnemonic for *stationery* could be *matching stationery has envelopes*. The letter *e* in *envelopes* matches the *e* in *stationery*. Remember to exercise your creativity!

CHAPTER SUMMARY: A QUICK REFERENCE GUIDE

PAGE	TOPIC	KEY POINTS	EXAMPLES
25	**Spelling with prefixes**	1. **Learn to recognize the two parts of any word containing a prefix and a root. Adding a prefix usually does not change the spelling of the root word.** Knowing common prefixes helps avoid spelling errors.	mis + treat = mistreat; un + pleasant = unpleasant

PAGE	TOPIC	KEY POINTS	EXAMPLES
26	**Spelling with suffixes**	2. **If a one-syllable word ends in a consonant-vowel-consonant, double the final consonant before adding a suffix that begins with a vowel.** This does not apply if the final consonant is *x* or if it is silent.	plan + er = planner; step + ed = stepped
		3. **If a two-syllable word ends in a consonant-vowel-consonant and is accented on the second syllable, double the final consonant before adding a suffix that begins with a vowel.** Guideline 3 has exceptions.	control + ed = controlled; occur + ed = occurred
		4. **If a word ends in *e* preceded by a consonant, drop the *e* before adding a suffix that begins with a vowel.** Guideline 4 has exceptions.	come + ing = coming; grieve + ing = grieving
		5. **If a word ends with *y* preceded by a consonant, generally change the *y* to an *i* before adding a suffix other than one beginning with *i*.** Guideline 5 has exceptions.	marry + ed = married; city + es = cities
28	**Miscellaneous spelling guidelines**	6. **After a soft *c*, write the *e* before the *i* when these two letters appear in sequence.** Guideline 6 has exceptions.	ceiling; deceive
		7. **Create a memory device (a mnemonic) for each word that is difficult to spell.**	argument: The word *argument* has gum in the middle.
		8. **Keep a good dictionary handy and use it whenever you question the spelling of a word.**	
31	**A list of 200 commonly misspelled words**	Study the words on the list; create a mnemonic for those you find difficult.	Matching *stationery* has *envelopes*.

QUALITY EDITING

The following paragraphs (adapted from Brusaw, Alred, and Oliu, *Handbook of Technical Writing*, Third Edition) contain 12 errors. In the spaces provided below, first list each word as it is misspelled. Then, write the correct spelling next to the misspelled word.

Learning to spell requires a systematic effort. The following system will help you learn to spell correctly.

1. Keep your dictionary handy and use it regulerly. If you are unsure about the speling of a word, do not rely on memery or guess work.

2. After you look up the spelling, write the word from memory severel times. Then check the ackuracy of your spelling. Practice is esential.

3. Keep a list of words you comonly misspel. Concentate on common words like calender, maintenence, and unnccessary.

1. _____   _____
2. _____   _____
3. _____   _____
4. _____   _____
5. _____   _____
6. _____   _____
7. _____   _____
8. _____   _____
9. _____   _____
10. _____   _____
11. _____   _____
12. _____   _____

APPLIED LANGUAGE

A. In the following list, one or two letters in each word should be doubled letters. Write the correct spelling of each word in the space provided.

1. _____   _____

2. _____   _____

3. _____   _____

4. _____   _____

5. _____   _____

6. _____   _____

7. _____   _____

8. _____   _____

9. _____   _____

10. _____   _____

11. _____   _____

12. _____   _____

B. Review the list of commonly misspelled words in Table 2-2. What words would you add to the list? Try to list at least ten.

1. _____

2. _____

3. _____

4. _____

5. _____

6. _____

7. _____

8. _____

9. _____

10. _____

ORKSHEET 2

SPELLING GUIDELINES

Part A. Prefixes

In the space provided, write the word that has the same meaning as the two words in italics. Use your dictionary, if necessary.

EXAMPLE: Your plan is *not practical*.   <u>impractical</u>

1. The new procedure is *not regular*.   _____

2. The truth was *not escapable*.   _____

3. The action taken by Ned is *not legal*.   _____

4. Her testimony was *not material*.   _____

5. The two books are *not fiction*.   _____

6. He was *not patient*.   _____

7. Erik is *not secure*.   _____

8. The members are *not active*.   _____

Part B. Suffixes

In the space provided, build two new words with each root by adding the suffixes indicated.

EXAMPLE: great   er, est   <u>greater</u>   <u>greatest</u>

	Root	Suffixes	New Words	
1.	defer	ed, ment	_____	_____
2.	transmit	er, al	_____	_____
3.	occur	ed, ence	_____	_____
4.	begin	er, ing	_____	_____
5.	repel	ed, ing	_____	_____
6.	allot	ed, ment	_____	_____
7.	boast	ed, ing	_____	_____
8.	chat	ed, ing	_____	_____

9.   extol   ed, ing   _____   _____

10. offer   ed, ing   _____   _____

11. ship   ed, ing   _____   _____

12. refer   ed, ence   _____   _____

Part C. More Suffixes

In the space provided, combine the root and suffix to build a new word.

	Root	Suffix	New Word
1.	delete	ing	_____
2.	manage	able	_____
3.	compose	ing	_____
4.	encourage	ment	_____
5.	compare	able	_____
6.	revise	ion	_____
7.	service	able	_____
8.	trace	able	_____
9.	surprise	ing	_____
10.	erase	able	_____
11.	peace	able	_____
12.	remote	ly	_____
13.	describe	ing	_____
14.	devote	ing	_____
15.	grieve	ous	_____
16.	decline	ing	_____
17.	forgive	ness	_____
18.	profane	ity	_____

Part D. More Suffixes

In the space provided, combine the root and suffix to build a new word.

	Root	Suffix	New Word
1.	wordy	ness	_____
2.	carry	er	_____
3.	hurry	ing	_____
4.	wealthy	est	_____
5.	survey	or	_____
6.	stray	ed	_____

Part E. *ie* OR *ei*

Fill in the blank to complete each of the words below. Insert *ie* or *ei* to create a correctly spelled word.

1. unbel _____ vable

2. fr _____ ght

3. dec _____ ving

4. n _____ ghbor

5. gr _____ f

6. rec _____ pt

7. rel _____ ve

8. r _____ gn

9. br _____ f

10. misch _____ vous

11. n _____ ce

12. perc _____ ve

CHAPTER 3

CLASSIFYING WORDS AS PARTS OF SPEECH

For the most part, your grammar and usage will go unnoticed—
until you make a mistake, that is.

—Maryann V. Piotrowski, *Re: Writing*

LEARNING UNIT GOALS

LEARNING UNIT 3-1: STATEMENT WORDS: NOUNS, PRONOUNS, AND VERBS

- Define and recognize nouns, pronouns, and verbs. (pp. 42–46)
- Use statement words effectively in your writing. (pp. 42–46)

LEARNING UNIT 3-2: MODIFYING WORDS: ADJECTIVES AND ADVERBS

- Define and recognize adjectives and adverbs. (pp. 47–50)
- Use modifying words effectively in your writing. (pp. 47–50)

LEARNING UNIT 3-3: CONNECTING WORDS: PREPOSITIONS AND CONJUNCTIONS

- Define and recognize prepositions and conjunctions. (pp. 51–53)
- Use connecting words effectively in your writing. (pp. 51–53)

LEARNING UNIT 3-4: INDEPENDENT WORDS: INTERJECTIONS

- Define and recognize interjections. (p. 53)

To express thoughts, you must group and connect words into sentences. For most people, the ability to do this has become an unconscious habit. Sometimes, however, you may have difficulty "saying what you mean." In business writing, you must *know* how to say what you mean.

Learning how to say what you mean involves a conscious knowledge of grammar. You must know how to arrange words in sentences so that they make sense. This includes knowing what grammatical constructions are available. Using an incorrect verb form could result in a lost business transaction. A knowledge of correct grammar also ensures respect as you speak to friends and business associates.

If you understand grammar, you will know

1. When you have written an incomplete sentence

2. How to recognize subjects and verbs to avoid making errors in agreement

3. How to recognize main clauses to avoid the common error known as a comma splice. (This occurs when the writer runs two sentences together with only a comma between them.)

4. How to use commas correctly because you can recognize independent and dependent clauses, phrases, and compound subjects and predicates

This chapter gives a preview of the parts of speech. You may wonder, "Why a preview?" If you have taken a reading course or a how-to-study course, you probably understand the importance of previews. A preview of a subject gives the mind an opportunity to see the whole picture and understand the relationship of the various ideas. Later chapters of this text discuss the parts of speech in detail.

The eight parts of speech in the English language are *nouns, pronouns, verbs, adjectives, adverbs, prepositions, conjunctions,* and *interjections.* These terms are only labels. A basic rule of grammar is that the label given to a particular word depends on how the word is used in the sentence, or the function of a word in a sentence. The same word can function as more than one part of speech. Since words are labeled according to their sentence function, we have used the general functions of the eight parts of speech to organize this chapter into four learning units: Learning Unit 3-1 explains statement words (nouns, pronouns, and verbs). In Learning Unit 3-2 you will learn about modifying words (adjectives and adverbs). Learning Unit 3-3 discusses connecting words (prepositions and conjunctions) and Learning Unit 3-4 covers independent words (interjections).

LEARNING UNIT 3-1
STATEMENT WORDS: NOUNS, PRONOUNS, AND VERBS

Nouns and verbs are the two most important parts of speech in the English language. Without nouns and verbs, you could not have sentences. Nouns—or their substitutes, pronouns—and verbs are the core of all your statements.

NOUNS

A noun is a word that names something. Look around you. What do you see? A *wall*? A *ceiling*? A *door*? A *pen*? A *person*? A *desk*? A *window*? The words in italics are all nouns because they name the things we see. Nouns also can name things or qualities that we do not see, such as *patience, courage, luck, intelligence*, and *skill*.

Nouns, then, are name words that name anything—persons, places, things, activities, qualities, and concepts.

Persons: *Alyssa* needs a loan to help finance her business.

Places: A new shopping mall opened in *Maplewood*.

Things: An incorrect accounting *entry* caused an error in the *ledger*.

Activities: James, the new employee, worked weekends as a *lifeguard*.

Qualities: The new supervisor lacked *patience*.

Concept: This management gives employees the *freedom* to express themselves.

How to Recognize Nouns

1. You probably have a noun when it makes sense to add *a, an*, or *the* (adjectives called *articles*) before the word.

a computer	an effect	the departments
a decorator	an informant	the stockbroker

   You can also add *the* before the word and *of* after the word.

the account of	the building of	the coat of
the arrival of	the bureau of	the freight of

2. Many nouns add *s* or *es* to show plurals.

communication, communications	organization, organizations
business, businesses	loss, losses
mechanic, mechanics	tax, taxes

3. Descriptive words (called adjectives) often appear before nouns.

difficult problem	*large* budget	*tall* box
efficient employee	*several* computers	*young* consultants
heavy desk	*similar* report	*zealous* worker

Nouns have several functions in sentences. These functions are discussed in Chapter 6.

PRONOUNS

Frequently, we use pronouns instead of nouns to make statements. A **pronoun** is a word that replaces a noun (or another pronoun). Note how pronouns can be substituted for nouns in the following sentences:

Paula (or *She*) deserves the promotion. (*She* is a pronoun.)

Shane congratulated *Jared* and *Monica* (or *them*). (*Them* is a pronoun.)

Juan (or *He*) recommends the use of voice mail. (*He* is a pronoun.)

When did you last telephone *Matthew* (or *him*)? (*Him* is a pronoun.)

In the following sentence, a pronoun can be substituted for two pronouns: *He* and *she* (or *They*) attended the meeting.

Pronouns help to make our language smooth and manageable. It would be awkward to say *Diane prepared Diane's own tax return.* Most people would prefer to say *Diane prepared her own tax return.* In this sentence, the word *her* is a pronoun, and *Diane* is the pronoun's antecedent. An **antecedent** is the word or group of words that a pronoun replaces. In each of the following sentences, the pronouns are in italics and the antecedents are in all-capital letters.

Justin is the PERSON *who* saw the accident.

The BOOK *that* Ms. Sachs requested is now in stock.

KATRINA has opened *her* own law office.

Note how the pronoun *everyone* substitutes for many nouns in the following sentence: *Everyone* took part in the discussion.

Chapters 8 and 9 discuss pronouns in more detail. Meanwhile, here is a short list of words that are often used as pronouns:

all	her	nobody	such	us
any	him	none	that	we
anybody	his	nothing	their	which
everybody	I	one	these	who
everyone	it	ours	they	whoever
everything	me	she	this	whom
he	mine	some	those	you

Since pronouns are substitutes for nouns, they have the same functions in a sentence as nouns. Pronouns are unlike nouns, however, in that many of them show possession with possessive pronouns such as *my, mine, our,* and *ours.* Nouns show possession with an apostrophe ('). You will learn more about this in Chapter 7.

VERBS

A **verb** is a word or group of words (called a *verb phrase*) that expresses (1) action or (2) state of being (existence). Verbs usually begin the part of a sentence that tells what someone or something is *doing* or *being*.

Action Verbs

To complete their action, some verbs must transfer this action to another word that follows the verb. The following sentence illustrates this type of action:

The president of the company *bought* a computer.

The action verb in this sentence is *bought*. If you stopped the sentence after the verb *bought*, the sentence would not make sense. The reader would ask, "Bought what?" The action verb *bought* transfers the action from the president (who did the buying) to a computer (which was bought and now belongs to the president). The word *computer* (a *direct object*) completes the action of the verb *bought*. Verbs that need direct objects to complete their action are called **transitive verbs**.

The action of some verbs stops with the verb. These verbs are **complete verbs** because they do not need help to complete their action. The following sentence illustrates the action of a complete verb:

The new company president *arrived* yesterday.

The action verb in this sentence is *arrived*. If you stopped the sentence after the verb *arrived*, the sentence would still make sense. The action verb *arrived* is complete in itself and does not need another word or object to complete its action. The word *yesterday* in the sentence gives added information that tells when the president arrived. A verb that does not need an object to complete its meaning is called an **intransitive verb**.

State-of-Being Verbs

Verbs can also express a state of being. To express a state of being, a verb, called a **linking verb**, joins a noun or pronoun at or near the beginning of a sentence (the subject) with another word or words (called a *subject complement*) that follows the verb. This word or words rename or describe the noun or pronoun in the subject and complete the meaning of the subject. Predicate words that rename the subject are *predicate nouns*. Predicate words that describe the subject are *predicate adjectives*. Since a linking verb does not need an object to complete its meaning, it is also an *intransitive verb*.

Common linking verbs are *am*, *is*, *are*, *was*, *were*, *be*, *being*, and *been* (forms of the verb *to be*). The sense verbs—*feel*, *taste*, *smell*, *sound*, and *look*—are also often linking verbs. Additional words often used as linking verbs are *appear*, *become*, and *seem*.

STUDY TIP

If it makes sense to substitute a form of the verb *to be*, such as *am*, *is*, or *are*, for a verb, you probably have a linking verb. This tip, however, has exceptions. Check your dictionary to see if it lists the verb you question as intransitive (*vi.*).

The young reporter *grew* two inches taller since he has worked here.

(You can say, *The young reporter is two inches taller*, so the word *grew* is a linking verb as used in this sentence. *Grew* also can be an action verb that needs an object [a transitive verb], as in the sentence, *The farmer grew wheat*.)

The dictionary classifies verbs as transitive or intransitive. This classification divides verbs according to their relation to objects. A transitive verb transfers its action from a subject to an object. An intransitive verb does not have an object. It is either complete in itself or it is a linking verb that needs a *subject complement* to complete its meaning. The following outline shows this relationship:

1. Transitive verbs (must have objects to complete their meaning)
2. Intransitive verbs (do not have objects to complete their meaning)
    a. Complete verbs (action stops with the verb)
    b. Linking verbs (link nouns or pronouns in the subject to a subject complement in the predicate)

Do not be concerned if the terms in the transitive/intransitive outline are unfamiliar to you. This discussion presents an overview of the transitive/intransitive method of verb classification. Chapters 4 and 5 also discuss transitive and intransitive verbs.

Since the dictionary uses the transitive/intransitive verb classification, this text also uses this classification. The reason is obvious: You can check your dictionary if you are not sure if a particular verb needs an object to complete its meaning. Some verbs can be either transitive or intransitive, depending on how they are used in a sentence. The definitions in the dictionary will give meanings, and sometimes examples, for verbs used as transitive verbs and intransitive verbs.

As indicated in the verb definition given at the beginning of this section, a group of verbs is called a **verb phrase**. Verb phrases function as a single verb. The following sentences contain verb phrases:

The new manager *has been waiting* for an hour.

Every employee *has received* a new desk.

Jack's Repair Shop *should have repaired* the dents in your car door.

The final verbs in verb phrases are the **main verbs**. They are preceded by **helping (auxiliary) verbs**. For example, in the above verb phrase *has been waiting*, the helping verbs are *has been* and the main verb is *waiting*. The main verb determines whether the phrase is transitive or intransitive. Helping verbs are discussed in Chapter 10.

How to Recognize Verbs

Verbs can show a change of time (or *tense*) with endings such as *s, ed, en,* and *ing*. Helping verbs also indicate a change in time. When you are looking for a verb in a sentence, add the words *yesterday* and *tomorrow* before the sentence and see which word in the sentence changes with the change in time. Note how the word *works* changes in the following sentences:

Every employee in the hotel *works* eight-hour shifts.

Yesterday, every employee in the hotel *worked* eight-hour shifts.

The word *works* in the first sentence changed to *worked* in the second sentence. This tells you that the verb in the first sentence is *works*.

Another way to recognize verbs is to see if it makes sense to put *to, he, she,* or *they* before a word. If so, you probably have a verb.

to repeat   he rescued   she sold   they shopped

YOU TRY IT

Find the nouns, pronouns, and verbs in the following paragraphs. Write N above the nouns, P above the pronouns, and V above the verbs. (Adapted from Kristin R. Woolever and Helen M. Loeb, *Writing for the Computer Industry,* p. 17.)

Trainers

Trainers conduct educational sessions for new customers or customers with new applications. They use customer documents as well as special training manuals designed specifically for them.

Field Engineers

Field engineers install, maintain, and repair computer hardware. Although they work with customers, field engineers are employed by the computer company, and they have manuals written exclusively for them.

Answers: *Nouns:* Trainers, Trainers, sessions, customers, customers, applications, documents, manuals, Engineers, engineers, hardware, customers, engineers, company, manuals. *Pronouns:* They, them, they, they, them. *Verbs:* conduct, use, designed, install, maintain, repair, work, are employed, have written.

LEARNING UNIT 3-2
MODIFYING WORDS: ADJECTIVES AND ADVERBS

Words that modify—adjectives and adverbs—make the meaning of other words more exact. The word *modify* means to describe, limit, or restrict. Adjectives and adverbs describe, limit, or restrict other words.

ADJECTIVES

An **adjective** is a descriptive or limiting word that modifies a noun or a pronoun. Single-word adjectives always come before the noun they modify. For example, look at the word *employee*. The word *employee* suggests a person who works for someone else. Note how much clearer your picture of that employee becomes when the writer adds adjectives. Here are several possibilities:

a conscientious employee   *a dissatisfied* employee
an indispensable employee   *a meticulous* employee
an efficient employee   *a talkative* employee

You probably remember *a, an,* and *the* as the adjectives (called *articles*) that you can use to decide if a word is a noun. These three words are always adjectives, regardless of how they are used in a sentence.

STUDY TIP

If you remember that adjectives often end in *ous, ious, able,* or *ible,* this will help you recognize some unusual adjectives. Examples are *marvelous, nervous,* and *vigorous; conscious, delicious,* and *desirous;* and *desirable, payable,* and *permissible.*

Adjectives answer the following questions about a noun or pronoun: *How many? What kind? Which one?*

How many? The manager hired *ten* employees.

What kind? Everyone likes a *helpful* person.

Which one? Randall does not want to sit in the *yellow* chair.

STUDY TIP

If you have difficulty remembering the questions that adjectives answer, a memory trick may help. You can use the mnemonic, *How many adjectives answer what kind and which one?*

Here is a list of words that are used frequently as adjectives:

angry	easy	high	real
bright	efficient	large	recent
busy	fast	little	simple
complex	happy	low	slow
difficult	hard	quick	soft
distant	helpful	rapid	tall

How to Recognize Adjectives

Adjectives usually precede the nouns and pronouns they modify. An adjective can, however, appear as a predicate adjective after a linking verb.

Written communications were sent to the *new* customers. (The adjective *written* answers *what kind?* about the noun *communications;* the adjective *new* answers *which one?* about the noun *customers.*)

The company president likes her *nine* employees. (The adjective *nine* answers *how many?* about the noun *employees.*)

All the company's sales managers are *young.* (*Young* is a predicate adjective that follows a linking verb and answers *what kind?* about the noun *managers.*)

Adjectives make comparisons by adding the endings *er* and *est* or the words *more* and *most.*

Use the *smaller* bookcase to store your books. (The adjective *smaller* modifies the noun *bookcase.*)

This is the *largest* box we have. (The adjective *largest* modifies the noun *box*. The ending *est* compares more than two persons or things.)

WRITING TIP

Avoid general adjectives, such as *good* or *nice*, and trite adjectives, such as *fond*. Use precise and colorful adjectives. Sometimes, writing is stronger without adjectives.

ADVERBS

An **adverb** modifies a verb, an adjective, or another adverb. Most adverbs answer the questions: *How? To what extent? Where? When?*

How?	All the new dictionaries *quickly* disappeared. (The adverb *quickly* modifies the verb *disappeared*.)
To what extent?	The *nearly* finished report was returned to the accounting department. (The adverb *nearly* modifies the adjective *finished*.)
Where?	I will meet you *here*. (The adverb *here* modifies the verb phrase *will meet*.)
When?	The new manager left *early*. (The adverb *early* modifies the verb *left*.)

STUDY TIP

If you have difficulty remembering the four questions that adverbs answers, you can use the mnemonic, *How and to what extent do adverbs go where and when?*

Adverbs that modify adjectives or another adverb usually come immediately before the word they modify. Adverbs that modify verbs can appear almost anywhere in the sentence. The following sentences illustrate how adverbs modify verbs, adjectives, and other adverbs. Each word in italics is an adverb; the word it modifies is in all-capital letters.

Adverbs used to modify verbs

Ms. Baxter *hurriedly* SIGNED the contract.

The proofreader DID his job *well*.

Adverbs used to modify adjectives

Our systems analyst is *too* SLOW.

You should replace this *really* OLD cabinet.

Adverbs used to modify other adverbs

The applicant accepted the position *somewhat* HESITANTLY.

This new laser printer performs *exceedingly* WELL.

How to Recognize Adverbs

Adverbs frequently end in *ly*. (Sometimes, however, a word ending in *ly* can be an adjective, such as a *costly* mistake.)

drove *carelessly*	ran *smoothly*	walked *briskly*
effectively presented	sang *frequently*	worked *willingly*

Like adjectives, adverbs make comparisons by adding the endings *er* and *est* or the words *more* and *most*. You must remember that adjectives make comparisons about nouns, while adverbs make comparisons about adjectives, verbs, and other adverbs. If you question the part of speech of a word, consult your dictionary.

Jane worked *harder* today than she worked yesterday. (The adverb *harder* modifies the verb *worked*, telling *how* Jane worked. Since *today* and *yesterday* are compared, *harder* is required.)

Jane worked the *hardest* of all the employees. (The adverb *hardest* modifies the verb *worked*, telling *how* Jane worked. Since Jane is compared with more than one person, *hardest* is required.)

YOU TRY IT

A. Find the adjectives and adverbs in the following paragraphs. Write Adj. above the adjectives and Adv. above the adverbs. (Adapted from Kristin R. Woolever and Helen M. Loeb, *Writing for the Computer Industry,* pp. 14-15.)

Who Uses Equipment

Workers who use computers regularly to perform various jobs may be partially or wholly

responsible for entering new information into the system.

Managers and other users may use a computer only occasionally. Usually they have personal

computers in their offices to send short messages quickly, schedule meetings efficiently, manipulate

financial and other data accurately, and prepare reports and other documents more economically.

B. Three groups of words and phrases are listed below. The groups are verbs, adjectives, and adverbs. Write an adverb in the space provided to modify each word listed. Exercise your creativity!

Verbs:

1. analyze _____

2. _____ programmed

3. solve _____

4. enter _____

Verbs:

5. _____ was generating

6. followed _____

7. explain _____

8. talk _____

Adjectives:

9. _____ fast

10. _____ positive

11. _____ competitive

Adverbs:

12. _____ effectively

13. _____ easily

14. _____ hastily

Answers: **A.** *Adjectives:* various, responsible, new, other, personal, their, short, financial, other, other. *Adverbs:* regularly, partially, wholly, only, occasionally, usually, quickly, efficiently, accurately, more, economically. **B.** 1. critically 2. painstakingly 3. creatively 4. quietly 5. actively 6. closely 7. carefully 8. loudly 9. surprisingly 10. very 11. extremely 12. more 13. somewhat 14. less

---

LEARNING UNIT 3-3
CONNECTING WORDS: PREPOSITIONS AND
CONJUNCTIONS

This learning unit discusses prepositions and conjunctions as connecting words. As mentioned earlier, the pronouns *who*, *which* and *what* can also be used to connect a word or words in a sentence.

PREPOSITIONS

Every preposition has an object. A **preposition** is a word that shows the *relation* of an object (a noun or pronoun) to another word in the sentence, usually a noun or a verb. The preposition combines with its object and any modifiers to form a **prepositional phrase.**

As modifiers, prepositional phrases usually limit or make the meaning of another word in the sentence more exact. Prepositional phrases usually function as adjectives or adverbs. Sometimes prepositional phrases function as nouns.

Note the following sentence:

I will finish the project *later.* (In this sentence, the adverb *later* modifies the verb phrase *will finish.*)

If you substitute the prepositional phrase *on Friday* for the adverb *later*, you have the following sentence:

I will finish the project *on Friday.* (Now the prepositional phrase *on Friday* modifies and limits the verb phrase *will finish.*)

In each of the following prepositional phrases, the preposition appears in italics:

in the meantime	*concerning* the merger	*on* Friday
from Chicago	*without* delay	*of* the errors
under his breath	*by* a prominent writer	*at* the end

You will learn more about prepositions in Chapter 14. Before you reach that chapter, however, you should learn to recognize prepositional phrases as modifiers. As you will see in Chapter 5, the ability to recognize prepositional phrases simplifies the study of sentence construction.

Here are some common prepositions:

about	behind	for	since
above	below	from	through
across	beneath	in	throughout
after	beside	into	to
against	between	like	toward
along	beyond	near	under
among	by	of	underneath
around	concerning	off	until
as	down	on	up
at	during	over	with
before	except	past	within

Did you notice that many prepositions in the above list show direction, motion, or position? If some words on the list are unfamiliar to you as prepositions, memorize them. This will save you time later when you analyze sentences.

STUDY TIP

If you write the prepositions on a 3 x 5 index card, you can refer to the card when you look for prepositional phrases in sentences. Checking the index card when you are not sure if a word is a preposition will help you remember the various prepositions.

How to Recognize Prepositions

A preposition must always be followed by an object, which must be a noun or pronoun. To decide if a word is used as a preposition, look for the noun that functions as the object of the preposition. It will answer the question *what?* or *whom?* Chapter 8 will tell you what to do if the object is a pronoun.

CONJUNCTIONS

A **conjunction** is a connecting word that joins words or groups of words called clauses and phrases. You are probably most familiar with *coordinating conjunctions* (*and, but, or, nor, for, yet,* and *so*); they join words or groups of words of equal grammatical value (for example, two nouns and two verbs). Conjunctions such as *although, because, if, since, that, until, when, where,* and *while* are *subordinating conjunctions*. They join clauses of unequal value. Conjunctions are discussed in Chapter 15.

Note the italicized conjunctions in the following sentences:

Pay now *or* pay later.

We purchased the machine *because* it is economical.

Mr. Durham *and* Ms. Sandusky did not report for work.

The new appraiser is slow *but* accurate.

You will get the promotion *if* your excellent work continues.

Business has been slow *since* Mr. Crane left the company.

YOU TRY IT

Find the prepositions and conjunctions in the following paragraph. Underline the prepositional phrases. Write P above the prepositions and C above the conjunctions. (Adapted from Kristin R. Woolever and Helen M. Loeb, *Writing for the Computer Industry*, p. 149.)

Visual symbols, or "icons," have become essential for much of the software produced

for personal computers today, and the writers who design the manuals use them

in the documentation as well. These symbols decrease the number of words necessary

on the page because they graphically represent certain tasks or indicate possible user

actions. A notepad, a hand, and almost unlimited other possibilities can serve as helpful cues

to users throughout the documentation.

Answers: *Prepositions:* for, of, for, in, of, on, as, to, throughout. *Conjunctions:* or, and, because, or, and. *Prepositional phrases:* for much, of the software, for personal computers, in the documentation, of words, on the page, as helpful cues, to users, throughout the documentation.

LEARNING UNIT 3-4
INDEPENDENT WORDS: INTERJECTIONS

An **interjection** is a word or group of words used to show strong feeling or sudden emotion. Usually, an exclamation point or comma follows interjections. Often interjections are independent of the rest of the sentence.

A pure interjection does not add to the basic meaning of a sentence. It simply suggests surprise, fright, confusion, wonderment, or any other similar emotion. Note these examples:

Ouch! Something bit me!

Oh, do you really believe her story?

Hurray! Swenson has been promoted.

Wow! We all have new chairs.

YOU TRY IT

How many interjections can you think of? Make a list of at least ten. Be creative, but also try to use common words.

1. _____
2. _____
3. _____
4. _____
5. _____

6. _____
7. _____
8. _____
9. _____
10. _____

Possible Answers: Wow! Super! Ahah! My goodness! Yea! Golly! Great! No! Really! Well! Exactly!

CHAPTER SUMMARY: A QUICK REFERENCE GUIDE

PAGE	TOPIC	KEY POINTS	EXAMPLES
43	**Nouns**	**Definition:** Nouns are words that name persons, places, things, activities, qualities, or concepts. **How to recognize nouns:** **1. a.** Add *a*, *an*, or *the* before the word. **b.** Add *the* before the word and *of* after the word. **2.** Many nouns add *s* or *es* for plurals. **3.** Descriptive words often appear before nouns. Nouns have several functions.	Persons: *Janet* looked through every file for the missing letter. Places: *Chicago* has many opportunities for recreation. Things: Your *home* is beautiful. Activities: Many people enjoy *walking*. Qualities: Patience is a *virtue*. Concepts: Look for the *kindness* to increase.
43	**Pronouns**	**Definition:** Pronouns are words that replace a noun (or another pronoun)—a noun substitute. Pronouns make our language smooth and manageable. They perform the same functions as nouns.	The supervisor looked everywhere for *her* missing key. James has sold *his* house.
44	**Verbs**	**Definition:** A verb is a word or group of words that expresses action or state of being (existence). **Classification of verbs:** Verbs are classified as action/linking verbs and transitive/intransitive verbs. **Transitive/intransitive verbs:** **1.** Transitive verbs (need an object) **2.** Intransitive verbs (do not need an object) **a.** Complete verbs **b.** Linking verbs	**1.** Lynn *found* a watch. **2. a.** Tom *laughed* loudly. **b.** Jim *seems* tired.

PAGE	TOPIC	KEY POINTS	EXAMPLES
		How to recognize verbs: **1.** Find the word that changes when the words *yesterday* and *tomorrow* are added before the sentence. **2.** Look for words that make sense when *to, he, she,* or *they* are added before the word.	
47	**Adjectives**	**Definition:** An adjective is a descriptive or limiting word that modifies a noun or pronoun. Adjectives answer the following questions about a noun or pronoun: *How many? What kind? Which one?* **How to recognize adjectives:** **1.** Adjectives usually precede the nouns and pronouns they modify, but they can appear as a predicate adjective after a linking verb. **2.** Adjectives show comparisons by adding *er* or *est* or the words *more* or *most*.	This is a *beautiful* house. The *ten* chairs arrived. Please look for *green* pencils.
49	**Adverbs**	**Definition:** Adverbs modify a verb, an adjective, or another adverb. Adverbs answer one of these questions: *How? To what extent? Where? When?* **How to recognize adverbs:** **1.** Adverbs frequently end in *ly*. (Sometimes, however, a word ending in *ly* can be an adjective, such as *costly* mistake.) **2.** Adverbs make comparisons by adding *er* and *est* or the words *most* and *more*.	He talked *slowly* and *carefully*. Note that *talked* is an intransitive complete verb.
51	**Prepositions**	**Definition:** Prepositions show the relation of an object (a noun or pronoun) to another word in the sentence. **How to recognize prepositions:** Prepositions are always followed by an object (noun or pronoun). Look for the noun or pronoun to decide if a word is a preposition. It will answer the question *what?* or *whom?*	You will find the door *at* the left.

PAGE	TOPIC	KEY POINTS	EXAMPLES
52	**Conjunctions**	**Definition:** Conjunctions join words or groups of words called phrases or clauses. Coordinating conjunctions join words or groups of words of equal grammatical value. Subordinating conjunctions join clauses of unequal value.	Sukie *and* Curt were not at home. Coordinating conjunctions: *and, but, or, nor, for, yet, so.* Subordinating conjunctions: *although, because, if, since, that, until, when, where, while,* and so on.
53	**Interjections**	**Definition:** Interjections show strong feeling or sudden emotion.	*Hurray*! Our team won the game.

QUALITY EDITING

Imagine that you are an office manager writing a memo to your supervisor requesting electronic equipment. Use proofreader's marks to edit the following two paragraphs of your memo. Look for errors in the parts of speech. Also watch for spelling errors. (Adapted from the Visual Education Corporation's *Proofreading Skills for Business.*)

Office equipment have changed in the last few decades. The traditional office was paper based. This means that information was recorded (usually typed) in paper and stored in paper. Once a document has typed, minor corrections can be made in the final copy. If major correction are made, the hole document must be retyped.

Word processing and other information processing equipment can changed this situation in many offices. Now you may type at a key board that resembles a typewriter key board. Then you view the typed-in materiel on a screen. The letters, numbers, and symbols are formed on the screen by a series of lighted dot. The document is not printed at the same time it is keyboard. Instead, the information is electronically stored. When you have finished keyboarding the information, the machine can tell a printer to print out the document.

APPLIED LANGUAGE

1. The following sentence contains two nouns and two verbs. Every other part of speech is used once. Identify the part of speech of each word and write it in the space provided.

    Goodness! In June he unexpectedly sold both stores and retired.

    a. Goodness   _____

    b. In   _____

    c. June   _____

    d. he   _____

    e. unexpectedly _____

    f. sold   _____

    g. both   _____

    h. stores   _____

    i. and   _____

    j. retired   _____

2. In this chapter you learned that the same word may be used as different parts of speech in different sentences or even in the same sentence. You can understand the different meanings of a word because you understand its usage in the context of the sentence. Identify how the word *light* is used in each of the following sentences.

    a. I am glad you can shed some *light* on the problem.   _____

    b. You have a *light*, airy room.   _____

    c. The rockets *light* up the sky.   _____

    d. Do not take these changes *lightly*. They are serious.   _____

3. Write a one-page, double-spaced paper about an occupation that interests you or about your present occupation if you are employed. Give some interesting facts about the occupation. What type of person enjoys this occupation? Considering the present economy, what is the future in this occupation? Will advancement be readily available? When you are finished with your paper, go back and write N above each noun, V above each verb, Adj. above each adjective, and Adv. above each adverb.

CLASSIFYING WORDS AS PARTS OF SPEECH

Part A. Complete Sentences

Each of the following statements is either complete or incomplete. In the space provided, write the letter C or I, depending on whether the sentence is a complete or incomplete statement. If the item is incomplete, also indicate whether the subject or the verb is missing.

_____**I/verb**_____   **EXAMPLE:** The controversial report that was given to the executive committee on Thursday.

_____   **1.** Since last Thursday was a very busy morning.

_____   **2.** The business manager, a young woman who received her degree in accounting from Ohio State University.

_____   **3.** Sam, please answer the phone.

_____   **4.** Cancel the order for four program bulletins.

_____   **5.** Canceled the order for four program upgrades.

Part B. What the Parts of Speech Do

Each of the clues below refers to one of the eight parts of speech. Write the part of speech suggested by each clue in the space provided. Each part of speech should be listed only once.

1. This one appears in every sentence.   _____

2. This one is a connector.   _____

3. This one always takes an object.   _____

4. This one may describe how you write.   _____

5. This one simply expresses strong feeling.   _____

6. This one is merely a substitute..   _____

7. This one may describe your computer.   _____

8. This one may be your signature.   _____

Part C. Identifying Parts of Speech

Indicate the part of speech (noun, pronoun, verb, adjective, adverb, preposition, conjunction, or interjection) for each italicized word in the following sentences. Write your answers in the spaces provided after the paragraph.

Most businesses have a strategy that guides them in competing with other firms. Sometimes, planners (1) *develop* the strategy (2) *through* a formal process that includes evaluations, interviews, and meetings with (3) *key* people. Other times, the strategy is created by the person who owns (4) *or* controls the business or by (5) *someone* responsible for a division or department. When people create these plans, they ask questions about the (6) *future* of the company. They also try to predict (7) *accurately* what their competitors (8) *will be doing* in the future. When Kim Lee's (9) *predictions* about the future of her company's competitors came true, her boss said, (10) *"Incredible!* How did you know that?"

1. _____

2. _____

3. _____

4. _____

5. _____

6. _____

7. _____

8. _____

9. _____

10. _____

Part D. Prepositions and Conjunctions

Insert either a preposition or a conjunction in the spaces provided below. Write the letter P or C after each one to indicate whether it is a preposition or a conjunction.

1. _____ mastering the use _____ page-layout programs may take time and effort, the results are valuable.

2. Many designers _____ the world _____ graphics were reluctant to use these programs at first, _____ now they find them indispensable.

3. Small firms can now create many _____ their own brochures _____ other materials _____ they look professional _____ have impact.

4. _____ the features _____ learners _____ these programs, is an on-line tutorial _____ learners can learn _____ practice page-layout techniques.

5. Many new page-layout users are insecure _____ using the programs _____ they have many hours _____ hands-on experience.

CHAPTER 4

GROUPING WORDS INTO SENTENCES

In effective professional [business] writing, the only "bad" sentence is one that has to be reread. If you stumble over a sentence, the structure may be at fault.

—Blair Spencer Ray, *Introduction to Professional Communication*

LEARNING UNIT GOALS

LEARNING UNIT 4-1: IDENTIFYING THE MAJOR SENTENCE PARTS

- Recognize the simple subject, complete subject, simple predicate, and complete predicate of a sentence. (pp. 62–65)
- Know how to recognize the skeleton of a sentence. (pp. 64–65)
- Distinguish between direct and indirect objects; subject and object complements. (pp. 65–69)
- Use the six major sentence elements correctly in your writing. (pp. 62–69)

LEARNING UNIT 4-2: IDENTIFYING CLAUSES AND PHRASES

- Recognize independent and dependent clauses. (p. 71)
- Understand the difference between clauses and phrases. (pp. 71–73)
- Use clauses and phrases correctly in business writing. (pp. 71–73)

The overview of the parts of speech in Chapter 3 has given you the vocabulary necessary to understand how words are grouped into sentences. Obviously, you cannot make a sentence by arbitrarily and mechanically manipulating the eight parts of speech. Sentences are formed by joining parts of speech in an order that creates an organic whole, reproducing in words a thought in the mind. As a photograph reproduces a scene, a sentence reproduces a complete thought.

Did you take geometry in high school? If so, you will recall this well-known axiom: The whole is equal to the sum of all its parts and is greater than any of its parts. This is what writing a sentence is all about—arranging parts to arrive at the best possible whole.

This chapter begins a study of the major sentence parts. Understanding the major sentence parts provides the necessary background to begin analyzing sentences. Learning Unit 4-1 explains subjects and predicates, direct and indirect objects, and subject and object complements. Clauses and phrases are discussed in Learning Unit 4-2. You will learn more about the topics in Learning Units 4-1 and 4-2 in later chapters. Now, let's start with Learning Unit 4-1.

LEARNING UNIT 4-1
IDENTIFYING THE MAJOR SENTENCE PARTS

In this learning unit, you will begin a study of the basic principles of sentence structure. These principles involve a clear relationship of simple subjects and simple predicates with direct objects, indirect objects, subject complements, and object complements. This learning unit explains these terms and their sentence function.

SUBJECTS AND PREDICATES

A **sentence** is a group of related words (can be a single word) that contains a complete thought and at least one subject and one predicate (in a command or a request, the subject is usually implied). The subject tells who or what is doing something or being something. **Note:** In the example sentences a vertical line separates the subject from the predicate.

In addition to defining the terms used with subjects and predicates, this section shows you how to find the simple subject and simple predicate. This is the skeleton, or core, of a sentence. Knowing how to recognize the skeleton, that is, to recognize the simple subject and simple predicate, is the beginning of understanding sentence construction. To help you recognize the simple subject and the simple predicate in the example sentences, we have underlined the simple subject once and the simple predicate twice.

Simple Subjects and Complete Subjects

Simple Subject.   The **simple subject** of a sentence is a noun(s), pronoun(s), or group of words acting as a noun; it has no **modifiers** (words that describe or limit the simple subject). Note the following sentence:

The <u>cashier</u> | <u><u>reported</u></u> the missing coin purse.
Cashier is the simple subject because it has no modifiers.

Complete Subject.   The **complete subject** is the simple subject and its modifiers. The following sentence shows the complete subject to the left of the vertical line:

┌──── Complete subject ────┐
My <u>friend</u> in Chicago | <u>bought</u> a new car.
Friend is the simple subject because it has no modifiers. The complete subject is *My friend in Chicago*—the simple subject and its modifiers.

When the subject of a sentence has only one word, it is the simple subject because it has no modifiers. It is also the complete subject. In grammar, the entire subject, regardless of the number of words, is called the complete subject.

<u>John</u> | <u>left</u> the meeting.
John is the simple subject because it has no modifiers. *John* is also the complete subject.

Pronouns are frequently used as one-word simple subjects and complete subjects.

<u>She</u> | <u>left</u> the meeting.
The simple subject is *she*; it has no modifiers. *She* is also the complete subject.

Simple Predicates and Complete Predicates

Simple Predicate.   The **simple predicate** of a sentence is the verb or verb phrase that tells something about the action or state of being of the simple subject. A verb phrase, you recall from Chapter 3, consists of a main verb and helping verbs. For now, we will use a single transitive or intransitive verb, which is the main verb in the verb phrase.

Our <u>director</u> | <u>purchased</u> a laser printer.
Purchased is the simple predicate because it is the verb that tells something about the subject.

Complete Predicate.   The **complete predicate** is the verb and any words that complete its meaning or modify it.

┌──────── Complete predicate ────────
The <u>manager</u> of this department | <u>purchases</u> the equipment for the
└────────────────┐
entire company.
Purchases is the verb, or simple predicate. The complete predicate is the verb and the words that complete its meaning.

Let's repeat the transitive/intransitive verb outline from Chapter 3 and relate the complete predicate to this outline.

1. Transitive verbs (must have objects to complete their meaning)

The insurance <u>company</u> | <u>provides</u> many excellent options.

2. Intransitive verbs (do not have objects to complete their meaning)
   a. Complete verbs (action stops with the verb)

   My <u>friends</u> | <u>work</u> for Chase Manhattan.

   b. Linking verbs (link nouns or pronouns in the subject to a subject complement in the predicate)

   The <u>cause</u> of the machine accident | <u>is</u> unknown.

From what you learned in Chapter 3 about verbs and from this outline, you can conclude that the complete predicate of a transitive verb sentence is (1) the transitive verb, (2) its objects, and (3) any modifiers. The complete predicate of an intransitive complete verb sentence is (1) the complete verb and (2) any modifiers. The complete predicate of an intransitive linking verb sentence is (1) the linking verb, (2) the subject complement that completes the meaning of the subject, and (3) any modifiers. The terms used in these sentences and the transitive/intransitive verb outline will be illustrated frequently. If necessary, reread the verb discussion in Chapter 3.

You can have a sentence in which the complete predicate is no more than the simple predicate, or verb. In grammar, the entire predicate, regardless of the number of words, is called the complete predicate. Note the following two sentences:

<u>Chen</u> | <u>ran</u>.

<u>Chen</u> | <u>ran</u> to the building.

In the first sentence, *ran* is the simple predicate because it has no modifiers. *Ran* is also the complete predicate. *Ran* is also the simple predicate in the second sentence, but because it has modifiers, *ran* is not the complete predicate. *Ran to the building* is the complete predicate.

Finding the Simple Subject and Simple Predicate

How do you find the simple subject and simple predicate in a sentence? First, you look for the verb or verb phrase (the simple predicate). Use the hints for finding the verb given in Chapter 3. Usually, adding the words *yesterday* or *tomorrow* before a sentence will help you find the verb quickly. Let's try it.

The president *rides* the bus to work.

Yesterday, the president *rode* the bus to work.

The word *rides* changed to *rode* when the time of the sentence was changed to the past (yesterday), so *rides* is the verb, or simple predicate.

To find the simple subject, insert *who?* or *what?* before the verb *rides* and ask, *Who rides?* The answer is *president*, the simple subject, which is always a noun or pronoun.

With practice, you can quickly find the skeleton of a sentence and understand how it is related to the other sentence parts. Study the sentences that follow. Under the simple subject is the key question word you should ask to find the subject.

The quality control <u>inspector</u> | <u>objects</u> to these cost-cutting changes.
who
Who objects? Inspector. *Inspector* is the simple subject.

<u>Leadership</u> | <u>affects</u> employee morale.
what
What affects? Leadership. *Leadership* is the simple subject. It is also the complete subject because it has no modifiers.

A <u>contract</u> | <u>is</u> an agreement that the parties intend to be legally
what
enforceable.
What is? Contract. *Contract* is the simple subject.

| <u>Help</u> me.
Who should help? In this sentence, the subject *you* is implied.

Compound Subjects and Compound Predicates

A **compound subject** contains two or more simple subjects. A **compound predicate** contains two or more simple predicates. These elements are joined by coordinating conjunctions such as the word *and* or the word *or.* Compound subjects and compound predicates may occur in the same sentence, as shown in the third example below.

Compound Subject:	<u>Lory</u> and <u>James</u>	<u>hired</u> two people for their word processing business.
Compound Predicate:	<u>Clark</u>	<u>washed</u> and <u>waxed</u> the company car.
Compound Subject and Compound Predicate:	The company <u>president</u> and the <u>accountant</u>	<u>bought</u> a new computer and <u>sold</u> the old computer.

You are now ready to learn about direct objects, indirect objects, subject complements, and object complements. Future chapters frequently refer to these sentence elements.

STUDY TIP
Remember that nouns in prepositional phrases are *not* simple subjects, since prepositional phrases function as *modifiers* (adjectives and adverbs).

DIRECT AND INDIRECT OBJECTS

Chapter 3 explained that some action verbs—the *transitive verbs*—must have a **direct object** to receive the action named by the verb. Direct objects are *always* nouns and pronouns. In this section, we first discuss direct objects. Then, you learn about indirect objects.

Direct Objects

The three steps used to find the direct object follow. Note that the first two steps are similar to the steps used to find the simple subject.

1. Find the verb or verb phrase. Does it need another word to receive the action? If so, the verb is transitive and will need a direct object.

2. Insert *who?* or *what?* before the verb to find the simple subject. (Ask *who?* or *what?* + verb.)

3. Now you must find the direct object. You can do this by asking *what?* or *whom?* after the verb. For example, if the verb is *sold* and the subject is *James*, you would say:

   James | sold *what?* (The answer is the direct object; it will be a noun or pronoun.)

   James | sold computers. (*Computers* is the direct object.)

Now let's try these steps on a sentence. The direct object is labeled DO.

$$\underset{\textbf{who}}{\text{The bakery shop \underline{owner}}} \mid \overset{DO}{\underset{\textbf{what}}{\underline{\text{purchases}} \text{ flour}}} \text{ daily.}$$

1. Find the verb or verb phrase. Remember to add *yesterday* or *tomorrow* to the sentence to see if a word changes. It will be the verb. (Yesterday, the owner *purchased*. The verb is *purchases*.)

2. Find the simple subject. *Who* or *what* does the action *purchases?* Owner. *Owner* is the simple subject.

3. See if the verb has a direct object. The direct object finishes the sentence in answer to the question *what?* or *whom?* Do we know what the owner purchases? Yes, flour. The direct object in this sentence is *flour*. (And now we also know that in this sentence, the verb *purchases* is transitive.)

A direct object *must* be a noun or pronoun. Direct objects *cannot* be an adverb or the object of a preposition. The direct object must finish the sentence in answer to the question *what?* or *whom?* (If a word or phrase in the sentence answers a different question, such as *how? to what extent? where?* or *when?*, then it is not a direct object.) Look at the next example.

$$\underset{\textbf{who}}{\underline{\text{Loren}}} \mid \overset{DO}{\underset{\textbf{what}}{\underline{\text{studied}} \text{ French.}}}$$

What did Loren study? French. *French* is the direct object. *French* is what Loren studied (In this sentence, the verb *studied* is transitive; it has a direct object, *French.*)

$$\underset{\textbf{who}}{\underline{\text{Loren}}} \mid \underline{\text{studied}} \text{ quietly.}$$

What did Loren study? We do not know. The sentence has no direct object. *Quietly* is an adverb that answers the question *how?* (In this sentence, the verb *studied* is intransitive complete; it does not have a direct object answering the question *what?*)

┌─ *prep. phrase* ─┐
<u>Loren</u> | <u>studied</u> in the barn.
who

What did Loren study? Again, we do not know, so the sentence has no direct object. *In the barn* is a prepositional phrase (used as an adverb) that answers the question *where?* (In this sentence, like the one above, the verb *studied* is intransitive complete; it does not have a direct object answering the question *what?*)

Mistaking an adverb or the object of a preposition for a direct object is the result of not recognizing a verb as transitive (must have an object) or as intransitive complete (does not have an object).

Compound Direct Objects.   Sentences may also have a **compound direct object**, that is, two or more nouns or pronouns that receive the action of the same transitive verb. Note the compound direct object in the following sentence:

┌────────────── Compound DO ──────────────┐
                DO   *DO*
The <u>supervisor</u> | <u>found</u> new pencils and pens on his desk.
      who   **what**   **what**

What did the supervisor find? Pencils and pens. *Pencils* and *pens* receive the action of the transitive verb *found.* They answer the question, Supervisor found *what?*

Indirect Objects

Occasionally, an indirect object appears between the verb and the direct object. **Indirect objects** are nouns or pronouns that answer the questions *to whom?* or *for whom?* about the direct object. Indirect objects can also be compound.

To have an indirect object (*IO*), you *must* have a direct object (*DO*). Note the following sentence.

                      IO   *DO*
The <u>salesperson</u> | <u>sold</u> her customer the last blue suit.
      who   **to whom**   **what**

What did the salesperson sell? Suit. *To whom* did the salesperson sell the suit? Customer. *Customer* is the indirect object.

You can expand indirect objects into prepositional phrases. For example, you can expand *customer* in the above sentence and say, *The salesperson sold the last blue suit to her customer.* When *customer* is expanded in this way, it is no longer the indirect object. *To her customer* is a prepositional phrase because it begins with a preposition and ends with an object (noun). An indirect object never follows the preposition *to* or *for.* The following sentences illustrate this:

                      IO   *DO*
The new <u>owner</u> | <u>wrote</u> (*to*) each customer a letter.
        who   **to whom**   **what**

In this sentence, the *to* is implied, but it is not written before the indirect object.

DO   ┌──── prep. phrase ────┐
The new <u>owner</u> | <u>wrote</u> a letter to each customer.
    who   what

This sentence does not have an indirect object. Instead, it has a prepositional phrase. *Customer* is the object of the preposition *to*. When the indirect object becomes a prepositional phrase, the *to* is the preposition of the prepositional phrase.)

Compound Indirect Object.   A **compound indirect object** is an indirect object that consists of two or more nouns or pronouns. Like all indirect objects, the compound indirect object follows a transitive verb and occurs before a direct object. It names the persons or things *to whom* and *for whom* something is given or done. Note the following sentence:

┌──── Compound IO ────┐
IO   IO   DO
<u>Loren</u> | <u>gave</u> Alexander and Macaulay a new file cabinet.
who   to whom   to whom   what

The compound indirect object is *Alexander* and *Macaulay*. This compound indirect object can also be expanded into a prepositional phrase: *Loren gave a new file cabinet to Alexander and Macaulay.* When the preposition *to* is used, a compound indirect object such as *Alexander* and *Macaulay* become objects of a preposition and are *no longer indirect objects.*

SUBJECT AND OBJECT COMPLEMENTS

A *linking verb*, as you know from Chapter 3, is an intransitive verb. Its purpose is to link a subject noun or pronoun with a predicate word that gives meaning to the sentence. Chapter 3 gave the common linking verbs. Forms of the verb *to be* are *always* linking verbs when they are used *alone* as main verbs.

This section first discusses subject complements. The second topic is object complements.

Subject Complements

Linking verbs always need a subject complement to make a complete statement. **Subject complements** rename or describe the subject. If you say, *My supervisor is,* you have not made a complete statement. Subject complements are *always* predicate nouns, predicate pronouns, or predicate adjectives.

A **predicate noun** or **predicate pronoun** completes the linking verb by renaming or identifying the subject. A **predicate adjective** completes the linking verb by describing or modifying the subject. The word that answers *who?* or *what?* after the linking verb is the subject complement—a predicate noun, predicate pronoun, or predicate adjective.

Note how subject complements (*SC*) are used in the following sentences:

Predicate Noun (PN): Rachel <u>Billingsley</u> | <u>is</u> the next shop manager. $\overset{SC,PN}{}$ (The subject, *Billingsley*, is renamed by its complement, *manager*, a predicate noun. You could say, Billingsley, is *who*? Billingsley is the manager.)

Predicate Pronoun (PP): The two <u>managers</u> | <u>are</u> they. $\overset{SC,PP}{}$ (The word *they* is a predicate pronoun that identifies the subject *managers*. *They* are the *managers.*)

Predicate Adjective (PA): Her <u>report</u> | <u>was</u> highly technical. $\overset{SC,PA}{}$ (The subject *report* is modified by its subject complement *technical*, a predicate adjective. You could say, Report is *what*? Report is technical).

Compound Subject Complement.   A **compound subject complement** is a subject complement that contains two or more predicate nouns, pronouns, or adjectives. Note the following sentence:

All new <u>employees</u> | <u>were</u> optimistic and energetic.

The compound subject complements *optimistic* and *energetic* are predicate adjectives that describe the subject *employees*.

Object Complements

A direct object can be followed by a word that renames or describes it. This is called an **object complement**. The object complement may be a predicate noun (*PN*) or predicate adjective (*PA*), but not a predicate pronoun (*PP*). Object complements help to complete the meaning of transitive verbs such as *call, consider, find, elect, make,* and *name*. The object complements in the sentences that follow are labeled *OC*. Following this label is a *PN* or *PA*, indicating whether the object complement is a predicate noun or predicate adjective.

The <u>court</u> | <u>appointed</u> Jane *guardian*. (*Guardian* renames *Jane.*)

<u>Marta</u> | <u>painted</u> the old chair *blue*. (*Blue* describes *chair.*)

Compound Object Complement.   A **compound object complement** is an object complement that contains two or more predicate nouns or predicate adjectives. In the sentence that follows, *long* and *dull* are predicate adjectives that modify the direct object *books*.

<u>Management</u> | <u>found</u> the books long and dull.

Object complements are not needed in most sentences. However, their use can add interest and color to your writing. They are a major sentence part so you should be able to recognize them.

YOU TRY IT

1. For each of the following sentences, underline the simple subject once and the simple predicate twice.

Health Record: The patient's name is Hector Moreno. Mr. Moreno sees his physician regularly. A prolonged and severe case of influenza was his only reported illness during the past ten years. His family physician, Dr. Susan Yee, recently reported a slight hearing loss for the patient. She recommended a review of Mr. Moreno's hearing by a specialist.

2. For each of the following sentences, underline the complete subject once and the complete predicate twice.

Report of Mr. Moreno's Hearing Test: The tests of Hector Moreno's hearing indicate a slight hearing loss in the left ear. The loss probably resulted from years of operating very noisy machinery. The patient's auditory discrimination falls at the low end of the normal range. I recommend no hearing devices for this patient at the present time.

3. For each of the following sentences, write DO above each direct object and IO above each indirect object.

Letter to Mr. Moreno: Dear Mr. Moreno: A specialist examined your hearing and found a slight impairment in the left ear. Your years of working in a noisy environment probably caused this impairment. We will send you a reminder in one year to have your hearing checked again. Please call Dr. Yee if you experience any further problems.

4. For each of the following sentences, write SC above each subject complement. Then write PN, PP, or PA to indicate whether the complement is a predicate noun, predicate pronoun, or predicate adjective. Write OC above each object complement. Then write PN or PA to indicate whether the object complement is a predicate noun or predicate adjective.

Letter to Dr. Yee: Dear Dr. Yee: Last year's tests of my hearing were thorough and efficient. The technician who conducted all the tests classified my hearing "low normal." Now, however, my hearing loss is worse. I am an older person, Dr. Yee, and I worry that my hearing will disappear entirely. I would like to schedule an appointment for another test.

Answers: **1.** *Simple subjects:* name, Mr. Moreno, case, physician, She; *Simple predicates:* is, sees, was, reported, recommended. **2.** *Complete subjects:* The tests of Hector Moreno's hearing, The loss, The patient's auditory discrimination, I; *Complete predicates:* indicate a slight hearing loss in the left ear, probably resulted from years of operating very noisy machinery, falls at the low end of the normal range, recommend no hearing devices for this patient at the present time. **3.** *DO:* hearing, impairment, impairment, reminder, Dr. Yee; *IO:* you. **4.** *SC/PA:* thorough, efficient, worse; *OC/PA:* "low normal"; *SC/PN:* person.

LEARNING UNIT 4-2
IDENTIFYING CLAUSES AND PHRASES

So far in this chapter, you have concentrated on grouping words into sentences. Within sentences, words can be grouped into *clauses* and *phrases*. You will learn about these two groups in this learning unit.

CLAUSES

A **clause** is a group of related words that contains both a verb and its subject. The two types of clauses are independent clauses and dependent clauses.

An **independent clause** (also called a *main clause*) can stand alone as a simple sentence because it contains a complete thought. A **dependent clause** (also called a *subordinate clause*) does not contain a complete thought and cannot stand alone because it depends on another clause for its meaning. In the examples below, the *independent* clauses are also simple sentences. The *dependent* clauses have a subject and verb but not a complete thought.

┌────── Independent clause ──────┐ ┌────────── Dependent clause ──────────┐
All the employees worked *as if their supervisor were in the office.*

The dependent clause in this sentence is an adverb clause that modifies the intransitive complete verb *worked* and answers the question, *How* did the employers work?

┌────────── Independent clause ──────────┐ ┌────── Dependent clause ──────┐
Our supervisor has the new computer *that we ordered a month ago.*

The dependent clause in this sentence is an adjective clause that modifies the noun *computer* and answers the question, *Which* new computer?

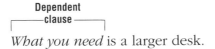

What you need is a larger desk.

The dependent clause in this sentence is a noun clause that functions as the subject of the intransitive linking verb *is*.

Each dependent clause functions as a *single* part of speech. You can see from the above examples of dependent clauses that they can function as adverbs (adverb clauses), adjectives (adjective clauses), or nouns (noun clauses). The first word of a clause is often a clue to the function of the clause. These three types of dependent clauses are discussed in later chapters.

PHRASES

A **phrase** is a group of related words that does *not* contain both a subject and a verb. Each phrase functions as a single part of speech, that is, as a noun, verb, adjective, or adverb.

In this discussion on phrases, we concentrate on the two most common phrases: prepositional phrases and verb phrases. Later chapters discuss appositive phrases, participle phrases, gerund phrases, and infinitive phrases.

Prepositional Phrases

A *prepositional phrase*, you recall from Chapter 3, begins with a preposition and ends with an object—a noun or pronoun. Adjectives often appear between the preposition and the object as shown in the following prepositional phrases:

in a *large* box	on the *green* chair	of *building* standards
by the *new* owner	before the *old* bridge	after the *long* day

Prepositional phrases function as adjectives and adverbs. To decide if a prepositional phrase functions as an adjective, ask if the phrase answers the questions *how many? what kind?* or *which one?* about a noun or pronoun. If the prepositional phrase doesn't answer any of these questions, it must function as an adverb and answer the questions *how? to what extent? where?* or *when?*

Note how the prepositional phrases function in the following two sentences:

> ┌────── *prep. phrase* ──────┐
>
> Some <u>members</u> of the accounting profession | <u>have</u> separate insurance programs.

> ┌────── *prep. phrase* ──────┐
>
> New <u>employees</u> | <u>receive</u> no benefits during their trial period.

In the first sentence, the prepositional phrase *of the accounting profession* functions as an adjective modifying the noun *members* and answering the question, *Which* members? In the second sentence, the prepositional phrase *during their trial period* functions as an adverb modifying the transitive verb *receive* and answering the question, *When* do employees receive no benefits?

The preposition in a prepositional phrase shows a relationship between the object (a noun or pronoun) in the phrase and another word in the sentence. In the first sentence above, the preposition *of* relates the noun in the prepositional phrase (*profession*) to the noun *members*. In the second sentence above, the preposition *during* relates the noun in the prepositional phrase (*period*) to the verb *receive*. Some prepositions also function as other parts of speech. From Chapter 3, you know that if a word is used as a preposition, you will find a relationship between the object that follows the preposition and another word in the sentence. Here are two additional examples:

> ┌──── *prep. phrase* ────┐
>
> The <u>manager</u> of the department | <u>was</u> late.

The prepositional phrase modifies *manager* and functions as an adjective. The preposition *of* relates *department* to *manager.*

┌─ *prep.* ─┐
└ *phrase* ┘

<u>She</u> | <u>completes</u> her work in a day.

The prepositional phrase functions as an adverb that modifies the verb *completes*. The preposition *in* relates *day* to *completes*.

Verb Phrases

As explained in Chapter 3, verb phrases consist of main verbs and helping (auxiliary) verbs. Main verbs can express action or state of being without any help from other words. Helping verbs can change the time (tense) of main verbs; they precede main verbs.

The English language has 23 helping verbs. Nine of these verbs are always helping verbs (*can, could, may, might, must, shall, should, will,* and *would*). The verbs *am, are, is, was, were, be, being, been, have, has, had, do, does,* and *did* are either helping verbs or main verbs. Helping verbs are often combined in verb phrases such as *have been* or *should have*.

Following are some examples of verb phrases:

are made	could have gone	might leave	should read
can go	had planned	must type	will run
could talk	may work	shall find	would sit

The parts of verb phrases may be separated by other words, such as the simple subject or words like *not* or *almost*.

<u>Did</u> <u>you</u> <u>find</u> her?

The <u>president</u> | <u>did</u> not <u>authorize</u> the purchase.

<u>Jonah</u> | <u>has</u> almost <u>finished</u> the report.

WRITING TIP

An understanding of how clauses and phrases function in a sentence is extremely important to business writers. When you use clauses and phrases to modify words, place them so the meaning is perfectly clear. In general, do not place the modifier too far from the word modified. Try to spot the weakness in the following sentence:

Beth met a friend just this morning who wants her old car.

Because the dependent clause *who wants her old car* modifies the word *friend*, the sentence should read: *Just this morning, Beth met a friend who wants her old car.* In this wording, *friend* is modified by the clause immediately following.

This chapter concludes with a reminder that one of the most serious errors you can make in writing is to write a sentence fragment. **Sentence fragments** are dependent clauses and phrases written as complete sentences. To avoid sentence fragments, be sure the word group has a subject, a verb, and makes sense. Also look for a dependent clause that begins with a subordinating conjunction. A final test that can be used to detect a sentence fragment is to read the sentence aloud. You will immediately hear that the fragment does not make sense.

FRAGMENT: Looked for the missing report. (Subject is missing.)
REVISED: Alvin looked for the missing report.
FRAGMENT: The company accountant received a raise. Because she presented excellent reports at the board meeting. (The dependent clause does not make sense standing alone. It must be attached to an independent clause.)
REVISED: The company accountant received a raise because she presented excellent reports at the board meeting.

YOU TRY IT

1. Underline the independent clause in each of the following sentences once. Underline the dependent clauses twice.

Text from Advertisement for Value-Plus Drugstores

Value-Plus Drugs provides the best service that you can find today. Our pharmacists are the experts that you can trust. The next time you need a prescription, call Value-Plus.

2. Underline the verb phrases and the prepositional phrases in each of the following sentences. Write P above the prepositional phrases and V above the verb phrases.

Notice to Customers

The advertised A-plus multi-vitamin special for this week has been sold out. In its place, we are offering the same size bottle of MegaVites for the special A-plus vitamin price. If you would prefer to wait for a new shipment of A-plus multi-vitamins, please ask the store clerk for a rain check.

Answers: 1. *Independent clauses:* Value-Plus Drugs provides the best service, Our pharmacists are the experts, call Value-Plus; *Dependent clauses:* that you can find today, that you can trust, The next time you need a prescription. 2. *Prepositional phrases:* for this week, In its place, of MegaVites, for the special A-plus vitamin price, for a new shipment, of A-plus multi-vitamins, for a rain check; *Verb phrases:* has been sold out, are offering, would prefer.

CHAPTER SUMMARY: A QUICK REFERENCE GUIDE

PAGE	TOPIC	KEY POINTS	EXAMPLES
62	**Sentence**	A group of related words containing a complete thought and at least one subject and one predicate. The subject tells who or what is being discussed; the predicate says something about the subject.	Every person in the room read the long report.

PAGE	TOPIC	KEY POINTS	EXAMPLES
62	**Simple subject**	A noun(s), pronoun(s), or group of words acting as a noun; it has no modifiers. The simple subject is found by locating the verb and asking *who?* or *what?* before the verb. A compound subject contains two or more simple subjects. A simple subject can also be a complete subject.	In the above sentence, *person* is the simple subject.
63	**Complete subject**	The simple subject and its modifiers.	*Every person in the room* is the complete subject of the sentence above.
63	**Simple predicate**	A verb or verb phrase that tells something about the action or state of being of the simple subject. A compound predicate contains two or more simple predicates. A simple predicate can also be a complete predicate.	In the above sentence, *read* is the simple predicate.
63	**Complete predicate**	The verb (simple predicate) and any words that complete its meaning or modify it.	*read the long report* is the complete predicate of the sentence above.
64	**Finding simple subject and simple predicate**	1. Find the verb by putting the word *yesterday* or *tomorrow* before the sentence. The word that changes with the change in time is the verb (simple predicate). 2. Insert *who?* or *what?* before the verb. The answer is the simple subject (noun or pronoun).	Gretchen *gives* her old clothes to charity. Yesterday, Gretchen *gave* her old clothes to charity. *Gives* changed to *gave*, so it is the verb. *Who* gives? Gretchen. Simple subject is *Gretchen.*
65	**Direct object**	A noun or pronoun that receives the action of a transitive verb. Find the direct object by stating the subject and the verb and asking *what?* or *whom?* Sentences may have compound direct objects.	James *hit* the *champ* in the jaw. *Champ* is the direct object. *In the jaw* is a prepositional phrase.
67	**Indirect object**	A sentence must have a direct object to have an indirect object. Indirect objects: 1. Appear between the verb and the direct object 2. Are nouns or pronouns that answer *to whom?* or *for whom?* about the direct object 3. Can be expanded into prepositional phrases 4. never follow the preposition *to* or *for.* Sentences can have compound indirect objects.	Jane sold *John* her old computer. *John* is the indirect object. Jane sold her old computer *to John. To John* is the prepositional phrase.

PAGE	TOPIC	KEY POINTS	EXAMPLES
68	**Subject complement**	A predicate noun, pronoun, or adjective that follows a linking verb and answers *whom?* or *what?* about the subject. A predicate noun or pronoun renames the subject. A predicate adjective modifies or defines the subject. A compound subject complement contains two or more predicate nouns, pronouns, or adjectives.	Harold was the only good *pitcher* on the team. *Pitcher* is a predicate noun that renames the subject.
69	**Object complement**	Follows a direct object and renames (predicate noun) or describes (predicate adjective) the direct object. Sentences can have compound object complements.	The vice president appointed Suki *manager. Manager* is a predicate noun that renames the direct object.
71	**Clause**	A group of related words containing a verb and its subject. An *independent (main) clause* can stand alone as a simple sentence. A *dependent clause* cannot stand alone as a simple sentence. It depends on another clause for its meaning and functions as adjectives, adverbs, or nouns.	*Independent clause*: When I stand near the window, *I can hear the ocean.* *Dependent clause: When I stand near the window*, I can hear the ocean.
71	**Phrase**	A group of related words that do not contain both a subject and verb. It functions as nouns, verbs, adjectives, and adverbs. *Prepositional phrases* function as adjectives and adverbs. *Verb phrases* function as simple predicates. They include main verbs and helping verbs.	*Verb phrase*: The dog's collar *has been lost.* *Prepositional phrase*: The woman *in the corner* is our president.

QUALITY EDITING

Your supervisor is a nutrition adviser who has prepared some information on eating for good health. He has asked you to check the following paragraphs from his writing. Edit his writing. Look for obvious errors such as sentence fragments, inappropriate prepositions, or spelling errors.

Vitamins cannot be store by our bodies are called water-soluble. Fat-soluble vitamins stored in body tissues. Vitamin A is a fat-soluble regulates many important body functions. This vitamin aides vision and helps with healing. Milk and eggs that are good soarces of Vitamin A.

B-complex vitamins is water-soluble. Niacin is the name of a B-complex vitamin that some experts believe that lowers cholesterol. Good sources of niacin are different than the sources of vitamin A. Some niacin sources that readily available include fish, chicken, and tukrey.

APPLIED LANGUAGE

A. In each of the following sentences, underline the complete subject once and the complete predicate twice. Then write DO above each direct object and IO above each indirect object.

1. The health care assistant sent Mr. Johnson a copy of his health records.

2. He gave the new insurance company the records.

3. The new insurance company sold Mr. Johnson a health management policy.

B. Complete each of the following sentences by adding a dependent clause.

1. The X-ray technician will complete her work when _____

_____

2. The paramedics worked as if _____

_____

3. Be sure to take all of the medication that _____

_____

C. Write a half-page letter to the Just-Right Vitamin Company to tell them why you do or do not take vitamins.

WORKSHEET 4

GROUPING WORDS INTO SENTENCES

Part A. Simple Subject

In the space provided, write the simple subject of the verb shown in italics. If the sentence is worded as a command, write the word *you* to indicate that you is understood as the subject. All prepositional phrases have been underlined.

_____ 1. Some parts of the instruments *are* very delicate.

_____ 2. A number of doctors and researchers *have disagreed* with the findings.

_____ 3. Both of those drugs *are listed* in the Guide to Prescription Drugs.

_____ 4. Many health management organizations now *employ* nurse practitioners.

_____ 5. Much discussion *has been held* about the new treatments.

_____ 6. Anyone without health insurance *is taking* serious risk.

_____ 7. Most patients *must complete* a long questionnaire.

_____ 8. Laboratory technicians, like many other hospital employees, *work* different shifts.

_____ 9. Some of the drugs *are available* only in generic form.

_____ 10. Families with children *benefit* from this health plan.

_____ 11. The time spent by each member of the surgical team is *shown* on the patient's invoice.

_____ 12. Certain treatments *are not covered.*

_____ 13. Any one of these doctors *will answer* your questions.

_____ 14. A patient, by refusing to follow directions, *may endanger* his or her health.

_____ 15. The purpose of this form, of course, *is* to provide a health history.

_____ 16. Keep careful *records* of all the exercise bicycles you sell this month.

_____ 17. Certain health-care related expenses for travel, meals, and lodging *may be deducted* from income.

_____ **18.** A donation <u>to the hospital's children's fund</u> *will help fund* important research.

_____ **19.** The dental assistant *emphasized* brushing and flossing.

_____ **20.** The amount <u>of the invoice</u> *depends* <u>on the services provided</u>.

Part B. Direct Object

Each sentence below contains one direct object. Write the direct object in the space provided. Note that every verb appears in italics.

_____ **1.** The pharmacist *should have* your records.

_____ **2.** Our basic policy *includes* dental and eye care.

_____ **3.** This medical facility *has* the best service in the state.

_____ **4.** Most clients *appreciate* our prompt service.

_____ **5.** All business owners *want* profits.

_____ **6.** The home health-services coordinator *assessed* the elderly woman's physical strength.

_____ **7.** Several states *demand* basic health insurance coverage for all employees.

_____ **8.** The agent *knows* the exact rate.

_____ **9.** Her physical therapist *spent* several years in training.

_____ **10.** Clark *gave* the client a copy of the policy.

_____ **11.** He *put* the document in a safe place.

_____ **12.** Her parents *prepared* a living will.

Part C. Clauses and Phrases

Classify each of the elements below by writing the appropriate letter in the space provided.

a. phrase b. dependent clause c. independent clause or sentence

_____ **1.** Without the needed medicines

_____ **2.** After the examination

_____ **3.** She breathed

_____ **4.** According to the health insurance policy

_____ **5.** Failing to get a response

_____ **6.** The test is positive

_____ **7.** Although the treatment is expensive

_____ **8.** Since January

_____ **9.** Because she provided services

_____ **10.** When the computer lost the data

_____ **11.** The laser printer is working

_____ **12.** She listened to her patient

_____ **13.** On the third shelf

_____ **14.** If the physical therapy technician is available

_____ **15.** They may help us

_____ **16.** While using the centrifuge

_____ **17.** In about one hour

_____ **18.** Our money was wasted

_____ **19.** Before we make a decision

_____ **20.** Whenever he complains about pain

_____ **21.** Cutting very precisely

_____ **22.** Everyone cooperated

_____ **23.** To fill the prescription

_____ **24.** On November 15

_____ **25.** The diagnosis was confirmed

Part D. Complete Sentences

For each pair of items below, write the letter *a* or *b* in the space provided to indicate which one is the complete sentence.

_____ **1. a.** Sells a machine that diabetics use to check their blood sugar level.
   b. Buy four dozen thermometers.

_____ **2. a.** The purchasing clerk ordered six cases of pain relievers.
   b. Order of six cases of pain relievers placed by the purchasing clerk.

_____ **3. a.** We may purchase a new vaporizer.
   b. If the register does not recognize the bar codes on the dental floss.

_____ **4. a.** Many of the ointments in stock.
   b. Weigh yourself.

_____ **5. a.** The technician drew blood from the boy's arm.
   b. A computer program that records the prescription and prints the label.

CHAPTER 5

SENTENCE PATTERNS, SENTENCE TYPES, AND SENTENCE ANALYSIS

A sentence should read as if its author, had he held a plough instead of a pen, could have drawn a furrow deep and straight to the end.

—Henry David Thoreau

LEARNING UNIT GOALS

LEARNING UNIT 5-1: SENTENCE PATTERNS

- Explain the common sentence patterns. (pp. 86–87)
- Use these sentence patterns effectively in writing. (pp. 86–87)

LEARNING UNIT 5-2: SENTENCE TYPES

- Distinguish four sentence types—statement, question, command, and exclamatory. (pp. 88–89)
- Use these sentence types effectively in writing. (pp. 88–89)

LEARNING UNIT 5-3: SENTENCE ANALYSIS

- Recognize simple subjects, simple predicates, direct objects, indirect objects, and object complements in transitive verbs. (pp. 91–94)
- Recognize simple subjects, simple predicates, and subject complements in intransitive verbs. (pp. 94–95)

If you were asked to use an adjective that most completely describes effective business writing, what adjective would you use? Would it be concise, complete, correct, direct, or precise? All these qualities are important, of course, and they are all related to *clearness*, or *clarity*. Clarity in writing is the result of clarity in thinking. To write clearly, you must think clearly. Then you must organize your thoughts, use clear words, and choose an effective sentence structure.

This chapter continues the study of sentence structure—a study that will help you write clear business documents. Learning Unit 5-1 describes five common sentence patterns. In Learning Unit 5-2, we explain the four sentence types. Learning Unit 5-3 organizes the knowledge you learned in Chapter 4 so that you can recognize the major sentence elements. Knowing these elements will simplify the mastery of the grammar mechanics in this text.

Before you begin Learning Unit 5-1, look at Figure 5-1. It reviews how to find the simple subject, direct and indirect objects, and subject and object complements.

Figure 5-1 Finding Simple Subject, Direct and Indirect Objects, and Subject and Object Complements

1. Simple subject:
   a. Find the verb or verb phrase (the simple predicate). You can do this by adding the words *yesterday* or *tomorrow* before the sentence. The word that changes form when you change the time of the sentence is the verb.
   b. Ask *who?* or *what?* before the verb.
   c. The answer is the simple subject (a noun or pronoun).

   Lena <u>is</u> the club president. (*Who* is? Lena. *Lena* is the simple subject.)

2. Direct object:
   a. Find the verb or verb phrase. Does another word receive the action? If so, the verb is transitive and it has a direct object.
   b. Ask *who?* or *what?* before the verb to find the simple subject.
   c. To find the direct object, ask: Simple subject + verb + *what?* or *whom?*
   d. The answer is the direct object (a noun or pronoun).

   DO
   A word <u>processor</u> <u>solved</u> his problem. (*Who* or *what* solved? Processor. *Processor* is the simple subject. Processor solved *what?* Problem. *Problem* is the direct object.)

3. Indirect object:
   a. Find the direct object. (Ask: Simple subject + verb + *what?* or *whom?*)
   b. Ask *to whom?* or *for whom?* about the direct object. (Ask: Simple subject + verb + direct object + *to whom?* or *for whom?*)
   c. The answer is the indirect object (a noun or pronoun).

$$\overset{ID}{} \qquad \overset{DO}{}$$

She gave her supervisor a desk lamp. (*What* did she give? Lamp. She gave the lamp *to whom?* Supervisor. *Supervisor* is the indirect object.)

4. Subject complement:
   a. A sentence with a linking verb must have a subject complement. A linking verb connects the simple subject with a word that renames or describes it—the subject complement.
   b. The subject complement is part of the complete predicate. If the subject complement renames the subject, it is either a noun, called a predicate noun, or a pronoun, called a predicate pronoun. If the subject complement describes the subject, it is an adjective, called a predicate adjective. The labels for the predicate words are predicate noun (*PN*), predicate pronoun (*PP*), and predicate adjective (*PA*).

$$\overset{SC,PN}{} \quad \overset{\text{prep. phrase}}{\lceil \qquad \qquad \rceil}$$

Thomas is a clerk in the municipal court. (The word *clerk* is a predicate noun that renames *Thomas.*)

5. Object complement:
   a. You must have a direct object to have an object complement.
   b. The object complement follows the direct object and renames (predicate noun) or describes (predicate adjective) it.
   c. Repeat the direct object and ask what word renames or describes it.

$$\overset{DO}{} \qquad \overset{OC,PN}{}$$

The president named Stacy chairperson. (*Chairperson* renames Stacy.)

Figure 5-1 Finding Simple Subject, Direct and Indirect Objects, and Subject and Object Complements (Continued)

This chapter adds sentence labels for the three types of verbs: transitive, intransitive complete, and intransitive linking. The labels for transitive (*vt.*) and intransitive (*vi.*) are the same as in the dictionary. Since the intransitive verbs are divided into complete verbs and linking verbs, we use *vi.c.* for the complete verb and *vi.l.* for the linking verb. (**Note:** Do not confuse the term *complete verb* with the term *complete predicate*. The *intransitive complete verb* is a simple predicate that does not have an object to complete its action; the verb is complete in itself.)

When the sentence has a verb phrase, the verb label is placed above the last word in the phrase (main verb), since this word determines whether the verb phrase is transitive or intransitive. Note the following examples:

$$\overset{vt.}{} \qquad\qquad\qquad\qquad \overset{vi.c.}{}$$

Donna closed the door.   The glass broke.

$$\overset{vi.l.}{} \qquad\qquad\qquad\qquad \overset{vi.c.}{} \quad \overset{\text{prep.}}{\overset{\lceil phrase \rceil}{}}$$

Wong is a worker.   Corbin will walk with you.

Labels Used in This Chapter

Simple subject = one underline	Direct object = *DO*
Simple predicate (verb or verb phrase) = two underlines	Indirect object = *IO*
	Subject complement = *SC*
Transitive verb = *vt.*	Predicate noun = *PN*
Intransitive complete verb = *vi.c.*	Predicate adjective = *PA*
Intransitive linking verb = *vi.l.*	Object complement = *OC*

LEARNING UNIT 5-1
SENTENCE PATTERNS

The English language offers various ways to express ideas. This learning unit explains the five sentence patterns used most often in modern business writing. Three of the five patterns use the transitive verb. The other two patterns use the intransitive verb. You will notice that most of the differences in the sentence patterns occur in the predicate.

TRANSITIVE VERB SENTENCE PATTERNS

Subject–Transitive Verb–Direct Object

The transitive verb must have a direct object to complete its meaning.

        vt.   *DO*
Scotty built a beautiful table.

         vt.   *DO*
Francis prepared the agenda.

Subject–Transitive Verb–Indirect Object–Direct Object

The indirect object always comes before the direct object.

        vt.   *IO*   *DO*
The company offered Montgomery a new automobile.

       vt.   *IO*   *DO*
Randall handed the person an application.

Subject–Transitive Verb–Direct Object–Object Complement

The object complement renames (predicate noun) or describes (predicate adjective) the direct object. Since you must have a transitive verb to have a direct object, you must have a transitive verb to have an object complement.

        vt.   *DO*   *OC,PN*
The manager appointed Elaine an assistant.

         vt.   *DO*   *OC,PA*
The new customer made the store owner angry.

INTRANSITIVE VERB SENTENCE PATTERNS

Subject–Intransitive Complete Verb

The complete action of the verb in this sentence pattern makes it possible to have short sentences of two or three words. These short sentences are not used very often in business writing, but they add variety and interest to your writing. Note that the sentences contain subjects that are both simple and complete and predicates that are likewise both simple and complete.

vi.c. *vi.c.*
Clocks tick.   Sales end.

vi.c *vi.c.*
Employees work.   Students study.

Subject–Intransitive Linking Verb–Subject Complement

Intransitive linking verbs must be followed by subject complements that either rename (predicate noun, or *PN*) or describe (predicate adjective, or *PA*) the subject.

vi.l. *SC,PN*
Those two companies are our principal competitors.

vi.l. *SC,PA* ⎾*prep. phrase*⏋
All the computer programmers are sensitive to your needs.

Y O U   T R Y   I T

A. Each of the sentences below uses a different sentence pattern. Match the appropriate pattern to its sentence. Write the letter of the pattern in the space provided. **Note:** An example has been provided with each pattern.

a. Subject—Intransitive Complete Verb
The companies negotiated.

b. Subject—Intransitive Linking Verb—Subject Complement
The managers were knowledgeable.

c. Subject—Transitive Verb—Direct Object
The printer's client checked the page proofs.

d. Subject—Transitive Verb—Indirect Object—Direct Object
Eric sent the bindery the book pages.

e. Subject—Transitive Verb—Direct Object—Object Complement
The finished books made the client happy.

_____ **1.** The press supervisor is very experienced.

_____ **2.** The customer laughed.

_____ **3.** Julia gave the customer the cost estimate.

_____ **4.** The artist finished the cover design.

_____ **5.** The new manager considered the current procedures inefficient.

LEARNING UNIT 5-2
SENTENCE TYPES

When you speak and write, you usually do not consciously decide what sentence type to use. Instead, you automatically make a statement, ask a question, give a command, or express sudden or strong emotion. Your voice and body language often indicate your sentence type when you speak. What about when you write? Let's look at these four sentence types from a writer's viewpoint.

SENTENCES THAT MAKE STATEMENTS

Most sentences make **statements** (called **declarative sentences**) that say or assert something. Usually, subjects come *before* verbs in sentences that make statements. They end with a period.

          vt.   *DO*   ┌── *prep. phrase* ──┐
The <u>computer</u> <u>keeps</u> the sales records for sales managers.

            vi.c.   ┌── *prep. phrase* ──┐
This <u>explanation</u> <u>will be printed</u> on customer statements.

Writers sometimes place the verb before the subject for emphasis. **Note:** In both of the following sentences, the president did the running. Even though *president* appears after the verb in the second sentence, *president* is the subject.

        vi.c. ┌─*prep. phrase*─┐
Our <u>president</u> <u>ran</u> down the hall.

┌─*prep. phrase*─┐ *vi.c.*
Down the hall <u>ran</u> our <u>president</u>.

SENTENCES THAT ASK QUESTIONS

From an early age, you received much of your learning by asking questions. Questions do not assert anything; they make inquiries about facts. When writing in a conversational style, **questions** (called **interrogative sentences**) involve the reader in the writing.

In sentences that ask questions, the subject often follows the verb or is in the verb phrase. These sentences end with a question mark. In the following two questions, the subject is in the middle of the verb phrase:

vi.c.

How <u>does</u> the <u>manager</u> <u>know</u> which items must be reordered?

vi.c. ┌────── *prep. phrase* ──────┐

What <u>did</u> the repair <u>person</u> <u>say</u> about the broken wheel?

Often, questions can be expressed as statements:

vi.l.   ┌ *prep. phrase* ┐   *vi.l.*   ┌ *prep. phrase* ┐

<u>Were</u> <u>you</u> in the room?   (<u>You</u> <u>were</u> in the room.)

vi.l.   ┌── *prep. phrase* ──┐   *vi.l.* ┌── *prep. phrase* ──┐

<u>Are</u> the <u>records</u> in the drawer?   (The <u>records</u> <u>are</u> in the drawer.)

You should also recognize the sentence pattern of a question introduced with a direct object or a helping verb, as in the following examples:

DO   *vi.c.*   *vi.c.*

What color <u>did</u> the <u>customer</u> <u>choose</u>? (The <u>customer</u> <u>did choose</u> what

DO

color?)

vi.c.   *vi.c.*

<u>Does</u> the <u>supervisor</u> <u>agree</u> with you? (The <u>supervisor</u> <u>does agree</u> with you.)

SENTENCES THAT GIVE COMMANDS

Think back to your early memories of spoken commands or requests. These commands probably were orders such as *Clean your plate* or *Be quiet*. Sentences that give **commands** (called **imperative sentences**) have verbs, but the subjects are not expressed. The speaker or writer directs the command or request to somebody—*you*. These sentences end with periods or exclamation points.

vt.   *DO*   *vi.c.*

(<u>You</u>) <u>Prepare</u> a trial transcript.   (<u>You</u>) <u>Stop</u>!

vt.   *DO*   *vi.c.*

(<u>You</u>) <u>Find</u> that lost voucher.   (<u>You</u>) Don't <u>touch</u> that!

SENTENCES THAT MAKE EXCLAMATIONS

Any sentence may be spoken or written with strong emotion. A sentence that makes an **exclamation** (called **exclamatory sentences**) states a fact as an exclamation. These sentences end with an exclamation point.

vi.c.   *vi.l.*

The <u>computers</u> <u>have been stolen</u>!   Look! The hair <u>color</u> <u>is</u> beautiful.

YOU TRY IT

A. In the space provided, indicate the type of sentence by writing S for statement, Q for question, C for command, and E for exclamation. Then, add the correct punctuation mark (period, question mark, or exclamation point) at the end of each sentence.

1. We distributed the new brochure you printed to our clients last week   _____

2. When will the next one be ready   _____

3. Your work is wonderful   _____

4. Keep it up   _____

5. Our printing presses always run best when we use your superb inks   _____

6. Unfortunately, we have had some difficulty obtaining the blue ink to match our logo

color   _____

7. Can you use the enclosed sample to match our logo color   _____

8. Let us know as soon as possible   _____

B. You would like to advertise your computer repair services. Your local copying shop prints copies of brochures. Write the shop a note inquiring about their service. Try to use all four sentence types.

Answers: 1. S, period 2. Q, question mark 3. E, exclamation point 4. C, period 5. E, exclamation point 6. S, period 7. Q, question mark 8. C, period

LEARNING UNIT 5-3
SENTENCE ANALYSIS

You may wonder why analyzing a sentence is important. Did you ever decide to take apart a mechanical gadget that did not work correctly? If you know how the internal parts of the gadget are supposed to work, you will have a good chance of figuring out why the gadget is not working. However, if you have no idea how the gadget's internal parts work, your chances of fixing it are not very good. The same is true of a sentence. If you take apart a sentence that performs as it should, you will learn how to fix sentences that do not perform well. As with the gadget, the key is to understand how the parts of a sentence *should* work together.

The ability to recognize the major parts of a sentence will help you determine the correct usage of words. The major sentence parts are the *simple subject, simple predicate, direct object, indirect object, subject complement* (predicate noun, pronoun, and adjective), and *object complement* (predicate noun and adjective). Modifiers—adjectives and adverbs—are not major sentence parts and are not included in our sentence analysis. Once you learn how to find the major parts, you will know that the other words in the sentence are modifiers. (Appositives, which you will study later, are also not included in our sentence analysis because they function as modifiers.)

This learning unit divides the analysis of sentences into transitive and intransitive verb sentences. Examples are given for both types of intransitive verbs: the complete verb and the linking verb. The last topic in this unit discusses how to show compound elements in your sentence analysis.

ANALYZING TRANSITIVE VERB SENTENCES

First, look at the sentence analysis flowchart in Figure 5-2. This flowchart shows the flow of the various steps used in the analysis of sentences. Note that Steps 1, 2, and 3 are the same for analyzing transitive and intransitive verb sentences. When the sentence verb is transitive, you follow the flowchart steps to the left. When the sentence is intransitive, you follow the flowchart steps to the right.

To help you understand the transitive verb and intransitive verb flow through the chart, we have grouped the example sentences as transitive and intransitive. In your analysis, you may not know whether your verb is transitive or intransitive. When you apply the knowledge you learned in Chapters 3 and 4 and adapt the information given below for each group, you will quickly learn to determine whether a verb is transitive or intransitive. Always remember that a transitive verb must have a direct object.

Before you follow the flowchart steps, you *must* look for prepositional phrases in your sentence. If your sentence has any prepositional phrases, cross them out. This will prevent you from mistaking a modifier (prepositional phrases are always modifiers) for a major sentence part. In the example

Figure 5-2 Sentence Analysis Flowchart

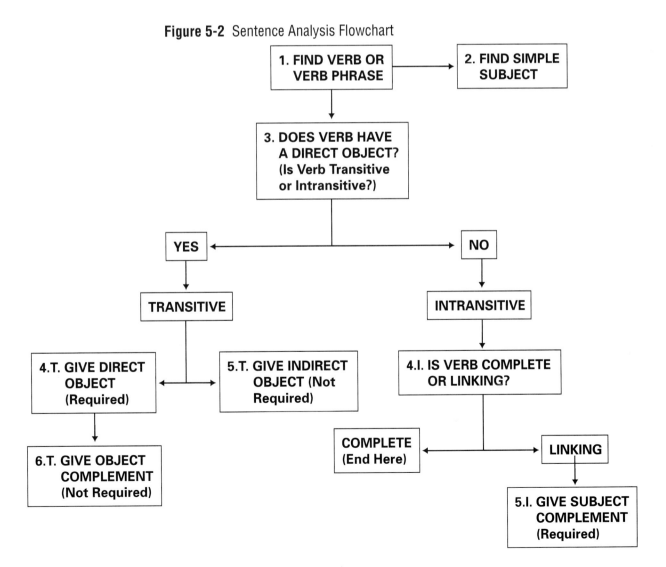

sentences, the prepositional phrases are crossed out. (To find a prepositional phrase, you recall, look for a preposition and its object, a noun or
pronoun.)

Transitive Verb Sentence 1

The members ~~of the board~~ received the message.

Step 1.   **Find the verb or verb phrase.** The verb is the word that
changes with time when you add *yesterday* or *tomorrow* before a sentence. Since the time of our example sentence is in
the past, we use *tomorrow* and say,

Tomorrow, the members *will receive* the message.

The word *received* in the example sentence changed to *will receive.*
Received is the verb.

Step 2.   **Find the simple subject.** Insert *who?* or *what?* before the
verb to find the subject. *Who* received? Members. *Members* is
the simple subject.

Step 3.   **Does the sentence have a direct object?** Repeat the simple subject and the verb and ask *what?* or *whom?* (The direct
object must be a noun or pronoun.) Members received *what?*
Members received message. This sentence has a direct object. *The verb is transitive.*

Step 4.T. What is the direct object? *Message* is the direct object.

Step 5.T. Does the sentence have an indirect object? No.

Step 6.T. Does the sentence have an object complement? No.

After we have analyzed each sentence, we will label it.

┌─*prep. phrase*─┐   *vt.*   *DO*
The <u>members</u> of the board <u>received</u> the message.

Transitive Verb Sentence 2

The cashier ~~by the window~~ gave Kora the money.

Step 1.   **Find the verb or verb phrase.** The word that changes with
time when you add the word *tomorrow* before the sentence
is *gave*, the verb.

Step 2.   **Find the simple subject.** *Who* gave? Cashier. *Cashier* is the
simple subject.

Step 3.   **Does the sentence have a direct object?** Cashier gave
what? Cashier gave money. This sentence has a direct object.
The verb is transitive.

Step 4.T. What is the direct object? *Money* is the direct object.

Step 5.T. Does the sentence have an indirect object? Ask *to whom?* or *for whom?* about the direct object. Cashier gave money *to whom?* Kora. *Kora* is the indirect object.

Step 6.T. Does the sentence have an object complement? No.

$$\overbrace{\text{prep. phrase}}\quad vt.\quad IO\quad DO$$

The <u>cashier</u> by the window <u>gave</u> Kora the money.

Transitive Verb Sentence 3

The salesperson ~~at the door~~ made Joseph angry.

Step 1. **Find the verb or verb phrase.** Adding *tomorrow* before this sentence changes *made* to *will make. Made* is the verb.

Step 2. **Find the simple subject.** *Who* made? Salesperson. *Salesperson* is the simple subject.

Step 3. **Does the sentence have a direct object?** Salesperson made *whom?* Salesperson made Joseph. This sentence has a direct object. *The verb is transitive.*

Step 4.T. What is the direct object? *Joseph* is the direct object.

Step 5.T. Does the sentence have an indirect object. No.

Step 6.T. Does the sentence have an object complement? The object complement follows a direct object and renames (predicate noun) or describes (predicate adjective) the direct object. Joseph is *angry. Angry* is the object complement. Since *angry* describes Joseph, it is a predicate adjective.

$$\overbrace{\text{prep. phrase}}\quad vt.\quad DO\quad OC,PA$$

The <u>salesperson</u> at the door <u>made</u> Joseph angry.

Transitive Verb Sentence 4

What room ~~of furniture~~ did you sell?

To analyze a question, turn it around into a statement if possible.

You did sell what room ~~of furniture~~.

Step 1. **Find the verb or verb phrase.** Adding *tomorrow* before this sentence changes *did sell* to *will sell. Did sell* is the verb.

Step 2. **Find the simple subject.** *Who* did sell? You. *You* is the simple subject.

Step 3.   Does the sentence have a direct object? You did sell *what*? You did sell room. This sentence has a direct object. *The verb is transitive.*

Step 4.T. What is the direct object? *Room* is the direct object.

Step 5.T. Does the sentence have an indirect object? No.

Step 6.T. Does the sentence have an object complement? No.

$$\overset{DO}{}\quad \overset{\ulcorner prep.\ phrase \urcorner}{}\quad \overset{vt.}{}$$
What room of furniture <u>did</u> <u>you</u> <u>sell</u>?

ANALYZING INTRANSITIVE VERB SENTENCES

Intransitive Complete Verb Sentence

Now we go to the right of the sentence analysis flowchart (Figure 5-2). Steps 1, 2, and 3 are the same as the transitive verb steps. You have an intransitive verb when (1) the action is complete with the verb or (2) the verb is a linking verb. We begin with an intransitive complete verb sentence.

The employee walked slowly ~~to the door~~.

Step 1.   Find the verb or verb phrase. When you add *tomorrow* before this sentence, *walked* changes to *will walk*. *Walked* is the verb.

Step 2.   Find the simple subject. *Who* walked? Employee. *Employee* is the simple subject.

Step 3.   Does the sentence have a direct object? Employee walked *what*? or *whom*? You are looking for a noun or pronoun that answers the *what*? or *whom*? The sentence does not answer this question. We do not have a direct object. *The verb is intransitive.*

Step 4.I. Does the sentence have a complete or linking verb? Employee walked *slowly*. This tells *how* the employee walked. *Slowly* is a modifier (adverb). In this sentence, *walked* is an intransitive complete verb. We could stop the sentence after *walked*, and it would make sense. The analysis of this sentence is finished. If we had a linking verb, we would be looking for a subject complement—a noun, pronoun, or adjective in the predicate that renamed or described the simple subject.

$$\overset{vi.c.}{}\qquad\qquad \overset{\ulcorner prep.\ phrase \urcorner}{}$$
The <u>employee</u> <u>walked</u> slowly to the door.

Intransitive Linking Verb Sentence

The director ~~of human resources~~ is an extremely intelligent person.

Step 1. **Find the verb or verb phrase.** When you add *yesterday* before this sentence, *is* changed to *was*, a past form of the verb *to be*.

Step 2. **Find the simple subject.** *Who* is? Director. *Director* is the simple subject.

Step 3. **Does the sentence have a direct object?** We know that when the verb *to be* (*is, are, were,* etc.) stands alone, it never has a direct object. So the answer here is no. *The verb is intransitive.*

Step 4.I. **Does the sentence have a complete or linking verb?** The sentence does not have a complete verb. You cannot stop the sentence after *director is.* You know that *is* used alone is a linking verb, so this sentence has a linking verb. Now you must go to Step 5.I.

Step 5.I. **Give the subject complement.** You are looking for a noun, pronoun, or adjective in the predicate that renames or describes the director. The director is a person. *Person* is a predicate noun that renames the director. *Person* is the subject complement.

$$\overbrace{\qquad\qquad}^{prep.\ phrase}\ vi.l, \qquad\qquad\qquad\qquad SC,PN$$

The <u>director</u> of human resources <u>is</u> an extremely intelligent person.

Remember that if you have difficulty determining whether a verb is transitive or intransitive, you can check the dictionary. Some verbs can be either transitive or intransitive, depending on how they are used in a sentence. The definitions in the dictionary will give meanings, and sometimes examples, for verbs used as both transitive verbs and intransitive verbs.

ANALYZING COMPOUND ELEMENTS

Recall from Chapter 4 that simple subjects, simple predicates, direct objects, indirect objects, subject complements, and object complements can be compound, that is, contain two or more parts. The compound parts of a sentence are usually joined by the conjunctions *and*, *but*, *or*, or *nor*. When you analyze a sentence with compound parts, you include the conjunction with the parts, as we have done in the following example. This eliminates the conjunction from the sentence. As connectors of compound parts, conjunctions should not cause you any difficulty in analyzing sentences.

The job applicants and former employees waited ~~for an hour~~.

Step 1. **Find the verb or verb phrase.** When you add *tomorrow* before this sentence, *waited* changes to *will wait*. *Wait* is the verb.

Step 2. **Find the simple subject.** *Who* waited? Applicants and employees. The simple subject is compound—*applicants and employees.*

Step 3. **Does the sentence have a direct object?** Applicants and employees waited *what?* or *whom?* You are looking for a noun or pronoun that answers the *what?* or *whom?* The sentence does not answer this question. We do not have a direct object. *The verb is intransitive.*

Step 4.I. **Does the sentence have a complete or linking verb?** The verb *waited* is followed by a prepositional phrase. Prepositional phrases are always modifiers. Complete verbs are followed by modifiers. *Waited* is a complete verb. The analysis of this sentence is finished.

The job <u>applicants</u> <u>and</u> former <u>employees</u> <u>waited</u> for an hour.

You can do the same as we did above for other major sentence parts. Just remember to look for compounds when you analyze sentences.

YOU TRY IT

Apply the sentence analysis steps to the eight sentences below. In the space provided, write vt. for transitive verb, vi.c. for intransitive complete verb, and vi.l. for an intransitive linking verb. For all sentences, cross out prepositional phrases, underline the simple subject once and the simple predicate twice. For transitive verb sentences, label the direct object DO and indirect object IO. For intransitive linking verb sentences, label the subject complement SC,PN (predicate noun) or SC,PA (predicate adjective).

_____ **1.** The designer of this brochure gave us a very colorful illustration.

_____ **2.** The committee has decided!

_____ **3.** The original pages are electronic instead of paper.

_____ **4.** The writer of the text provided us a lively story.

_____ **5.** [You] Please print us 5,000 copies of the pamphlet in full color.

_____ **6.** Jennifer arrived late.

_____ **7.** This new press is the latest product in printing technology.

_____ **8.** We are amazed at the machine's speed!

Answers: 1. vt. The designer of this brochure gave us a very colorful illustration. 2. vi.c. The committee has decided! 3. vi.l. The original pages are electronic instead of paper. 4. vt. The writer of the text provided us a lively story. 5. vt. [You] Please print us 5,000 copies of the pamphlet in full color. 6. vi.c. Jennifer arrived late. 7. vi.l. This new press is the latest product in printing technology. 8. We are amazed at the machine's speed!

CHAPTER SUMMARY: A QUICK REFERENCE GUIDE

PAGE	TOPIC	KEY POINTS	EXAMPLES
86	**Sentence patterns**	Transitive verb sentence patterns: **1.** Subject—transitive verb—direct object **2.** Subject—transitive verb—indirect object—direct object **3.** Subject—transitive verb—direct object—object complement Intransitive verb sentence patterns: **1.** Subject—intransitive complete verb **2.** Subject—intransitive linking verb—subject complement	$\quad\quad$ *vt.* $\quad\quad$ *DO* Sharon <u>found</u> the ledger. $\quad\quad$ *vt.* $\quad\quad$ *IO* Bob <u>gave</u> the employee a new *DO* job. $\quad\quad\quad$ *vt.* $\quad\quad$ *DO* The <u>president</u> <u>appointed</u> Loren *OC, PN* manager. $\quad\quad$ *vi.c.* Chris <u>sang</u>. $\quad\quad\quad$ *vi.l.* That <u>manager</u> <u>is</u> a wonderful *SC, PN* person.
88	**Sentence types**	**Statements:** Say or assert something. Usually the subject comes before the verb. These sentences end with a period.	$\quad\quad\quad$ *vt.* $\quad$ *DO* All the <u>workers</u> <u>went</u> home.
		Questions: Make inquiries. The subject comes after the verb or in the verb phrase. These sentences end with a question mark. **Commands:** Are directed to an implied *you.*. They have verbs but the subjects are not expressed. These sentences end with a period or exclamation point. **Exclamations:** State a fact as an exclamation to add emotion to a sentence. They end with an exclamation point.	$\quad\quad$ *vt.* $\quad\quad$ *DO* <u>Did</u> <u>you</u> <u>see</u> all those errors? *vt.* *DO* <u>Go</u> <u>home</u>! Oh! What a wonderful idea.
91	**Sentence Analysis**	Steps are given in the text.	

QUALITY EDITING

You work at ThriftyPrints, a print shop that specializes in business cards and letterhead. Your boss has prepared the draft of a new flyer and has asked you to check the following paragraphs. Edit the draft advertisement. Look for obvious errors such as incorrect end-of-sentence punctuation and misspelled words.

Business cards and letterheads are our specialty? ThriftyPrints aims to give you to the best possible service that availalbe today. We will help you design a logo and select the collors that will makes your business cards and letters distinctive.

Their's more. We guarantee that your job that will be done in three days And, ThriftyPrints will do all this for the lowest prices that ever encountered?

Bring your ideas for your new cards and letterheads to ThriftyPrints today. When use the coupon that shown below, you will receive 10 percent off your entire order

APPLIED LANGUAGE

A. Complete the following sentences by adding a dependent clause.

1. Pay the invoice for the paper when _____

_____

2. The person who receives the reward will be whoever _____

_____

3. We specified the ink that _____

_____

B. Complete the following sentences by adding a subject complement to each. Indicate whether the subject complement is a predicate noun or a predicate adjective by writing PN or PA above it.

1. The latest technology in printing is _____

_____

2. The customer will be _____

_____

3. The document is a(n) _____

_____

C. Pretend that you work for Simpson Children's Bookstore. Write a half-page letter to the Personalized Greeting Card Company to tell them why your customers do or do not like their new line of greeting cards for children. Try to use all four sentence types: statement, question, command, and exclamation.

SENTENCE PATTERNS AND TYPES; SENTENCE ANALYSIS

Part A. Sentence Patterns

Each of the sentences below uses a different sentence pattern. Write the letter of the appropriate pattern in the space provided next to each sentence.

a. Subject—Intransitive Complete Verb
b. Subject—Intransitive Linking Verb—Subject Complement
c. Subject—Transitive Verb—Direct Object
d. Subject—Transitive Verb—Indirect Object—Direct Object
e. Subject—Transitive Verb—Direct Object—Object Complement

_____ **1.** The office manager misplaced the purchase order.

_____ **2.** The account executive is very productive.

_____ **3.** [You] Tell me the amount of the invoice.

_____ **4.** The supervisor considered Terry's performance average.

_____ **5.** Janice drove.

_____ **6.** The client is not satisfied.

_____ **7.** Mr. Lee gave the cameraperson the new artwork.

_____ **8.** They labeled the machine nonoperational.

_____ **9.** Ray ordered 500 business cards.

_____ **10.** The president is charismatic.

Part B. Sentence Types

Add the appropriate punctuation to each of the following sentences. Then, in the space provided, write S, Q, C, or E to indicate whether the sentence is a statement, question, command, or exclamation.

_____ **1.** Make certain that the printer agrees to provide three-day service

_____ **2.** What a superb layout

_____ **3.** Does your ink supplier produce metallic inks

_____ **4.** We plan to print our annual report on recycled paper

———— **5.** Will you hire at least one female manager

———— **6.** Hooray for the new page layout program

———— **7.** Will we need additional paper to complete the job

———— **8.** Were the reports delivered in time for the meeting

———— **9.** You should learn about creating electronic pages

———— **10.** What a wonderful job this team has accomplished

———— **11.** Try a little harder

———— **12.** When did you acquire the new image setter

———— **13.** Use the paper that has the visible fibers

———— **14.** We created the report with transparent overlays

———— **15.** The graphic artist has not sent us the new designs

———— **16.** Rats

———— **17.** When we recycle old newspapers, what happens to the ink that comes out

———— **18.** Be sure to use a large font so that our customers can easily read the information

———— **19.** Get three estimates for that print job

———— **20.** Jeff Gray will be the new account executive for ThriftyPrints

Part C. Sentence Analysis

Apply the sentence analysis steps to the eight sentences below. In the space provided, write vt. for a transitive verb, vi.c. for an intransitive complete verb, and vi.l. for an intransitive linking verb. For all sentences, cross out prepositional phrases, underline the simple subject once and the simple predicate twice. For transitive verb sentences, label the direct object DO and indirect object IO. For intransitive linking verb sentences, label the subject complement SC,PN (predicate noun) or SC,PA (predicate adjective).

———— **1.** She gave the payroll department the records.

———— **2.** The noise from the old printing press gave Jim a headache.

———— **3.** Carlos practiced all day.

———— **4.** The president's speech motivated the employees.

———— **5.** The negotiating teams of both firms agreed.

———— **6.** We sent the printer the master copy of the report.

———— **7.** The new manager of accounting is a Harvard graduate.

———— **8.** The final cost of each card was enormous.

SECTION 2

WORDS THAT NAME

CHAPTER 6

NOUNS: CLASSES, FUNCTIONS, AND PROPERTIES

Knowing what we do about names and the power they command we can surmise that nouns are important. They are, in fact, the most important. . . .Of all parts of speech, only nouns are independent.

—John Fairfax and John Moat, *The Way to Write*

LEARNING UNIT GOALS

LEARNING UNIT 6-1: CLASSIFYING NOUNS

- Recognize proper nouns and three types of common nouns. (pp. 105–7)

LEARNING UNIT 6-2: NOUN FUNCTIONS

- Recognize nouns functioning as major sentence elements. (pp. 108–10)
- Recognize nouns as objects of prepositions, appositives, modifiers, nouns of direct address, and clauses. (pp. 110–13)

LEARNING UNIT 6-3: NOUN PROPERTIES

- Explain nominative, objective, and possessive case as a noun property. (p. 114)
- Understand person and number as noun properties. (pp. 114–15)
- Define gender; recognize gender bias in your writing. (pp. 115–16)

Every word in the English language expresses some idea. Verbs express the idea of action or existence. Adjectives and adverbs express the idea of modifying. Prepositions and conjunctions express the idea of connecting. The noun alone, however, *names* its idea. Nouns name persons, places, things, activities, qualities, and concepts.

As stated in the chapter quote, nouns are a unique part of speech because they are the only independent part of speech. Other parts of speech, either directly or indirectly, depend on nouns for their existence. Nouns, however, depend on no other part of speech; they stand alone.

In Learning Unit 6-1, we discuss the classes of nouns. Learning Unit 6-2 lists the functions of nouns, and Learning Unit 6-3 lists the attributes of nouns.

LEARNING UNIT 6-1
CLASSIFYING NOUNS

All nouns are classified into two basic types: proper nouns and common nouns. Most people recognize proper nouns. The different groups of common nouns can be more difficult to identify. This learning unit will help you understand three different groups of common nouns.

PROPER NOUNS

The word *proper* comes from a Latin word meaning "one's own." **Proper nouns** name *specific* persons, places, things, activities, qualities, and concepts that are important enough to have a separate name. These nouns always begin with a capital letter.

New York	Purdue University	Fourth of July	England
Kathy Ramos	St. Paul's Cathedral	January	Honeywell

COMMON NOUNS

Common nouns name a general class of persons, places, things, activities, and concepts. A common noun is not capitalized. We discuss three groups of common nouns in this section: concrete nouns, abstract nouns, and collective nouns.

When proper nouns are used extensively, sometimes they are treated as common nouns and not capitalized. Following are examples of proper nouns that have become common nouns:

French fries, french fries	India ink, india ink
Venetian blinds, venetian blinds	Bohemia, bohemia

If your dictionary does not use a capital letter for a word but says the word is "often cap," you can be sure that the common noun usage is generally accepted.

WRITING TIP

Remember that some nouns can be both proper and common nouns; for example, Turkey is a country and turkey is a bird.

Concrete Nouns

A **concrete noun** names anything physical—something that you can perceive through one or more of the five senses (sight, hearing, smell, taste, or touch). Concrete nouns bring pictures to your mind. For example, when a friend tells you about his or her new blue pickup truck, you mentally see a new blue pickup truck. Other examples of concrete nouns follow.

newspaper	pencil	contract	computer	desk	book
television	box	paper	file	lamp	printer

Abstract Nouns

An **abstract noun** does not name things in the physical world, that is, things perceived through one or more of the five senses. Abstract nouns name qualities, ideas, conditions, acts, or relationships that are formed in the mind and are separate from their objects. Many abstract nouns end in *ness* and *ty*.

kindness	goodness	loyalty	personality	fear
wickedness	liberty	charity	intelligence	pride

Abstract nouns often mean different things to different people. When you hear that a person has achieved success, you may think the person has a large sum of money. However, to this person, success may mean that he or she climbed a mountain.

WRITING TIP

In writing, prefer concrete nouns over abstract nouns because they are more precise and forceful.

Collective Nouns

A **collective noun** names a group or collection of persons, places, things, activities, qualities, and concepts. Recognizing collective nouns is important when you study noun and verb agreement. Collective nouns are plural in meaning. The meaning of a collective noun in a particular sentence, however, determines whether it takes a singular or plural verb. This is discussed in detail in Chapter 12. Following are some examples of collective nouns:

| committee | team | audience | jury | club | crew |
| company | class | family | group | band | school |

WRITING TIP

Although collective nouns are usually common nouns, they become proper nouns when used to name a particular group or company.
the band, the County Creek Jazz Band
the company, the Electronic Security Company

YOU TRY IT

A. Write each of the nouns in the list below under the appropriate heading—concrete, abstract, or collective.

hotel	committee	politeness	efficiency
thought	staff	manager	group
desk	captain	crew	service

Concrete: Abstract: Collective:

1. _____   5. _____   9. _____

2. _____   6. _____   10. _____

3. _____   7. _____   11. _____

4. _____   8. _____   12. _____

B. Identify the six proper nouns in the list below and write them in the spaces provided. Be sure to capitalize each proper noun correctly.

hotel riverside	television	dr. jefferson
first-class service	fig tree restaurant	fig tree
alex mettler	mrs. rivera	north dakota
reservations manager	doctor	travel agency

1. _____

2. _____

3. _____

4. _____

5. _____

6. _____

LEARNING UNIT 6-2
NOUN FUNCTIONS

Nouns are versatile words. They have several functions in sentences. When we focused on the major sentence parts in Chapters 4 and 5, you learned that some of these major sentence elements must always be nouns or pronouns. In this learning unit, we focus first on the function of nouns as major sentence elements. Then, we discuss nouns as objects of prepositions, as appositives and modifiers, as nouns of direct address, and as clauses. Pronouns, as substitutes for nouns, have the same functions as nouns.

NOUNS AS MAJOR SENTENCE ELEMENTS

The major sentence parts, you recall, are the *simple subject, simple predicate, direct object, indirect object, subject complement* (predicate noun, pronoun, and adjective), and *object complement* (predicate noun and adjective). Nouns function as subjects of verbs, as objects of verbs, and as subject and object complements.

Nouns as Subjects of Verbs

As subjects of verbs in sentences and clauses, nouns usually occur at the beginning of a sentence or clause. From Chapter 5, you know that subject nouns may also be located after the verb. Often, more than one noun is the subject of a verb.

The nouns used as subjects in the example sentences that follow are in italics. The verbs are underlined twice. The prepositional phrases are indicated so you will remember that major sentence parts are never in prepositional phrases.

Nouns as Subjects of Transitive Verbs.
Since transitive verbs must have direct objects, we have labeled the direct objects in the example sentences DO. You know from Chapters 4 and 5 that to find the subject of a verb, you insert *who?* or *what?* before the verb.

DO

The office *secretary* records all office appointments. (*Who* records? Secretary. *Secretary* is the subject of the transitive verb *records*.)

DO   ┌──prep. phrase──┐

Elena and *Samuel* sent letters to all stockholders. (*Who* sent? Elena and Samuel. *Elena and Samuel* is the compound subject of the transitive verb *sent*.)

Nouns as Subjects of Intransitive Verbs.   Intransitive verbs, you recall, are (1) complete verbs that need no other words to complete their action and (2) linking verbs, which must have subject complements. In the first two example sentences that follow, nouns are subjects of intransitive complete verbs. In the third example sentence, a noun is the subject of an intransitive linking verb.

The maintenance *engineer* <u>complained</u> constantly. (*Who* complained? Engineer. *Engineer* is the subject of the intransitive complete verb *complained*.)

$\overline{\qquad prep.\ phrase \qquad}$

The *supervisor* and the new *salesperson* <u>talked</u> about the products. (*Who* talked? Supervisor and salesperson. *Supervisor and salesperson* is the compound subject of the intransitive complete verb *talked*.)

Workers <u>are</u> more educated today. (*Who* are? Workers. *Workers* is the subject of the intransitive linking verb *are*.)

Nouns as Objects of Verbs

In Chapters 4 and 5, you learned that transitive verbs must have direct objects and they can also have indirect objects. Nouns and pronouns function as direct objects and indirect objects of transitive verbs. In the example sentences that follow, the nouns are in italics, the verbs are underlined twice, and the prepositional phrases are labeled.

Nouns as Direct Objects.   The direct object completes the action of the verb by answering *what?* or *whom?* In the following sentences, the nouns in italics function as direct objects:

Most legal offices <u>maintain</u> a court *calendar.* (Offices maintain *what?* Calendar. *Calendar* is the direct object.)

$\overline{\qquad prep.\ phrase \qquad}$

The accountant for Morse Company <u>developed</u> a *chart* of accounts. (Morse Company developed *what?* Chart. *Chart* is the direct object.)

Nouns as Indirect Objects.   To have an indirect object in a sentence, you must have a direct object. The indirect object comes after the verb and before the direct object. To find the indirect object, you ask *to whom?* or *for whom?* about the direct object. In the following sentences, the nouns functioning as indirect objects are in italics.

DO
<u>Hand</u> the *accountant* the ledger. (*To whom* did the implied subject *you* hand the ledger? Accountant. *Accountant* is the indirect object.)

DO
The store <u>offered</u> the *customer* a discount. (*To whom* did the store offer a discount? Customer. *Customer* is the indirect object.)

Nouns as Subject and Object Complements

Subject complements are predicate nouns, pronouns, and adjectives that complete the meaning of the subject. Object complements are predicate nouns and adjectives that follow direct objects. To have a subject complement, you must have an intransitive linking verb. To have an object complement, you must have a direct object.

In the example sentences that follow, the subject is underlined once, the verb is underlined twice, and the predicate noun is in italics. Prepositional phrases are also indicated.

Nouns as Subject Complements.   A predicate noun functioning as a subject complement *(SC)* renames the simple subject.

> SC
>
> The stranger was the new *salesperson.* (*Salesperson* is a predicate noun that renames the subject *stranger.*)

> SC   ┌─────prep. phrase─────┐
>
> Jacob has been the *winner* of every sales contest. (*Winner* is a predicate noun that renames the subject *Jacob.*)

> ┌──────prep. phrase──────┐   SC
>
> The future president of financial operations will be *Francisco.* (*Francisco* is a predicate noun that renames the subject *president.*)

Nouns as Object Complements.   A predicate noun functioning as an object complement (*OC*) renames the direct object.

> DO   OC
>
> The computer operators call Lee a *genius.* (*Genius* is a predicate noun that renames the direct object *Lee.*)

> DO   OC
>
> All the employees named Doris *manager.* (*Manager* is a predicate noun that renames the direct object *Doris.*)

> DO   OC
>
> The budget committee appointed Brad *spokesperson.* (*Spokesperson* is a predicate noun that renames the direct object *Brad.*)

NOUNS AS OBJECTS OF PREPOSITIONS

Prepositional phrases, you recall, begin with a preposition and end with an object that is a noun or pronoun. Modifiers can appear between the preposition and object. Prepositional phrases function as adjectives and adverbs. Because prepositional phrases are modifiers, they are not a major part of a sentence. In the following sentences, the prepositional phrases are indicated. The nouns in the prepositional phrases are in italics.

> ┌──────────prep. phrase──────────
>
> Many large offices require the services of a computerized accounting
>
> ──────┐
>
> *program.*

> ┌──prep. phrase──┐
>
> The house on the *corner* has many windows.

> ┌──prep. phrase──┐
>
> The chair behind the *desk* is oak.

NOUNS AS APPOSITIVES

An **appositive** is a noun or noun phrase that immediately follows another noun or noun phrase. The appositive renames or identifies the noun or noun phrase (appositives make you positive). Note the italicized appositives in the following sentences:

Mr. Caruso, *our present treasurer,* does not favor this proposal.

The company clerk consulted an expert, *her accountant.*

As stated in Chapter 5, since appositives rename (or modify) nouns or noun phrases that already function as a major sentence part, they do not fit in the sentence analysis flowchart.

NOUNS AS MODIFIERS

Nouns can act as adjectives and adverbs. As adjectives, nouns modify other nouns or pronouns. As adverbs, nouns modify verbs, adjectives, or other adverbs. Study the following sentences:

Effective supervisors pay attention to *company* policy. (The noun *company* functions as an adjective and answers the question *which one?* about the noun *policy.*)

The new dental assistant went *home.* (The noun *home* functions as an adverb and answers the question *where?* about the verb *went.*)

Ming's computer needs a new hard disk. (The noun *Ming*'s functions as an adjective and answers the question *which one?* about the noun *computer.*)

WRITING TIP

Writers sometimes become lazy and use a string of nouns as adjectives. This affects the clarity of the writing. A problem occurs with terms such as *city traffic control ordinances* or *college dormitory planning committee.* To avoid these awkward terms, change a noun to an adjective, use the possessive case, or change a noun to a prepositional phrase. You could say *city ordinances for traffic control* or *planning committee for college dormitories.*

NOUNS OF DIRECT ADDRESS

A **noun of direct address** occurs when the person spoken to or written to is explicitly named. Nouns used as direct address are not too common. However, you should know about them so you do not consider a noun of address as the simple subject.

When you use a noun of address, you name whomever or whatever you are speaking to in the sentence. The noun of address is set off by a comma(s). For example, note this sentence:

Mr. Jordan, where did you say you found the petty cash box? (The noun of address is *Mr. Jordan.* To find the simple subject, you ask *Who* found? The answer is *You* found. *You* is the simple subject, not *Mr. Jordan.*)

Nouns of address do not always occur at the beginning of a sentence. You can find them in the middle or at the end of the sentence.

Come here, *Joe,* and study these reports. (*You* is the implied subject of two independent clauses. The verb in the first clause is *come,* an intransitive complete verb. The verb in the second clause is *study,* a transitive verb.)

You can leave now, *Beth.*

NOUN CLAUSES

In Chapter 4, you learned to recognize independent clauses and dependent clauses. The first word of a dependent clause is often a clue to the function of the clause. **Noun clauses** are dependent clauses that usually begin with the words *that, what, which, who, whom,* and *whose.* (*Ever* can be added to most of these words, such as *whatever* and *whoever.*) Noun clauses are important in sentence analysis because they are not modifiers. Like nouns, they can function as a major sentence part—subjects, objects, or complements. (Noun clauses can also be appositives.)

Use the following steps to determine how a noun clause functions in a sentence:

1. Find the clause by looking for the clue words and a subject and verb. Include the modifiers in the clause. Noun clauses are part of the independent clause.

2. Think of the clause as a single noun. To do this, you can run the words of the clause together.

3. Use the sentence analysis flowchart to decide how the noun functions in the sentence.

Note the following example:

What this salesperson promises is always dependable.

Step 1. Find the clause. Look for the clue word, a subject, and a verb. In this sentence, the clause is *what this salesperson promises.*

Step 2. Think of the clause as a one-word noun: Whatthissalespersonpromises is always dependable.

Step 3. Use the sentence analysis flowchart steps:

  a. Find the verb or verb phrase. The verb in this sentence is the linking verb *is.*

  b. Find the subject. *Who* or *what* is? Whatthissalespersonpromises is. The noun clause functions as the subject of the sentence.

  c. Does the sentence have a direct object? No. Linking verbs do not have direct objects. *The verb is intransitive.*

    d. Does the sentence have a complete or linking verb? We know that *is* used alone is always a linking verb. The sentence has a linking verb.

    e. Give the subject complement. *Dependable* is a predicate adjective that describes the noun clause subject.

YOU TRY IT

In each of the sentences below, a noun is italicized. From the list of noun functions, determine the function the italicized noun performs in the sentence. Then, write the letter of the function in the space provided. **Note:** Use each function only once.

Noun Functions:

a. subject of transitive verb
b. subject of intransitive verb
c. direct object of verb
d. indirect object of verb
e. object of preposition

_____ **1.** The manager at the front desk gave *Ms. Carson* her mail.

_____ **2.** *Professor Zimmer* ordered his meal from room service.

_____ **3.** The guest of the hotel left in a *taxi*.

_____ **4.** The young *child* in Room 354 slept all morning.

_____ **5.** Mrs. Reynolds called her *office* from the hotel lobby.

Noun Functions:

a. subject complement
b. object complement
c. appositive
d. modifier
e. direct address

_____ **6.** The hotel named Yolanda *"Employee of the Month."*

_____ **7.** *Kevin*, please come to the front desk.

_____ **8.** Janice, the night *manager*, called the elevator repair service at midnight.

_____ **9.** Dr. Kelly asked for her *room* key.

_____ **10.** Maru Machida is the new public relations *director*.

Answers: 1. d 2. a 3. e 4. b 5. c 6. b 7. e 8. c 9. d 10. a

LEARNING UNIT 6-3
NOUN PROPERTIES

Nouns have four properties or characteristics: case, person, number, and gender. Understanding the noun properties can help you compare their change in form with that of pronouns.

CASE

From your study of sentence analysis, you know that certain sentence elements are related. For example, when a sentence has an intransitive linking verb, the sentence must have a predicate noun or predicate adjective that is related to the subject. **Case** shows how nouns and pronouns are related to other words in the sentence. We must understand case to know when to change the form of nouns and pronouns.

The English language has three cases: nominative, objective, and possessive. The **nominative case** (also called the subjective case) indicates the person or thing that is doing the action. Subjects and predicate nouns are in the nominative case. The **objective case** indicates the person or thing receiving the action of the verb. Direct or indirect objects of verbs, objects of prepositions, and object complements are in the objective case. The **possessive case** indicates ownership or a relationship.

Nouns usually do not cause case problems. They only change their form in the possessive case. They do not change their form in the nominative or objective case. Note the following sentences:

The *door* to the safe is open. (Since *door* is the subject, it is in the nominative case.)

James closed the *door* to the safe. (Since *door* is a direct object, it is in the objective case.)

Although in the first sentence *door* is in the nominative case and in the second sentence *door* is in the objective case, the word *door* has not changed its form. *Door* is still *door*.

Nouns in the possessive case, however, change their form. The apostrophe is used to show noun possession. Note the following examples:

From	To
the book of the professor	the professor's book
the desk of the president	the president's desk
the computer of the secretary	the secretary's computer

Chapter 7 discusses singular and plural possessive nouns.

Unlike nouns, some pronouns change their form in the nominative and objective cases. Pronouns also change form in the possessive case.

PERSON

Grammar has three persons: the person speaking (first person), the person spoken to (second person), and the person spoken of (third person). **Person** is a property of nouns, pronouns, and verbs. Nouns *do not* change form for person. The person of a noun is shown by its relationship to other words in the sentence. The only words that change form for person are personal pronouns and verbs.

NUMBER

In the English language, **number** is the grammatical property of nouns, pronouns, and verbs that indicates whether something is singular or plural.

Nouns, pronouns, and verbs usually change form to indicate singular or plural. Since these changes can cause difficulty for writers, they are discussed in detail in later chapters.

GENDER

The **gender** of nouns and pronouns is determined by sex. The masculine gender denotes the male sex, feminine gender denotes the female sex, and the neuter gender denotes things of no sex, that is, neither masculine or feminine.

In the past, it was common practice to use the masculine gender when the gender of a group was unknown or included both males and females. Today, this is no longer true. Business writers avoid sexual bias by (1) using plural subjects and pronouns, (2) rewording a sentence, or (3) referring to both males and females. In the sentences that follow, first we give an example of the traditional method of using the masculine pronoun to denote a common gender. Then, we give examples of the three alternatives.

Traditional:	A message from the president said that an employee should park his car in the rear parking lot.
Plural Subjects and Pronouns:	A message from the president said that all employees should park their cars in the rear parking lot.
Reword Sentence:	A message from the president said that all employees should use the rear parking lot.
Use Both Sexes:	A message from the president said that an employee should park his or her car in the rear parking lot.

The last method—using his and her—can become awkward. First try to use plural subjects and pronouns. Next, try to reword the sentence. Then, you may not have to refer to both sexes.

Many terms that refer definitely to males are no longer in standard use. Table 6-1 gives a list of gender-biased words and their nonsexist alternatives.

YOU TRY IT

In the space provided, complete each sentence by entering the missing word or words.

1. The subject of a sentence is considered to be in the _____ case.

2. The possessive case form of Jack is _____.

3. She greeted the guest. In this sentence, the word *guest* is in the _____ case because it is the direct _____.

4. The term *number* refers to whether a noun is singular or _____.

5. An alternative to the gender-biased term *salesman* is _____.

Answers: 1. nominative; 2. Jack's; 3. objective, object; 4. plural; 5. salesperson.

Table 6-1 Gender-Biased Words and Nonsexist Alternatives

Gender-Biased Words	Nonsexist Alternatives
airline stewardess, -steward	flight attendant
anchorman	news anchor, anchor
auto repairman	auto repairer
baggage man	baggage handler
busboy, -girl	dining room attendant
businessman, -men	business executive(s), business person(s), manager(s)
cameraman	camera operator, cameraperson
chairman, -men	chair(s), chairperson(s), director(s), department head(s), moderator(s)
clergyman	pastor, minister, clergy member, the clergy (plural)
congressman, -men	congressional representative(s), representative(s), member(s) of congress, senator(s)
craftsman	craftperson, craft worker, artisan
deliveryman, -boy	deliverer
draftsman	drafter
fireman, -men	firefighter(s)
foreman, -men	first-line manager/s, supervisor/s
garbage man	garbage collector
housewife	homemaker
insurance man	insurance agent
layman, -men	layperson, laypeople
mailman	mail carrier, letter carrier, postal worker
man, men	individual(s), human being(s), human race, humanity, human(s), men and women, people
man-hour	labor-hour, worker-hour
manpower	personnel, staff, workers, work force, human energy, labor force
meter maid	parking enforcement officer
middleman	intermediary, agent, dealer, reseller, marketing intermediary
salesman, -men	sales agent(s), sales force, sales representative(s), salespeople, salesperson(s)
serviceman, -men	service technician(s), service person
statesman	leader, public servant, diplomat, politician

CHAPTER SUMMARY: A QUICK REFERENCE GUIDE

PAGE	TOPIC	KEY POINTS	EXAMPLES
105	**Proper and common nouns**	**Proper nouns** name important specific persons, places, things, activities, qualities, and concepts; they begin with a capital letter.	*Jane, Roseville, General Motors*
		Common nouns name a general class of persons, places, things, activities, qualities, and concepts; they do not begin with a capital letter.	*daughters, towns, tables*
106	**Concrete and abstract nouns**	**Concrete nouns** are perceived through the senses.	*train, cake, shirt*
		Abstract nouns cannot be perceived through the senses; they name intangibles.	*memory, security*
106	**Collective nouns**	**Collective nouns** name a group or collection of persons, places, things, activities, qualities, and concepts. They are important for noun and verb agreement.	*flock, tribe, congregation*
108	**Noun functions**	**1. Nouns as major sentence elements**	
		a. Nouns as *subjects of verbs* (transitive and intransitive) usually occur at the beginning of a sentence or clause.	*Orville* cut the paper. The *accountant* is ill.
		b. Nouns as *objects of verbs* function as direct objects and indirect objects of transitive verbs.	Ginger found the missing *stapler.* Give the *messenger* a tip.
		c. Nouns as *subject and object complements* are predicate nouns that rename subjects and predicate nouns that follow direct objects and rename them.	Webster is a nice *boss.* Loren made Jim an *owner.*
		2. Nouns as objects of prepositions. The last word in a prepositional phrase is the object of the preposition, which is a noun or pronoun.	The lamp on the *table* is broken.
		3. Nouns as appositives. An appositive is a noun or noun phrase that immediately follows another noun or noun phrase.	Suzy, *our accountant,* left for the day.
		4. Nouns as modifiers. Nouns can act as adjectives and ad-	She has a *gold* coin. Wong has gone *home.*

PAGE	TOPIC	KEY POINTS	EXAMPLES
		5. Nouns of direct address. They name whomever or whatever you are speaking to in the sentence; they cannot be the simple subject.	*Sarah*, leave the light on.
114	**Noun properties**	**1. Case:** Shows how nouns and pronouns are related to other sentence parts. The three cases are (a) nominative, which indicates person doing action (subjects and predicate nouns); (b) objective, which indicates person receiving action (direct and indirect objects, objects of prepositions, object complements); and (c) possessive, which denotes ownership. Nouns only change form in possessive case. **2. Person:** Indicates person speaking, spoken to, and spoken of. Nouns do not change form for person. **3. Number:** Nouns, pronouns, and verbs change form to indicate singular or plural. **4. Gender:** It is improper to use the masculine gender when the gender of a group is unknown or includes males and females. To avoid gender bias: **a.** use plural subjects and pronouns **b.** reword a sentence. **c.** refer to both males and females.	

QUALITY EDITING

Your supervisor at the Riverside Hotel asked you to edit the following draft of a letter to the hotel's restaurant supply company. Watch for errors in spelling, end-of-sentence punctuation, capitalization, and gender-biased terms.

Dear mr. Casey:

The waitresses in the lotus restaurant, our fine dinning room, tell me that the custimers has complaned about the coffe that you provided last week. The businessman say that your coffe is bitter This is the thrid complaint that we have had about the bevrages you have supply. Can you explain this problem.

Unfortunately, Mr. casey, I must advice you that if we have another insident of this type we will be forsed to change supplirs.

Sincerely,

Rose Hopper

Bevridge Controler

APPLIED LANGUAGE

1. Replace each of the common nouns listed below with a proper noun.

Common Nouns: Proper Nouns:

a. city   _____

b. politician   _____

c. dentist   _____

d. hotel   _____

e. singer   _____

f. flower shop   _____

g. coffee   _____

h. dry cleaning service   _____

2. In the spaces below, list five gender-biased terms that are not listed in Table 6-1. Then, write a nonsexist alternative for each term.

Gender-Biased Terms: Nonsexist Terms:

a. _____   _____

b. _____   _____

c. _____   _____

d. _____   _____

e. _____   _____

3. Refer to the Quality Editing exercise above. Pretend that you are John Casey writing a note to Rose Hopper. Explain that your coffee was not up to its usual quality because you had tried a new, less expensive supplier. Assure her that you are going to go back to your previous supplier and that the problem she experienced will not recur. After your note is finished, circle all of the nouns you have used.

ORKSHEET 6

NOUNS: CLASSES, FUNCTIONS, AND PROPERTIES

Part A. Proper, Common, and Collective Nouns

Underline all of the nouns in the memo below. Then, write each proper noun and each collective noun in the spaces provided following the memo.

MEMO

TO: Gloria Lee

FROM: Stan Porter

SUBJECT: Sales Team

The summer sales campaign will begin in two weeks. This year, Prestige Hotels will develop a promotion for professional people. You will be asked to lead the team that will make this promotion a success for the Prestige Hotel in our city. The promotion will be entitled the "Prestige for Professionals Campaign." We have selected this group—professionals such as doctors and lawyers—because often they have clients or patients who come from other cities. Many of these guests are affluent and able to pay premium rates. If local professionals are familiar with our hotels, we expect them to reserve our rooms for their visitors.

Your team will need to form committees to plan the campaign here. The committees will identify the audiences for the promotion, the media that will be used to reach them, and the incentives that will generate the most sales.

Last year, the campaign for sales to groups and conventions was rated as the most creative. I hope that we take that honor again this year.

Proper Nouns:

1. _____

2. _____

3. _____

4. _____

Collective Nouns:

1. _____

2. _____

3. _____

4. _____

Part B. Noun Functions

Some of the nouns in the following sentences are italicized. Write SC above the italicized nouns that are subject complements and OC above those that are object complements.

1. Miss Faber, the new director, is a tourism *economist.*

2. Those hotels are our principal *competitors.*

3. They elected Mr. Duarte *chairperson.*

4. Louis is the hotel *concierge.*

5. The purchase of the hotel's stock made Ms. Eberhard the principal *stockholder.*

Part C. Nouns as Modifiers

The following list contains 20 nouns. Combine them so that you have 10 pairs of nouns with the first noun modifying the second one. For example, combine *desk* and *clerk.* The result is *desk clerk.* Combine the nouns as you wish, but the results must make sense!

food	dessert	telephone	lobby
service	furniture	check	garbage
tray	preparation	room	cooperation
convention	operator	sales	policy
removal	personnel	staff	safety

1. _____

2. _____

3. _____

4. _____

5. _____

6. _____

7. _____

8. _____

9. _____

10. _____

Part D. Noun Clauses

In each of the sentences below, double underline the simple predicate of the independent clause and single underline the noun clause. In the space provided, write whether the noun clause functions as subject, DO (direct object), or SC (subject complement).

_____   **1.** What you did is wrong.

_____   **2.** That is what she wants.

_____   **3.** Judy said that she sold the picture.

_____   **4.** Tell the president that you will be late.

_____   **5.** That she had become manager was not generally known.

_____   **6.** The rumor was that the fire began at midnight.

Part E. Nominative and Objective Case

In each of the following sentences, underline the nouns that are in the nominative case once and the nouns that are in the objective case twice.

1. The new mail carrier delivered the package to the hotel.

2. The elevator stopped on the ninth floor.

3. That hotel does not have an exercise room for its guests.

4. She is the chief accountant for the hotel chain.

5. Dr. Evian will be the on-call physician for May.

6. The public relations director advertised the new business services available to the guests.

7. The hotel has an engineer who conducts safety checks daily.

8. Room service is not profitable.

9. The desk clerk gave the guests their keys.

10. They named George Alopoulos the new manager.

Part F. Possessive Case

Write the possessive form of the italicized phrase in the space following each sentence.

1. The *name of the guest* is Garcia.

_____

2. The *invoice of the food supplier* was inaccurate.

_____

3. They measured the *girth of the refrigerator.*

_____

4. Give the clean towels to the *assistant to the lifeguard.*

_____

5. What is the *policy of the firm?*

_____

6. The *concession of Mr. Salluto* is not open.

_____

7. The *style of the pianist* is too loud for our lobby.

_____

8. Fiduciary matters are the *responsibility of the controller.*

_____

9. She objected to the manner *of the receptionist.*

_____

10. The *repair of the ceiling* would require *work of one month.*

_____

Part G. Gender-Biased Terms

Provide a nonsexist alternative for each of the gender-biased terms below.

1. congressman   _____

2. waitress   _____

3. chairman   _____

4. bell boy   _____

5. maid   _____

6. manpower   _____

7. sales girl   _____

8. deliveryman   _____

9. fireman   _____

10. mailman   _____

11. foreman   _____

12. housewife   _____

13. policeman   _____

14. repairman   _____

15. cameraman   _____

16. cowboy   _____

17. tradesman   _____

18. washerwoman   _____

19. weatherman   _____

20. nursemaid   _____

CHAPTER 7

NOUNS: PLURALS AND POSSESSIVES

When you are unsure about a plural, check the dictionary. Most college dictionaries provide the plural forms of all troublesome nouns.

—Maxine Hairston and John Rusckiewicz, *The Scott, Foresman Handbook for Writers, 2nd edition*

LEARNING UNIT GOALS

Did you ever wonder how we would communicate if there were no nouns? What happens if you take out all the nouns of a short newspaper article? Do you learn anything from the article? Without nouns, you would learn little, if anything, about what the article said. The independence of nouns as a part of speech and our complete dependence on them make nouns the center of our communication system.

Remembering the rules of noun plurals can be a challenge. To help you, Learning Units 7-1 and 7-2 group the noun plural rules into two major groups: regular noun plurals and irregular noun plurals. Within these two groups you will find several subgroups. Since you probably know some of the noun plural rules, you may be surprised how quickly you can move through them.

The difficulty with noun possessives occurs when in the ownership relationship, writers lose track of the possessor and the thing possessed. Learning Unit 7-3 will help you understand this relationship. Learning Unit 7-4 presents some special uses of possessive nouns. Now, let's begin by studying the regular noun plurals and the most common group of irregular noun plurals.

LEARNING UNIT 7-1
REGULAR NOUN PLURALS AND IRREGULAR
NOUN PLURALS ENDING IN *y*, *o*, AND *f* SOUND

In grammar, *number* indicates whether something is singular or plural. A **singular noun** refers to only one of anything (person, place, thing, quality, etc.). **Plural nouns** refer to more than one of anything. Words such as *machine* and *building*, therefore, are singular, while words such as *machines* and *buildings* are plural.

This learning unit discusses the regular noun plurals and the irregular noun plurals ending in the *y*, *o*, and *f* sound. Most college dictionaries do not list the regular plural forms. The plural forms of unusual, or *irregular*, nouns are, however, usually included. Consult your dictionary whenever you are in doubt about a plural form.

REGULAR NOUN PLURALS

For most nouns, plurals are formed by adding *s* or *es* to the singular. Often the sound of the plural noun determines whether you add *s* or *es* to the singular noun.

Nouns That Add *s*

Most common and proper nouns end in a sound that unites smoothly with the addition of *s*.

automobile, automobiles	cabinet, cabinets	Martin, Martins
employee, employees	computer, computers	Perry, Perrys
fountain, fountains	desk, desks	Smith, Smiths

Nouns That Add *es*

Common and proper nouns ending in a sound that does not unite smoothly with *s* add *es* to make the plural form pronounceable. Add *es* to nouns ending in *s, x, z, ch,* or *sh.*

business, businesses	quiz, quizzes	bush, bushes
tax, taxes	church, churches	Williams, Williamses

IRREGULAR NOUN PLURALS ENDING IN *y, o,* AND *f* SOUND

This group of irregular noun plurals includes exceptions to the basic rules of forming regular noun plurals.

Common Nouns Ending in *y*

Common nouns ending in *y* fall into two groups:

1. **Nouns ending in *y* preceded by a vowel (*a, e, i, o, u*) form the plural by adding *s* only.**

alloy, alloys	buoy, buoys	guy, guys
attorney, attorneys	fairway, fairways	turkey, turkeys

2. **Nouns ending in *y* preceded by a consonant (letters other than vowels) form the plural by changing the *y* to *i* and adding *es*.** Plurals of nouns ending in *quy* are also formed this way.

city, cities	colloquy, colloquies	duty, duties
quantity, quantities	secretary, secretaries	utility, utilities

Note: *All proper nouns* ending in *y* form the plural by adding *s*: February, Februarys; Mary, Marys; Henry, Henrys.

Nouns Ending in *o*

Use the following rules for forming the plural of nouns ending in *o*:

1. **Add *s* to nouns ending in *o* preceded by a vowel.**

cameo, cameos	folio, folios	radio, radios
curio, curios	portfolio, portfolios	trio, trios

2. **Add *s* or *es* to nouns ending in *o* preceded by a consonant.** These plural noun forms follow no rule. Some nouns have two plural forms, *s* and *es*, but *es* is usually preferred. Check your dictionary when you are in doubt.

Nouns that add *s*	Nouns that add *es*	Nouns that add *s* or *es*
albino, albinos	embargo, embargoes	echoes, echos
canto, cantos	hero, heroes	mementos, mementoes
dynamo, dynamos	potato, potatoes	mosquitoes, mosquitos

halo, halos	tomato, tomatoes	mottoes, mottos
memo, memos	veto, vetoes	zeros, zeroes

3. Add *s* to musical terms ending in *o*.

alto, altos	cello, cellos	solo, solos
banjo, banjos	piano, pianos	soprano, sopranos

WRITING TIP

When the dictionary gives two spellings, this usually means that the first spelling is preferred. *Merriam Webster's Collegiate Dictionary, Tenth Edition* **lists** *cargoes* **first so it is the preferred spelling.**

Nouns Ending in *f, fe,* or *ff*

Nouns ending in the *f* sound usually form the plural by adding *s*. In some nouns, however, the *f* or *fe* is changed to *v* and *es* is added. Other nouns use both forms—the added *s* and the change to *ves*.

Nouns that add *s*	Nouns that change to *ves*	Nouns using both forms
bailiff, bailiffs	half, halves	calves, calfs
chief, chiefs	knife, knives	dwarfs, dwarves
roof, roofs	leaf, leaves	hooves, hoofs
safe, safes	thief, thieves	scarves, scarfs
staff, staffs	wife, wives	wharves, wharfs

YOU TRY IT

In the space provided, write the plural of each of the following singular nouns:

1. law _____

2. court _____

3. dress _____

4. attorney _____

5. city _____

6. radio _____

7. chief _____

8. lawyer _____

9. exhibit _____

10. box _____

11. way _____

12. stationery _____

13. potato _____

14. leaf _____

Answers: 1. laws 2. courts 3. dresses 4. attorneys 5. cities 6. radios 7. chiefs 8. lawyers 9. exhibits 10. boxes 11. ways 12. stationeries 13. potatoes 14. leaves

LEARNING UNIT 7-2
MISCELLANEOUS IRREGULAR NOUN PLURALS

This learning unit gives nine miscellaneous irregular noun plural rules. You probably know some of these rules, so concentrate on learning the rules that are unfamiliar to you.

1. **Nouns that change spellings to form plurals.** This group includes nouns made plural by a vowel change and nouns that form plurals with *en*.

child, children	louse, lice	ox, oxen, or oxes
foot, feet	man, men	tooth, teeth
goose, geese	mouse, mice	woman, women

2. **Nouns that are either singular or plural, end in *s* but are singular in meaning, and have identical singulars and plurals.** Nouns that are either singular or plural are used only with plural verbs. Nouns ending in *s* but singular in meaning require singular verbs when used as subjects in sentences.

Either singular or plural	End in *s*—meaning is singular	Identical singulars and plurals
goods	civics	corps
pliers	economics	deer
riches	mathematics	moose
scissors	news	rendezvous
thanks	phonetics	series
trousers	semantics	sheep

Note: When *corps* is used in the singular, the *s* is silent.

3. **Numbers, capitalized letters (except *A, I, M,* and *U*), or words referred to as words.** The plurals of this group are formed by the addition of *s* or *es*.

Numbers	Letters and words	Words referred to as words
a few *7s*	*Ds* and *Fs*	many *yeses* and *noes*
the *1990s*	*ifs* and *buts*	too many *ands*

Note: You may also use an apostrophe if the expression would otherwise be easily misread (*so's, me's,* etc.).

4. **Lowercase letters and capitalized *A, I, M,* and *U.*** Usually, the plurals of this group are formed by the addition of an apostrophe and *s ('s)*.

four *e's*	five *l's*	several *A's*

5. **Abbreviations.** The plural of an abbreviation is usually formed by the addition of *s* to the singular form.

bbls.	hrs.	Ph.D.s	RNs.	YMCAs
CPAs	mos.	pks.	wks.	yrs.

Note: You may choose to insert an apostrophe to prevent a possible misreading (*rpm's*).

Some abbreviations have the same form for both the singular and the plural.

bu.	bushel *or* bushels	mi.	mile *or* miles
deg.	degree *or* degrees	min.	minute *or* minutes
ft.	foot *or* feet	oz.	ounce *or* ounces
lb.	pound *or* pounds	yd.	yard *or* yards

Note: When writing long reports, writers frequently refer to other pages in the report. If an abbreviation for the word *pages* or *and following* is desired, the following is standard usage:

page (p.), pages (pp.)
See pp. 4–15 (pages 4 through 15)

following (f., plural ff.)
See pp. 20 ff (page 20 and following pages)

6. **Plurals of contractions.** The plural of a contraction is formed by the addition of *s*.

don't   don'ts   can't   can'ts

Note: A contraction that already ends in *s* should not be used in the plural form.

7. **Nouns foreign in origin.** Many nouns that are foreign in origin are commonly used in formal, scientific, and technical matter. Some of these nouns have only their foreign plurals; others have been given an additional (English) plural. You will note that the nouns listed in Table 7-1 have been taken from Latin, Greek, and French. The plurals of foreign nouns are formed by changing the endings as indicated.

8. **Compound nouns. Compound nouns** consist of a combination of two or more words that are written in solid form as one word, as a hyphenated word, or as separate words.

   a. *To form the plural of a compound noun written as one word without a hyphen, make the final element plural.*

businessperson, businesspeople
cupful, cupfuls (also cupsful)
letterhead, letterheads

step child, stepchildren
stockholder, stockholders
weekday, weekdays

   b. *To form the plural of a hyphenated compound word that does not have a noun, make the final element plural.*

follow-up, follow-ups
trade-in, trade-ins

strike-over, strike-overs
write-up, write-ups

   c. *To form the plural of a compound word consisting of one or more nouns and an adjective or a preposition, make the principal noun plural.*

aide-de-camp, aides-de-camp
brother-in-law, brothers-in-law

bill of lading, bills of lading
looker-on, lookers-on

Table 7-1 Nouns Foreign In Origin

	Singular	Foreign Plural	English Plural
sis to *ses*	analysis basis diagnosis oasis	analyses bases diagnoses oases	
um to *a*	addendum curriculum datum memorandum	addenda curricula data memoranda	curriculums memorandums
us to *i*	alumnus (mas.) gladiolus nucleus stimulus	alumni gladioli nuclei stimuli	gladioluses nucleuses
ex or *ix* to *ices*	appendix index	appendices indices	appendixes indexes
a to *ae*	alumna (fem.) formula larva vertebra	alumnae formulae larvae vertebrae	formulas vertebras
on to *a*	criterion phenomenon	criteria phenomena	criterions phenomenons
eau to *eaux*	bureau tableau trousseau	bureaux tableaux trousseaux	bureaus tableaus trousseaus

Note: Some compound words form the plural by adding *s* either to the first word or the second word. Usually, writers follow the preferred choice given in the dictionary. In the following examples, the preferred spelling is given first.

> attorneys general, attorney generals
> courts-martial, court-martials
> notaries public, notary publics

9. **Plurals of personal names with titles.** The plurals of personal names accompanied by titles may be correctly expressed in more than one way. The tendency in business writing is to avoid the use of formal plural titles.

Singular	Formal plural	Preferred informal plural
Miss	Misses	Miss Noel and Miss Smith
Mr.	Messrs.	Mr. Wong and Mr. Lobe
Mrs.	Mmes.	Mrs. Hanson and Mrs. Stevens
Ms.	Mses. (seldom used)	Ms. Lang and Ms. Holt

Note: In most of today's business offices, correspondents have dropped the use of *Miss* and *Mrs.* These titles have been replaced with *Ms.*, which does not reveal a woman's marital status. Many business writers prefer to use no courtesy title.

WRITING TIP

Forming noun plurals can be tricky. Sometimes, the only safe procedure is to use your dictionary. Remember the following formula:

Doubt + Dictionary = Correct spelling

YOU TRY IT

A. Write the plural of each noun in the space provided.

1. child _____

2. mouse _____

3. scissors _____

4. A _____

5. tooth _____

6. foot _____

7. deer _____

8. YMCA _____

B. Write the plural of each noun in the space provided. These nouns are foreign in origin.

1. analysis _____

2. nucleus _____

3. datum _____

4. index _____

C. Write the plural of each compound noun in the space provided.

1. trade-in _____

2. sister-in-law _____

Answers: **A.** 1. children 2. mice 3. scissors 4. A's 5. teeth 6. feet 7. deer 8. YMCAs **B.** 1. analyses 2. nuclei or nucleuses 3. data 4. indices or indexes **C.** 1. trade-ins 2. sisters-in-law

---

LEARNING UNIT 7-3
SINGULAR AND PLURAL POSSESSIVE NOUNS

The *possessive case* is used to show possession, authorship, brand, kind, or origin: *John's* car, Roger *Keefe's* book, *Campbell's* soup, *dentists'* meeting, and *Darwin's* theory. A noun in the possessive case usually ends in an apostrophe and *s* (*'s*) and is followed by another noun. This noun can often be expanded into a phrase. Note the following examples:

A. J. *Cronin's* book   the book written by A. J. Cronin
Mr. *Caruso's* store   the store owned by Mr. Caruso
the *firm's* policy   the policy of the firm
today's program   the program for today

When a possessive noun is needed, *another noun is likely to follow* as in the preceding examples. An adjective, of course, may come between the two nouns, as in *Mr. Caruso's new store* or *today's expanded program*. You know the possessive case is needed if you can substitute a possessive pronoun (particularly *his, its,* or *their*) for the word in question. For example, *Joseph Santo's recent promotion* could be changed to read *his recent promotion.*

This learning unit gives the rules for singular and plural possessive noun forms. The last topic in the learning unit shows you how to ask the right questions about forming the possessive case of nouns.

WRITING TIP

Be careful not to confuse noun plurals with noun possessives. Noun plurals are more than one of anything and usually the noun itself changes form to show this. Noun possessives show ownership by adding an apostrophe and *s* (*'s*) or only an apostrophe.

SINGULAR POSSESSIVE NOUNS

Noun possessives are difficult for many writers because they forget the two important factors of the possessive relationship: the possessor and the thing possessed. The *possessor* is represented by a word written in the possessive case that functions as an adjective. For example, note the following sentence:

> Larry's plan to launch the advertising campaign is similar to last year's plan.

The *thing possessed* (plan) is a noun modified by the possessive nouns *Larry's* and *year's*. In working with possessives, always start by knowing *who possesses what*. Then look at the noun that "owns" something and decide whether it is singular or plural. If the noun is singular, apply Rule 1.

RULE 1.   Form the possessive case of a singular noun by adding an apostrophe and *s* (*'s*) after the noun.

a *fiduciary's* responsibility	a *month's* work
a *trustee's* background	*Bob's* notebook
Jess's reports	Ms. *Baker's* appointment
the *recruiter's* patience	your *coworker's* mail

Some writers apply Rule 1 whenever they make singular nouns possessive. Most writers, however, deviate from the basic rule when its application would result in a word that is awkward to pronounce. As a result, we have Rule 2.

RULE 2.   Form the possessive case of a singular noun that has two or more syllables and ends in an *s, x,* or *z* sound by adding only an apostrophe.

Henry Phillips' new book	Mr. Perkins' assignment
Ms. Rabinowitz' inquiry	the witness' manner

Many writers apply Rule 2 every time they work with nouns of two or more syllables that end in an *s* or *z* sound. Other writers apply Rule 2 only when they feel that the addition of *'s* would lead to an awkward pronunciation.

All the possessive forms in the following list are correct. If you pronounce each word carefully, you *may* decide that some of these forms are more acceptable than others.

Carole *Reynolds'* idea   Carole *Reynolds's* idea
Mr. *Fairless'* office   Mr. *Fairless's* office
Mr. *Lopez'* invention   Mr. *Lopez's* invention
Ms. *Carruthers'* report   Ms. *Carruthers's* report
Steinmetz' theory   *Steinmetz's* theory

Some professional writers add only an apostrophe to singular nouns of *one* syllable *(Betz', Ross', Leeds',* etc.). However, few, if any, handbooks of correct English usage recommend this practice.

PLURAL POSSESSIVE NOUNS

Before you make a plural noun possessive, be sure that you have the correct plural form. Then, apply whichever of the following rules is appropriate.

RULE 3. Form the possessive of a regular plural noun (one ending in *s*) by adding only an apostrophe after the *s*.

the *attorneys'* arguments   the *boys'* accounts
the *Briggses'* dilemma   the *Calderases'* home
the *horses'* bridles   the *Murphys'* indebtedness

RULE 4. Form the possessive of an irregular plural noun (one not ending in *s*) by adding an apostrophe and *s*.

children's shoes   *men's* trousers
salespeople's territories   *women's* organizations

ASK THE RIGHT QUESTIONS

Many writers find the use of nouns in the possessive case challenging simply because they fail to ask the right questions. Consider this sentence:

The (jurors, juror's, jurors') votes will be collected and counted.

Writers who are doubtful about the correct form of the second word in the preceding sentence should ask these questions:

1. Is the word used in the possessive case?

ANSWER: When a noun *(votes)* follows a word in question, the use of a noun suggests that the possessive case is needed. You can verify this by using the test for possession and saying *their votes.* Also, the word *votes* (like all possessive case forms) can be expanded into a phrase: the votes *of the jurors*. So we should use the possessive case.

2. Is the word in question singular or plural?

ANSWER: Note the use of the word *jurors* (not *juror*) in the phrase given in the answer above: votes *of the jurors* (usually 12 people). We definitely need the plural form.

3. How is the plural form *jurors* made possessive?

ANSWER: To make the possessive case of a regular plural noun (one ending in *s*), add only an apostrophe after the *s*.

The correct form, then, is *jurors'*.

YOU TRY IT

1. What are the two ways to form the possessive case of a singular noun?

a. _____

b. _____

2. How do plural nouns that end in *s* usually show ownership?

_____

3. Change the following phrases into possessive nouns.

a. the decision of the court   _____

b. the expertise of the attorney   _____

c. the history of Mr. Parsons   _____

d. the reports of the accountants _____

e. the businesses of the women   _____

Answers: 1. a. By adding an apostrophe and s ('s) **b.** By adding only an apostrophe. **2.** By adding an apostrophe. **3. a.** the court's decision **b.** the attorney's expertise **c.** Mr. Parsons' history **d.** the accountants' reports **e.** the women's businesses.

LEARNING UNIT 7-4
SPECIAL USES OF THE POSSESSIVE

Before you study the eight special uses of the possessive given in this learning unit, be aware that occasionally the possessive form is not followed by the noun it modifies. That noun is sometimes not expressed or it is located elsewhere in the sentence. Note the following examples:

I met your attorney at *Morton's*. (Morton's home)
That portfolio is *Darin's*. (Darin's portfolio)

The eight special uses that follow show how to express the possessive in joint ownership, separate ownership, nouns in apposition, compound words or phrases, possessive abbreviations, inanimate objects, time and measurement, and names of organizations.

1. **Joint ownership.** To show joint ownership, add the sign of the possessive (apostrophe or *'s*) to the last noun that names a possessor.

   Frank and *Carl's* delivery truck   Stacy and *Nelson's* store

2. **Separate ownership.** To show separate ownership, make each noun possessive.

   buyers' and *sellers'* points of view   *Leo's* and *Tom's* voices

3. **Nouns in apposition.** If nouns are used in apposition to the name of the possessor, add the sign of the possessive to the appositive or rewrite the sentence using a phrase.

   We appraised Dr. McGrath, the *dentist's*, property. (You may prefer this wording: We appraised the property of Dr. McGrath, the dentist.)

4. **Compound words or phrases.** In compound words or phrases, show possession by placing the sign of the possessive (apostrophe or *'s*) at the end.

   my son-in-law's car

   the secretary-treasurer's report

   the cookbooks' index

5. **Possessive abbreviations.** To make a singular abbreviation possessive, place an apostrophe and *s* after the period. If the abbreviation is plural, place an apostrophe after the *s* but not before it.

   Nissen, Jr.'s market   the Ph.D.s' theses
   the Randall Co.'s bargains   two M.D.s' opinions

6. **Inanimate objects.** Usually, you should not use the possessive case when referring to inanimate objects. The use of an *of* phrase (prepositional phrase beginning with *of*) is usually less awkward.

   the roof of the house (not *the house's roof*)

   the top of the desk (not *the desk's top* or *the desk top*)

   If people are in some way involved, the possessive case is usually acceptable, as in *this store's policies* or *my company's pension plan.*

7. **Time and measurement.** You can also state common expressions that refer to time and measurement in the possessive case.

   a minute's delay   at arm's length
   five cents' worth   one week's time
   ten minutes' delay   three weeks' time

8. **Names of organizations.** When you write the name of an organization, do not use an apostrophe unless an apostrophe is in its

official name. Follow the style used on the organization's letter-head.

Bankers Corporation   California Teachers Association

Hanson's Hardware   Fairview's Furniture Corporation

Y O U   T R Y   I T

Write the possessive of the following items in the space provided.

1. presence of a judge and jury   _____

2. votes of the men and the women   _____

3. the actions of the sergeant at arms   _____

4. the activities of two YMCAs   _____

5. the delay of a year   _____

Answers 1. judge and jury's presence 2. men's and women's votes 3. sergeant at arms' actions 4. two YMCAs' activities 5. a year's delay.

CHAPTER SUMMARY: A QUICK REFERENCE GUIDE

PAGE	TOPIC	KEY POINTS	EXAMPLES
126	**Singular and plural nouns**	**Singular nouns** refer to one of anything. **Plural nouns** refer to more than one of anything.	*basket, clock* *baskets, clocks*
126	**Regular noun plurals**	**1. Nouns ending in a sound uniting smoothly with the addition of *s*.** These nouns add *s* only. **2. Nouns ending in a sound that does not unite smoothly with the *s* sound.** Add *es* to nouns ending in *s, x, z, ch,* or *sh*.	*chair, chairs* *lunch, lunches*

127	**Irregular noun plurals ending in *y, o,* and *f* sound**	**1. Common nouns ending in *y*.**
		    a. Preceded by a vowel, add *s*.
		    b. Preceded by a consonant, change *y* to *i* and add *es*.
		    Note: Proper nouns ending in *y*, add *s*.
		 2. Nouns ending in *o*.
		    a. Preceded by a vowel, add *s*.
		    b. Preceded by a consonant, add *s* or *es*. These follow no rule.
		    c. Musical terms, add *s*.
			
			 valley, valleys
			 country, countries
			
			 ratio, ratios
			 domino, dominos
			 mottoes, mottos
			
			 stereo, stereos

PAGE	TOPIC	KEY POINTS	EXAMPLES
		3. Nouns ending in _f_, _fe_, or _ff_. Usually form the plural by adding _s_; _f_ or _fe_ sometimes changed to _v_ and _es_ added; some nouns use both forms.	_chief, chiefs_ _self, selves_ _scarves, scarfs_
129	**Miscellaneous irregular noun plurals**	**1. Nouns that change spellings.** Nouns with a vowel change and those using _en_.	_foot, feet_ _ox, oxen_
		2. Nouns that are either singular or plural, end in _s_ but have a singular meaning, and have identical singulars and plurals. Nouns singular or plural are used with plural verbs; nouns ending in _s_ but singular in meaning use singular verbs.	_clothes_ _news_ _deer_
		3. Plurals of numbers, capitalized letters (except _A_, _I_, _M_, _U_), or words referred to as words. Plurals are formed by the addition of _s_ or _es_.	_8s_ _Bs_ and _Cs_ so many _buts_
		4. Plurals of lowercase letters and capitalized _A_, _I_, _M_, and _U_. Usually the plurals of this group are formed by adding apostrophe and _s_ (_'s_)	six _z's_ all _A's_
		5. Plurals of abbreviations. Usually formed by adding _s_ to the singular form; some plurals are the same for the singular and plural.	_IRAs, CDs_ _oz._ (ounce or ounces)
		6. Plurals of contractions. They add _s_.	_can't, can'ts_
		7. Nouns foreign in origin. This group changes _sis_ to _ses_, _um_ to _a_, _us_ to _i_, _ex_ or _ix_ to _ices_, _a_ to _ae_, _on_ to _a_, and _eau_ to _eaux_.	_parenthesis, parentheses; bacterium, bacteria; bacillus, bacilli; index, indices, indexes; alga, algae, algas; phenomenon, phenomena, phenomenons; tableau, tableaux, tableaus_
		8. Compound nouns. Two or more words written solid, hyphenated, or separate. **a.** When a compound noun is written as one word without a hyphen, final element is plural.	_blackboard, blackboards_
		b. When no part of hyphenated compound word has a noun, final element is plural.	_go-between, go-betweens_
		c. When the compound has one or more nouns and an adjective or preposition, then principal noun is plural	_father-in-law, fathers-in-law_
		9. Plurals of personal names with titles. Informal preferred.	_Ms. Rogers_ and _Ms. Wong_ went home.

PAGE	TOPIC	KEY POINTS	EXAMPLES
133	**Singular possessive nouns**	1. **Possessive of singular noun.** Add an apostrophe and *s*. 2. **Possessive of singular noun with two or more syllables ending in *s*, *x*, or *z* sound.** Add only apostrophe.	*Mary's* computer *Milton Horowitz'* desk
134	**Plural possessive nouns**	3. **Possessive of plural noun ending in *s*.** Add only an apostrophe after *s*. 4. **Possessive of plural noun not ending in *s*.** Add an apostrophe and *s*.	the *students'* books *oxen's* horns
134	**Ask the right questions**	1. **Word used in possessive case?** Yes, when noun follows word; test with *his*, *its*, *their*; and word can be expanded into phrase. 2. **Word singular or plural?** When singular, use singular possessive; when plural, use plural possessive. 3. **How do you make noun possessive?** When plural ends in *s*, add only apostrophe.	The (dogs, dog's dogs') tails were short. Noun *tails* follows noun in question; *their tails*, and could be said as *tails of the dogs*. Noun *tails* is plural, so use *dogs'*.
135	**Special uses of the possessive**	A possessive is not always followed by the noun it modifies. Follow possessive rules anyway. 1. **Joint ownership.** Add a possessive sign to last noun naming possessor. 2. **Separate ownership.** Make each noun possessive. 3. **Nouns in apposition.** Add the sign of possessive to appositive or rewrite with phrase. 4. **Compound words or phrases.** Place the sign of possessive at end. 5. **Possessive abbreviations.** If singular, place an apostrophe and *s* after the period. If plural, place an apostrophe after *s* but not before. 6. **Inanimate objects.** Usually, you do not use the possessive; an *of* phrase is less awkward. 7. **Time and measurement.** Usually, this can be expressed in the possessive. 8. **Name of organization.** Use an apostrophe only if it is in the official name.	Elsie and *Lorenzo's* summer home *Joan's* and *Elaine's* pictures We sold the car of Dr. Schultz, the *professor*. *brother-in-law's* farm *R.N.s'* uniforms back of the chair (*not chair's back*) *two months'* time *Sears* *Johnson's Salvage Company*

QUALITY EDITING

Edit the following news article before it goes to press.

CITY DAILY TIMESES

At 2:00 p.m. yesterday, the courts decision was hand down in the case of *Phillips and Andersons' Printing Company* v. *Ms. Nora Jones*. The juries decision was in favor of the defendant, Ms. Jones . During the course of several days hearings, Ms. Joneses atternys presented exhibites including a series' of datas and analysises. Several memorandas written by Patricia Phillips and Sue Andersons' about Ms. Jones payment of her debtes showed that no amountes were owed to the womens' business's by Ms. Jones. After the hearing, Ms. Jones said her atornys expertise had saved her from finansial rune.

APPLIED LANGUAGE

A. Write a short sentence using the plural form of each word listed below.

1. memo

_____

2. bureau chief

_____

3. accountant

_____

B. Write a short sentence using the apostrophe or apostrophe plus *s* possessive forms of the item listed below.

1. contents of the memo

_____

2. decision of the bureau chief

_____

3. report of the accountant

_____

4. results of the three analyses

_____

5. actions of all the sheriffs

_____

6. meetings of the stockholders

_____

Name _____

Date _____

NOUNS: PLURALS AND POSSESSIVES

Part A. Regular Noun Plurals

1. Write the plural of each of the following nouns.

a. suit _____   **f.** jurist _____

b. court _____   **g.** clerk _____

c. case _____   **h.** defendant _____

d. firm _____   **i.** hearing _____

e. judge _____   **j.** objection _____

2. What did you do to make the words in item 1 plural?

_____

3. List one common noun that ends in each of these letters: *s*, *x*, *z*, *ch*, and *sh*. Next to the word, write its plural.

Common Noun   Plural

s   _____   _____

x   _____   _____

z   _____   _____

ch   _____   _____

sh   _____   _____

4. What did you do to make the words in item 3 plural?

_____

Part B. Irregular Noun Plurals

1. Write the plural of each of the nouns listed below.

a. attorney _____   **c.** way _____

b. key _____   **d.** joy _____

2. What did you do to make the words in item 1 plural?

_____

3. What kind of letter came before the y in each of these words?

_____

4. Write the plural of each of the nouns listed below.

    a. city _____   **c.** stationery _____

    b. company _____   **d.** boundary _____

5. What did you do to make the words in item 4 plural?

_____

6. What kind of letter came before the y in each of these words?

_____

7. Write the plural of each of the nouns listed below.

    a. radio _____   **e.** portfolio _____

    b. memo _____   **f.** potato _____

    c. hero _____   **g.** piano _____

    d. chief _____   **h.** leaf _____

Part C. Noun Plurals

Draw a line under any word that is not acceptable as a plural form. Then, write the correct plural form in the space provided. If the sentence has no errors, put a C in the space.

1. The lesson in civics emphasized the nation's legal foundation. _____

2. The CPA's audited our books in May. _____

3. The firm was established in the 1920's. _____

4. What were the basis of his conclusion? _____

5. The paper has four appendixes. _____

6. The memorandas were prepared in haste. _____

7. Five criterion were established for selection of the new partner. _____

8. The city's development department has two bureaus. _____

9. Did the officer interview the looker-ons at the scene of the accident? _____

10. His son-in-laws are partners in the firm. _____

Part D. Possessives

In the sentences below the words in italics indicate possession. Write the same possession in a form using apostrophe or apostrophe plus *s* in the space below each sentence.

EXAMPLE: The *four exhibits of the prosecutors* were missing.

        the prosecutors' four exhibits

1. The *manner of the witnesses* led the attorney to believe they had been coached.

_____

2. The *law courses of Mr. Watkins* are well attended.

_____

3. Take the legal brief to *the offices of Mr. Leibwitz.*

_____

4. The *arguments of the attorneys* did not change the *minds of the jurors.*

_____

5. The *policies of both companies* require preparation of time reports.

_____

6. The *signatures of both beneficiaries* are required on the form.

_____

7. The *legal research of Ms. Martinez* is thorough and efficient.

_____

8. The *decision of the bureau chiefs* was to prepare for litigation.

_____

9. The *staff of Representative Jones* in developing the chief arguments of the bill.

_____

10. The *arguments of the staff* will be used by Representative Jones.

_____

Part E. Special Uses of Possessives
Read the first sentence in each item below and then complete the second sentence by filling in the correct possessive form.

1. Maria and Antonia own a small business together. The company is

_____ business.

2. The buyer and the seller disagreed. Their opinions are different. The

_____ points of view differ.

3. Howard's Fashions, Inc., is filing a suit against the landlord. The action is

_____ suit.

4. This case belongs to Mr. Arness. It is _____ case.

5. Mr. Long and Ms. Ivory share an office. It is

_____ office.

6. Mr. Long and Ms. Ivory each have a desk. The desks are

_____ desks.

7. The defendant's attorney examined the exhibits of Ms. Shears, the prosecutor. The attorney examined _____ exhibits.

8. United Company and American Company share ownership of the building. It is _____ building.

9. Stanley has a portfolio. So does Maurice. The two portfolios are _____ portfolios.

10. The firm is owned jointly by the three partners, Messrs. Arbuthnot, Lee, and Sanchez. The firm is _____ firm.

CHAPTER 8

PERSONAL PRONOUNS AND PRONOUN TYPES

Pronouns are so frequently the keys or turning points of sentences that they cannot be too carefully studied or too thoroughly mastered.

—James G. Fernald and Cedric Gale,
English Grammar Simplified

LEARNING UNIT GOALS

LEARNING UNIT 8-1: PERSONAL PRONOUNS: BASIC CASE RULES

- Recognize the singular and plural personal pronoun forms for the nominative, objective, and possessive cases. (pp. 148–49)
- Know the personal pronoun case rules for nominative, objective, and possessive cases. (pp. 149–53)
- Use personal pronoun cases correctly in your writing. (pp. 148–53)

LEARNING UNIT 8-2: PERSONAL PRONOUNS: ADDITIONAL CASE RULES

- Recognize usage of the nominative case and objective case for conjunctions *than* and *as*, infinitives, and appositives. (pp. 154–57)
- Understand usage of compound personal pronouns. (pp. 157–58)

LEARNING UNIT 8-3: PRONOUN TYPES

- Recognize demonstrative, indefinite, relative, interrogative, and reciprocal pronouns. (pp. 158–62)
- Know how to use these five pronoun groups. (pp. 158–62)

Learning occurs most rapidly when you build new information on what you already know. Your knowledge of the noun has prepared you for this study of the pronoun—the noun substitute.

When you hear the word *pronoun*, which pronouns come to mind? Most people think first of the personal pronoun (*I, you, he, she, it, we,* and *they*). In Learning Units 8-1 and 8-2, we discuss the personal pronoun—the first of seven types of pronouns. Included in this discussion is the compound personal pronoun—the second type of pronoun. Learning Unit 8-3 discusses the remaining five types of pronouns.

LEARNING UNIT 8-1
PERSONAL PRONOUNS: BASIC CASE RULES

Have you ever stumbled over the use of the pronouns *I* and *me* or *he* and *him* when speaking or writing? If you were writing, you probably looked up the correct usage of these pronouns in a stylebook or asked a friend. Did you remember the usage rule the next time you questioned how to use these pronouns? You probably forgot the rule and had to look it up or ask your friend again. This learning unit will help you understand why personal pronouns have definite usage rules. Once you understand why, you will remember how to use the personal pronouns.

Since most pronouns are short substitutes for nouns or a group of words used as nouns, they have the same functions and characteristics as nouns. Like nouns, pronouns may be used in sentences as subjects, objects, or complements. Pronouns also have the same properties as nouns: case (nominative, objective, and possessive); person (first, second, and third); number (singular and plural); and gender (masculine, feminine, and neuter).

The study of the correct usage of the personal pronoun depends on understanding the nominative and subjective cases. Pronouns in the possessive form are usually modifiers.

From Chapter 3, you know that the word or group of words that a pronoun replaces is called its *antecedent.* An antecedent of a pronoun is usually a noun, but it could be another pronoun.

Brenda Clark deposited $2,000 in *her* checking account. (*Brenda Clark* is the antecedent of the pronoun *her.*)

WRITING TIP

Be sure that the antecedent of your pronoun is clear to your readers.

UNCLEAR: Amy and Beth returned with *her* new office furniture.

CLEAR: Amy and Beth returned with *their* new office furniture.

Amy and Beth returned with *Beth's* new office furniture.

DECLENSION OF PERSONAL PRONOUNS

In Chapter 6, you learned that nouns present few challenges in case, person, number, or gender. Of the three cases (nominative, objective, and possessive), nouns only change form in the possessive case. What makes pronouns so challenging is that they change their form when they function as subjects (nominative case), objects (objective case), or complements (nominative case). Pronouns also change their form for person, gender, and number. When you understand these changes, you will eliminate any problems you may have with pronouns.

Personal pronouns stand for persons or things (nouns). Table 8-1 shows the **declension** of personal pronouns. The word *declension* refers to the changes made in personal pronouns to express correctly their person, gender, number, and case as indicated by their use in a sentence.

As you study Table 8-1, note that the first column gives the person and gender of personal pronouns. The remaining columns give the singular and plural personal pronouns for the nominative case, objective case, and possessive case. You will be referring to this table frequently.

RULES FOR BASIC CASE FORMS

As explained in Chapter 6, case shows how nouns and pronouns are related to other words in the sentence. The nominative case indicates the doer of the action of the verb. The objective case indicates the person or thing receiving the action. The possessive case indicates ownership or a relationship. Since pronouns change their form for all cases, you must understand

Table 8-1 Declension of Personal Pronouns

	Nominative Case (Doer of the Action or Subject Form)		Objective Case (Receiver of the Action or Object Form)		Possessive Case (Ownership)	
	Singular	**Plural**	**Singular**	**Plural**	**Singular**	**Plural**
First person (person speaking)	I	we	me	us	my mine	our ours
Second person (person spoken to)	you	you	you	you	your yours	your yours
Third person (person or thing spoken of)						
Masculine gender	he	they	him	them	his	their theirs
Feminine gender	she	they	her	them	her hers	their theirs
Neuter gender	it	they	it	them	its	their theirs

the purpose of the three cases. Then, you will know when the function of a pronoun in a sentence requires that you change its form.

Nominative Case Personal Pronouns

The nominative case is used for (1) the subject of a verb and (2) the subject complement of a verb. These two fundamental uses of personal pronouns appear below as Nominative Case Rule 1 and Nominative Case Rule 2. From Table 8-1, you know that the nominative case pronouns are *I, you, he, she, it, we,* and *they*.

The first and most important step in using a personal pronoun correctly in a sentence is to determine how the pronoun functions in the sentence. You can determine how a pronoun functions by using the sentence analysis flowchart steps you learned in Chapter 5.

NOMINATIVE CASE RULE 1.   Use the nominative case form when the pronoun is the subject of a verb.

When following this rule, begin by making certain that the pronoun in the sentence is functioning as the subject of the verb. Then, look at Table 8-1 to be sure the subject pronoun is in the nominative case. Also check the table for the person, gender, and number of the pronoun. Note the following examples:

> *She* looked everywhere for the lost stock certificate. (*Looked* is the verb. *She* is the subject. Table 8-1 shows *she* as a third-person, feminine gender, singular nominative case pronoun.)

> *We* should have a cellular telephone. (*Should have* is the verb phrase. *We* is the subject. Table 8-1 shows *we* as a first-person, plural nominative case pronoun.)

The only difficulty you may have with this nominative case rule is if you are confused by a compound subject. The following sentence has a compound subject:

> The contractor and (*he/him*) will look over the plans. .

To make sure you have chosen the correct pronoun, use the following two steps:

1. Temporarily cross out all of the compound element *except* the pronoun you question.

> ~~The contractor and~~ (*he/him*) will look over the plans.

2. Find the verb or verb phrase and the subject. *Will look* is the verb phrase. *He* or *him* is the subject. Table 8-1 shows *he* as a singular nominative case pronoun. In the objective case columns, we see that *him* is a singular objective case pronoun. Since the subject must be a nominative case pronoun, *he* is the correct pronoun. The sentence should read as follows:

> The contractor and *he* will look over the plans.

This procedure also works when the elements of the compound subject are both pronouns.

> (*She/Her*) and (*I/me*) appeared before the judge who decided the labor dispute.

> **1.** Temporarily cross out one of the compound elements, *including* the conjunction.

>> (*She/Her*) ~~and~~ (*I/me*) appeared before the judge who decided the labor dispute.

> **2.** Find the verb or verb phrase and the subject. *Appeared* is the verb. *She* or *her* is the subject. Table 8-1 shows *she* as a singular nominative case pronoun and *her* as an objective case pronoun. Since the subject must be a nominative case pronoun, *she* is the correct pronoun. Now, repeat the procedure for the second pronoun choice.

>> ~~(*She/Her*) and~~ (*I/me*) appeared before the judge who decided the labor dispute.

> The subject must be a nominative case pronoun. *I* is a nominative case pronoun. *Me* is an objective case pronoun. The sentence should read as follows:

>> *She* and *I* appeared before the judge who decided the labor dispute.

NOMINATIVE CASE RULE 2. Use the nominative case form when the pronoun is a subject complement (predicate pronoun) that follows an intransitive linking verb.

To follow this rule, you must have an intransitive linking verb and know the subject of the verb. Linking verbs always have subject complements. To find the subject complement, look for a predicate pronoun that renames the subject. This predicate pronoun must be in the nominative case. Note the following sentences:

> It may be *she* at the door. (*She* renames the subject *it* and is in the nominative case.)

> Is it *he* who first raised the question? (Reverse the words *Is it* and you have *It is he*. *He* renames the subject *it* and is in the nominative case.)

> We found ourselves wishing that we were *they*. (*They* is the subject complement of *we*, which is part of a dependent clause. *They* is in the nominative case.)

In conversation, many authorities consider it acceptable to say *It's me*. To be technically correct, however, you should say *It's I*. You can avoid this problem by repeating the name and saying *It's Beth*. You should *never* use *It's me* in written correspondence.

Objective Case Personal Pronouns

As you might expect, personal pronouns are in the objective case when they function as objects. The objective case personal pronouns are *me, you, him, her, it, us,* and *them.*

WRITING TIP

You do not have to be concerned about the case usage of *you* and *it*. These pronouns never change form in the nominative or objective case.

OBJECTIVE CASE RULE 1.   Use the objective case form when the pronoun is the direct and indirect object of a verb and the object of a preposition.

To have a direct object and an indirect object, you must have a transitive verb. In the examples that follow, you know that the pronoun is the direct or indirect object of a verb. In other sentences, you may not be sure how a pronoun functions in a sentence. The flowchart steps for a transitive verb (see Chapter 5) will help you decide the function of a pronoun. Then, you can look at Table 8-1 to make certain you are using an objective case pronoun.

Pronoun is the direct object of a verb

Several of our members congratulated *her.*

A committee of five selected *us.*

Pronoun is the indirect object of a verb

We gave *him* an unusually large order. (Like all indirect objects, you can expand *him* into a prepositional phrase: *We gave an unusually large order to him.*)

Kurt showed *her* the new budget.

When a sentence contains a compound direct or indirect object, follow the same steps as those used earlier for a compound subject. The pronoun you choose must be in the objective case.

To have an object of a preposition, you must have a prepositional phrase. The pronoun object of a prepositional phrase must be in the objective case. Do not be confused when words such as *after, but, between, except, for, like,* and *to* are used as prepositions to introduce prepositional phrases. Note the following examples of pronouns as objects of prepositions:

┌─── *prep. phrase* ───┐

Mr. Whelan gave photocopies to Morgan and *me.*

┌── *prep.* ──┐
    phrase

Ms. Gage talked with *him* about the mortgage.

prep. phrase

The senior accountant divided the work between Wong and *me*.

WRITING TIP

If you are not sure if a pronoun is in the nominative or objective case, try to use it with a transitive verb such as *locked*. If you can use the pronoun before the verb, the pronoun is in the nominative case. If you can use the pronoun after the verb, the pronoun is in the objective case. Note the following examples:

> *He* locked the door on Jim. (nominative)
>
> Jim locked the door on *him*. (objective)

Possessive Case Personal Pronouns

The possessive case pronouns are *my*, *mine*, *your*, *yours*, *his*, *her*, *hers*, *its*, *our*, *ours*, *their*, and *theirs*. The possessive forms of personal pronouns such as *hers*, *its*, *yours*, *ours*, and *theirs* are always written without an apostrophe.

POSSESSIVE CASE RULE.   Use the possessive case form to show possession, authorship, brand, kind, origin, or any similar relation.

> The final decision will be *yours*.
>
> The Merton Company may lose *its* best customer. (Sometimes writers incorrectly use an apostrophe with the possessive form of *it*. Remember that *it's* is *always* a contraction of the words *it is*.)
>
> Mr. Stein is unaware of *their* new policy. (*Their* is used as an adjective to modify a noun. Pronouns used as adjectives are called possessive adjectives.)

YOU TRY IT

A. Write the appropriate personal pronoun in the spaces provided in the following paragraph. The pronoun will substitute for the noun that appears in italics under the answer rule.

During the past several months, the Learntech Company has worked diligently to develop and produce Learntech's new multimedia computer game, *The Qwizzard*. In the game, players

search for a magic window that will give _____ new perspectives on the
    the players

world. Along the way, _____ meet the Qwizzard. The Qwizzard presents
    the players

challenges to the players for _____ to solve. Once the players solve the
    the players

challenges, the Qwizzard gives _____ a clue that helps _____

the players the players

on _____ way to the next level of _____ search.

the players' the players'

B. Write the appropriate personal pronoun in the spaces provided in the following memo.

TO: Larry Cummings, Sales
FROM: Gina Jolsen, Programming
SUBJECT: Qwizzard Programming Standards

_____ have received _____ questions about the quality

control for Qwizzard. Since Qwizzard was created by a team of _____ best

programmers, _____ should have no concern about assuring

_____ customers that Qwizzard is highly reliable. _____ is

easy to install and _____ directions for play are clear. Furthermore,

_____ screens were developed by _____ most talented

graphic artist, Shelley Ferguson. _____ spent months on the final designs;

_____ attention to the smallest details produced the finest graphics of any

software product _____ have produced. Accept _____

personal assurance that _____ customers will be delighted with Qwizzard.

Answers: A. them, they, them, them, their, their. Possible answers: **B.** We, your, our, you, your, It, its, its, our, She, her, we, my, your.

LEARNING UNIT 8-2
PERSONAL PRONOUNS: ADDITIONAL CASE
RULES

This learning unit gives three additional nominative case rules and three additional objective case rules. The last topic in this learning unit discusses the group of pronouns known as the intensive/reflexive compound personal pronouns. Writers often have difficulty with these pronouns.

ADDITIONAL NOMINATIVE CASE RULES

NOMINATIVE CASE RULE 3. **Use the nominative case form when the conjunctions _than_ or _as_ precedes the pronoun and a following verb is implied.**

It will help you use the correct pronoun form with these conjunctions if you mentally repeat the implied verb.

No worker could be more sincere than *he* [meaning *he is*].

Paul Jackson is about as tall as *I* [meaning *I am*].

NOMINATIVE CASE RULE 4.   Use the nominative case form when the pronoun functions as an appositive identifying a subject or a subject complement.

Recall that an appositive is a noun or pronoun that usually follows another noun or pronoun and renames or identifies it. In the first example sentence that follows, the pronoun precedes the noun it renames.

We workers need a break.

The two supervisors, Albert and *I*, were questioned by the company president. (*Albert and I* rename the supervisors and are in apposition to the subject *supervisors*.)

The only dissenters were my fellow employees, *you* and *he*. (*You* and *he* are in apposition to the subject complement *employees*.)

NOMINATIVE CASE RULE 5.   Use the nominative case form for a pronoun following the infinitive *to be* when the infinitive has no subject.

You may already know that the infinitive is the word *to* and a present verb form, such as *to sing, to talk*, and *to type*. You will recognize the present verb forms without the *to* as transitive and intransitive verbs. Infinitives are a special type of verb form that you will study in Chapter 12.

When the infinitive *to be* has no subject, the pronoun following *to be* acts like a subject complement and must be in the nominative case. Note the following sentence:

It appeared to be *she* who is causing the disturbance about the long hours. (*It* is the subject of the intransitive linking verb *appeared*. The pronoun *she* immediately follows the verb *to be* and functions as the linking verb's subject complement [predicate pronoun].)

This usage of the nominative case personal pronoun can cause confusion because in spoken English the objective case personal pronoun is often used. For example, you might hear someone say:

It appeared to be *her* who is causing the disturbance about the long hours.

When you analyze this sentence, you can see that the predicate pronoun *her* should be in the nominative case (*she*) as stated in Nominative Case Rule 2. Although *her* might be acceptable in verbal communication, writers should use *she* in their written communication. In Objective Case Rule 3 you will learn to use the objective case when the infinitive *to be* has a subject.

S T U D Y   T I P

The preposition *to* must not be confused with the infinitive *to*. The infinitive *to* is always followed by a verb. The preposition *to* introduces a prepositional phrase and is followed by a noun and its modifiers.

She went to the store. (*To* and the noun *store* form a prepositional phrase).

She went to sleep. (*To* and the verb *sleep* form an infinitive.)

ADDITIONAL OBJECTIVE CASE RULES

OBJECTIVE CASE RULE 2.   Use the objective case form after the conjunctions *than* and *as* when the pronoun is the object of an understood verb.

When you mentally fill in the missing verb, you should not have a problem. Note these sentences.

Noise bothers Larry as much as [it bothers] *me.*

The city official likes Mr. Regus as much as [he likes] *us.*

OBJECTIVE CASE RULE 3.   Use the objective case form for subjects of infinitives and for pronouns following the infinitive *to be* that have a subject.

Note that this is the only *exception* to the rule that subjects of verb forms should be in the nominative case. Let's first study an example where the pronoun is the subject of an infinitive.

We taught *her* to look for every accounting error.

When you study this sentence, you see that *taught* is a transitive verb. Transitive verbs must have direct objects, and pronoun direct objects are in the objective case. Therefore, you must use the objective case pronoun *her.* In this sentence, the infinitive verb form follows the objective case *her.* So we can say that subjects of infinitives must be in the objective case. This is explained further in Chapter 12.

The following sentence has a subject before the infinitive *to be*:

I am surprised that you thought him to be *me.* (A noun or a pronoun that precedes the infinitive *to be* is considered its subject. That subject—*him* in this sentence—and any pronoun that follows the infinitive must both be in the objective case.)

OBJECTIVE CASE RULE 4.   Use the objective case form when the pronoun functions as an appositive following a direct object, an indirect object, or an object of a preposition.

He resented his benefactors, Walter and *me*. (The direct object is *bene-factors*; the appositives are *Walter* and *me*.)

I spoke with the authors, Roy Chapman and *her*. (*Authors* serves as the object of a preposition; therefore, the appositive pronoun that follows it must be in the objective case.)

COMPOUND PERSONAL PRONOUNS

A few of the personal pronouns are made compound by adding the suffix *self* (singular) or *selves* (plural) to the personal pronoun form. These pronouns are the same in the nominative and objective case, and they have no possessive case.

Person	**Singular**	**Plural**
First:	myself	ourselves
Second:	yourself	yourselves
Third:	himself, herself, itself	themselves

Compound personal pronouns, which have the properties of person, number, and sometimes gender, serve two main purposes: (1) they intensify, or add emphasis, to nouns **(intensive pronouns)** and (2) they reflect the action of the verb back on the subject **(reflexive pronouns)**.

As intensive pronouns, compound personal pronouns are often used as an appositive with a noun or pronoun. Intensive pronouns reinforce, or give emphasis to, a related word or group of words in the sentence.

The president *himself* replied to my question.

I will install the equipment *myself*. (*Myself* intensifies *I*, the subject.)

The distributor *himself* commented on a possible shortage. (*Himself* intensifies the subject *distributor*.)

Reflexive pronouns occur in the predicate of the sentence and refer back to the subject. When a compound personal pronoun functions as a reflexive pronoun, it may be used as the direct object of a transitive verb or as the object of a preposition where the object denotes the same person or thing as the subject of the sentence or clause.

Notice how the following compound personal pronouns function as reflexive pronouns and reflect the verb action back on the subject:

She worried *herself* into a state of frenzy. (*Herself* is used as the direct object of the verb telling *whom* or *what* received the action. As a reflexive compound personal pronoun, *herself* refers back to the same person—the subject.)

He knew at the time that he was endangering *himself*. (*Himself* reflects back on *he*, the subject of the noun clause.)

Do not use the reflexive form of the compound personal pronoun as a substitute for a personal pronoun or as the subject of a sentence.

CORRECT: Roger expected help from Bradford and *me*.

INCORRECT: Roger expected help from Bradford and *myself*.

CORRECT: Clayton and *I* will deliver the furniture.

INCORRECT: Clayton and *myself* will deliver the furniture.

YOU TRY IT

Insert one of the following pronouns into the numbered spaces in the paragraph below. **Note:** A pronoun can be used more than once.

he   us
itself   him
himself

Rudy Green, founder of Learntech, created the company's first product _____.

No designer was more dedicated than _____. For years, no one developed

a more sophisticated product than _____. Tough competition forced

_____ to produce only the best. The business of developing multimedia

software has grown increasingly complex. The design process _____

demands more skills, time, and financial resources than ever before. The company's managers,

Rudy's staff, and _____, must keep up with unprecedented technological

change. The result is that today's consumers still do not like the competitors as much as

_____.

Answers: himself, he, he, him, itself, he, us

LEARNING UNIT 8-3
PRONOUN TYPES

The English language has seven types of pronouns: (1) personal, (2) compound personal, (3) demonstrative, (4) indefinite, (5) relative, (6) interrogative, and (7) reciprocal. Personal and compound personal pronouns were discussed in Learning Units 8-1 and 8-2. This learning unit discusses the remaining five pronoun types.

DEMONSTRATIVE PRONOUNS

The word *demonstrative* means pointing out the one referred to. The four **demonstrative pronouns** that point out their antecedents are *this* (plural, *these*) and *that* (plural, *those*).

Like nouns, demonstrative pronouns function as subjects, objects, and complements. They can come before or after their antecedents. The antecedents may be a noun, phrase, clause, or a separate sentence. Note that demonstrative pronouns *do not* introduce clauses.

This is my new screwdriver. (A definite screwdriver is pointed out. The pronoun *this* functions as a noun subject.)

Please hand me *that*. (A definite object is pointed out. The pronoun *that* functions as a direct object since it follows a transitive verb.)

When demonstrative pronouns are used with nouns, they function as adjectives and are called **demonstrative adjectives**.

This ladder is not high enough to reach *that* shelf. (*This* and *that* function as demonstrative adjectives and modify the nouns *ladder* and *shelf*.)

This letter is well written; *that* is too formal. (*This* functions as a demonstrative adjective, modifying *letter*; *that* is a demonstrative pronoun functioning as a subject.)

WRITING TIP

When you use demonstrative adjectives with such nouns as *kind(s)*, *sort(s)*, and *type(s)*, remember that these words mean one class or two or more classes. The demonstrative adjective must agree with the usage of these words as singular or plural.

INCORRECT: *this* **kinds,** *that* **kinds,** *these* **kind,** *those* **kind**

CORRECT: *this* **kind,** *that* **kind,** *these* **kinds,** *those* **kinds**

INDEFINITE PRONOUNS

Indefinite pronouns point out persons, places, and things, but they are not as specific as demonstrative pronouns. Typically, as their name indicates, indefinite pronouns designate an unidentified or not immediately identifiable person or thing.

The indefinite pronouns include the following:

all	each	few	nobody	some
any	either	many	none	somebody
anybody	every	many a	nothing	someone
anyone	everybody	more	one	something
anything	everyone	most	others	
both	everything	neither	several	

Indefinite pronouns usually do not have antecedents, but they may have a group as an antecedent. They are the same in the nominative and objective cases; change form like nouns in the possessive case; function as nouns and adjectives; and are singular, plural, or singular and plural.

Note the following examples of the use of indefinite pronouns:

Somebody said that you should call a meeting. (Functioning as a noun, *somebody* is the subject of the sentence.)

The supervisor called Beth *many* times. (Functioning as an adjective, *many* modifies *times*.)

RELATIVE PRONOUNS

In contrast with indefinite pronouns, **relative pronouns** introduce certain dependent adjective and noun clauses and *relate* or connect them to an antecedent (word or phrase) in the independent clause. In dependent clauses, relative pronouns can function as subjects, objects of verbs, or objects of prepositions.

The relative pronouns are *who, whom, whose, which, that,* and *what.* Use *who* when the antecedent is a person. Use *which* to refer to anything except persons. Use *that* when the antecedent is a person or thing. (Chapter 9 discusses when the use of *that* is preferred over *which.*) Compound relative pronouns are formed by adding *ever* or *soever* to the relative pronoun.

> This is the person *who* will train you. (*Who* is the subject of the dependent adjective clause *who will train you* and modifies the noun *person* in the independent clause.)

> The employee *that* borrowed my jacket lost it. (*That* is the subject of the dependent adjective clause *that borrowed my jacket* and modifies *employee* in the independent clause *The employee lost it.*)

> Give to the charity fund *whatever* you can afford. (*Whatever* introduces a noun clause.)

WRITING TIP

The relative pronoun *what* is the same as *that which;* it has no antecedent.

Take *what* (*that which*) you want. (The relative *what* combines in itself an antecedent and relative that is equivalent to *that which.*)

The only relative pronoun that changes case form is *who.* The case of *who* depends on its use in the dependent clause. This is why so many people have difficulty with the use of *who* and *whom.* After you study Chapter 9, you should not stumble over the usage of *who* and *whom.*

INTERROGATIVE PRONOUNS

The pronouns *who, whom, whose, which,* and *what* function as **interrogative pronouns** when they are used to ask questions. Note that except for *that,* the interrogative pronouns are the same as the relative pronouns. Since interrogative pronouns ask for their antecedents, no antecedents precede them. The antecedents are in the answer to the question. Like relative pronouns, the only difficulty with these pronouns is the case difference between *who* and *whom* (explained in Chapter 9).

> *Who* do you want to answer your telephone calls?

> *Which* desk is yours? (Here, the interrogative pronoun is used as an *interrogative adjective* modifying *desk.*)

> *What* is your problem?

RECIPROCAL PRONOUNS

The word *reciprocal* means something shared, felt, or shown by both sides. **Reciprocal pronouns** are *each other* and *one another*. As you might expect, *each other* refers to two persons or things. *One another* refers to three or more persons or things.

The two new employees will work well with *each other*.

The shipping department employees work well with *one another*.

Table 8-2 gives a summary of the seven pronoun types. Once you understand the meaning of each pronoun group title, you will soon recognize the pronouns belonging to each group. Chapter 9 continues the discussion of how to use pronouns.

STUDY TIP

All pronouns in Table 8-2 function in some typical noun sentence positions, but they cannot all function in every noun sentence position.

Table 8-2 Summary of Pronoun Types

Pronoun Group	Explanation	Pronoun Forms	Examples
Personal pronouns	Personal pronouns change in form to show person(s) speaking, person(s) spoken to, person(s) or thing(s) spoken of. They must agree with their noun antecedent in person, gender, and number. Their case depends on sentence usage.	First person: *I, me, my, mine; we, us, our, ours.* Second person: *you, your, yours.* Third person: *he, him, his; she, her, hers; it, its; they, them, their, theirs.*	*We* visited the new loading dock. Did *you* find the sales contract? Let's look for *her* old insurance policy.
Compound personal pronouns	Compound personal pronouns are personal pronouns that add the suffixes *self* or *selves*. They (1) intensify or add emphasis (intensive pronouns) and (2) reflect verb action back on the subject (reflexive pronouns).	*myself, yourself, himself, herself, itself, oneself, ourselves, yourselves,* and *themselves,*	I asked the supervisor *myself.* He bought *himself* a new fax machine.
Demonstrative pronouns	Demonstrative pronouns point to persons, places, and things to which writer wants to call attention; contrast relation between two or more objects; and may be used as adjectives.	*this, these; that, those*	Who did *this?*
Indefinite pronouns	Indefinite pronouns point out indefinite persons, places, and things; usually do not have antecedents or change forms, but can have a group as an antecedent; and function as adjectives and nouns. Some indefinite pronouns are always singular, some are always plural, and others are singular or plural.	*all, any, anybody, anyone, anything, each, everybody, everyone, everything, few, many, most, much, neither, none, one, several, some, somebody, someone,* etc.	*Anyone* can enter the race. *Several* left the meeting early. *Most* workers have left the building.

Table 8-2 Summary of Pronoun Types (Continued)

Pronoun Group	Explanation	Pronoun Forms	Examples
Relative pronouns	Relative pronouns are connecting words that relate directly to an antecedent (word or phrase) and introduce dependent clauses describing a noun in the main clause. They function as subjects, objects of verb, or objects of preposition.	*who, whom, whose, which, what, that*	This is the salesperson *who* will win the award.
Interrogative pronouns	Interrogative pronouns ask questions and have no antecedents. The antecedents are in the answer to the question.	*who, whom, whose, which, what*	*Who* can operate this computer?
Reciprocal pronouns	Reciprocal pronouns indicate the relationship to another.	*one another, each other*	Donna and Lois gave gifts to *each other*.

YOU TRY IT

1. Name the four demonstrative pronouns.

   a. _____

   b. _____

   c. _____

   d. _____

2. Is the word *that* in the sentence below a demonstrative pronoun or a relative pronoun?

   _____

   Kevin said that he would program the new module.

3. What indefinite pronoun is used in the sentence below?

   _____

   Few people are as talented as he.

4. Write a sentence in which the pronoun *who* is used as a relative pronoun.

   _____

5. Write a sentence in which the pronoun *who* is used as an interrogative pronoun.

   _____

6. Complete the following sentence by adding a reciprocal pronoun.

   Ms. Ryan and Ms. Gee helped _____.

Answers: 1. a. this b. that c. these d. those; 2. relative; 3. few; 4. The sentence should use *who* as the subject of a dependent clause, as in *The man who wrote the book will receive royalties.* 5. The sentence should ask a question, as in *Who wrote the book?* 6. one another or each other

CHAPTER SUMMARY: A QUICK REFERENCE GUIDE

PAGE	TOPIC	KEY POINTS	EXAMPLES
149 ·	**Personal pronoun**	The personal pronouns (1) stand for persons or things (nouns) and (2) change forms for case, person, gender, and number.	See Table 8-1.
150	**Nominative case basic rules**	**Nominative Case Rules 1 and 2:** Use nominative case when pronoun is (1) the subject of the verb and (2) the subject complement.	*I* sold the new store. I am *she*.
152	**Objective case basic rule**	**Objective Case Rule 1:** Use objective case when a pronoun is the direct and indirect object of a verb and the object of a preposition.	The professor praised *him*. He gave *her* a new car. I talked to *her*.
153	**Possessive case rule**	Use possessive case to show possession, authorship, brand, kind, origin, or any similar relation.	The dog wagged *its* tail.
154	**Nominative case additional rules**	**Rule 3:** Use nominative case when the conjunction *than* or *as* precedes the pronoun and a following verb is implied. **Rule 4:** Use nominative case when a pronoun functions as an appositive identifying a subject or a subject complement. **Rule 5:** Use nominative case for a pronoun following the infinitive *to be* when infinitive has no subject.	No workers could be as good as *they* (meaning they are). We, Jack and *I*, played cards. I would like to be *she*.
156	**Objective case additional rules**	**Rule 2:** Use objective case after the conjunctions *than* and *as* when a pronoun is the object of an understood verb. **Rule 3:** Use objective case for subjects of infinitives and for pronouns following the infinitive *to be* that have a subject. **Rule 4:** Use objective case when pronoun functions as an appositive after direct object, indirect object, or object of preposition.	John bothers Larry as much as [John bothers] *me*. We taught *her* to type. Did you think her to be *me*? Jim asked *me* to come. Lynn called us, Curt and *me*.
157	**Compound personal pronouns**	Compound personal pronouns (1) intensify, or add emphasis, to nouns (intensive pronouns) and (2) reflect the action of the verb back on the subject (reflexive pronouns).	*Intensive:* The supervisor *herself* answered the question. *Reflexive:* Corey found *himself* without a job.
161	**Pronoun overview**	See Table 8-2.	

QUALITY EDITING

You work in the marketing department of Learntech. The new assistant has prepared this draft of an announcement. Edit the announcement.

```
NOTCIE


The marketing gruop, the staff and me, are pleased to

annunciate the promotion of Josefina Lopez to manger. I

am very please to make this announcement herself since

I have worked with Josefina for the paste five years.

Josefina has made herself valuble to the other managers

and I. We rely on his good judgment and creativiaty

dailly.


Josefina will works directly with John Bush in Sales

John and her will be responsable for the promotion of

Qwizzard. This represents a significant opportunity for

he and she to help one anouther to bild Qwizzard sales.


Let's all congradulate Josefina.
```

APPLIED LANGUAGE

1. Select an article from a newspaper or a magazine. Circle all of the pronouns. Then, make a list of the first six pronouns you found and identify the type of each pronoun.

2. Find the four pronouns in the following sentence. Write them in the space provided and indicate what type of pronoun each is.

   This is the manager who can help us to improve ourselves.

   a. _____   c. _____

   b. _____   d. _____

3. Compare the following two sentences.
   a. That is the machine that can help you to improve yourself.

b. Those are the courses that can help them to improve themselves.

How are the sentences alike? How are they different? On the line below, write another sentence that fits this pattern.

_____

_____

4. Write a short note to your supervisor in the graphic design department at Learntech to recommend Lucy Chang for a job. Try to use a relative pronoun, an indefinite pronoun, a demonstrative pronoun or adjective, and a reciprocal pronoun. (Ideas: Ms. Chang was the employee who . . .; few can . . .; everyone thinks . . .; this company is.)

PERSONAL PRONOUNS AND PRONOUN TYPES

Part A. Personal Pronouns

In the space provided, write the pronoun that should be substituted for the word or words in italics.

_____ 1. Ms. Martinez claims that you are better informed on the matter than *Ms. Martinez*.

_____ 2. Joyce said: "Please be patient with my staff and *Joyce*."

_____ 3. Learntech's competitors claimed that the idea was really *Learntech's competitors*.

_____ 4. Josefina is not yet aware that John and *Josefina* will be working together.

_____ 5. Mr. Green has asked that we refer all applicants to either Ms. Morgan or *Mr. Green*.

_____ 6. Michelle Lee said that the program was checked by *Michelle Lee* and her assistant.

_____ 7. According to Victor, the leaders will be *Victor* and Carlos.

_____ 8. We will buy new software programs if we can afford *new software programs*.

_____ 9. Gretchen told us that everyone except *Gretchen* and Olga had reviewed the document.

_____ 10. Maggie knew that *Maggie* and the others were next.

Part B. More Personal Pronouns

Circle the correct pronoun of the two or more pronouns shown in parentheses.

1. An investigation should be made by either you or (I, me).

2. My colleagues and (I, myself, me) may attend the digital media conference.

3. Were you present when (we, us) program designers presented our plan?

4. Many people thought the best programmer to be (he, him).

5. Neither (she, her) nor her employer could correct the problem.

6. If anyone in our department can fix it, it will be (she, her).

7. The director refused to give (we, us) engineers a chance to examine the program.

8. No one but you and (I, me, myself) know where the disks are stored.

9. Yesterday, I discussed the issue with both (she, her) and her colleagues.

10. (He, Him) and Latham will assist you with the books.

11. (They, Them) and Jan will attend the meeting.

12. Tomorrow, Sylvia will give the report to both (they, them) and her staff.

13. Kenneth and (I, me) will help with the project.

14. Ms. Tso said that it was not (she, her) who suggested the change.

15. (We, Us) would like you to keep better records.

16. The right to distribute the new program is (ours, our's).

17. It was (she, her) who created the files.

18. Both you and (he, him) should check your work more carefully.

19. Are you and (she, her) supervisors in your department?

20. Pamela will work faster than (he, him).

21. No manager could be better informed than (she, her).

22. Jenson is as accurate as (she, her).

23. (We, Us) editors clarify written information.

24. The only ones to finish were my colleagues, Karen and (he, him).

25. The two accountants, Alice and (I, me), prepared the budget.

26. The new policy affects Bob as much as (she, her).

27. Dr. Green spoke with the contributors, Salinas and (I, me).

28. Wilson accepted the award from Sam and (me, myself).

29. The animator and (I, myself) will leave soon.

30. They decided to keep the information to (theirselves, themselves).

Part C. Other Pronoun Types

Five pronoun types are used in the sentences below. Label the italicized pronouns by writing D (demonstrative), Ind. (indefinite), Rel. (relative), Int. (interrogative), or Rec. (reciprocal) above the italicized word(s).

1. *These* will work so well that they may replace our current ones.

2. The company made the new benefits available to *everyone*.

3. Ernest is the employee *who* shows the most improvement.

4. Hugh and Siga communicated with *each other* by electronic mail.

5. *Which* is the faster machine?

6. *What* will Nathan do after Qwizzard is launched?

7. Qwizzard is the product *that* will make our company profitable.

8. *Those* are not the specification sheets for our software.

9. *Several* of our designers have received awards for their work.

10. In Qwizzard, the players must cooperate with *each other* to earn clues.

CHAPTER 9

PRONOUN USAGE

Writers often find that by getting a good sense of pronouns, writing becomes clearer and more direct.

—Val Dumond, *Grammar for Grownups*

LEARNING UNIT GOALS

LEARNING UNIT 9-1: AGREEMENT OF ANTECEDENTS AND PERSONAL PRONOUNS

- Make antecedents and personal pronouns agree in number. (pp. 173–75)
- Make antecedents and personal pronouns agree in gender. (pp. 175–76)

LEARNING UNIT 9-2: COMPANY NAMES AND INDEFINITE PRONOUNS AS ANTECEDENTS

- Use company names correctly as antecedents. (pp. 176–77)
- Use indefinite pronouns correctly as antecedents. (pp. 177–79)

LEARNING UNIT 9-3: USAGE PROBLEMS OF MISCELLANEOUS PRONOUNS

- Know the correct use of *who* and *whom*; *whoever* and *whomever*. (pp. 180–82)
- Know the correct use of *whose*; *who, which,* and *that*. (pp. 182–83)

Have you ever read a sentence in which you were confused because the antecedent of a pronoun was not clear? Proper pronoun usage begins by avoiding an ambiguous pronoun reference (pronoun points to more than one antecedent) and a vague pronoun reference (pronoun refers to an implied word that is not in the sentence).

To correct ambiguous and vague pronoun references, you have three choices. You can (1) replace the pronoun with a noun, (2) move the pronoun closer to its antecedent, or (3) rewrite the sentence. Often, the best solution is to rewrite the sentence.

This chapter divides pronoun usage into three units. In Learning Unit 9-1, you will learn about the agreement of antecedents and personal pronouns. Learning Unit 9-2 discusses company names and indefinite pronouns as antecedents. In Learning Unit 9-3, you are taught how to solve the usage problems of *who* and *whom*; *whoever* and *whomever*; *whose*; and *who*, *which*, and *that*.

LEARNING UNIT 9-1
AGREEMENT OF ANTECEDENTS AND PERSONAL PRONOUNS

Since personal pronouns are substitutes for nouns, they should always refer to a specific person, place, or thing. Complete agreement must exist between pronouns and their antecedents. No reader should have to question the antecedent of a pronoun. In the following examples only one solution is given under *USE*. You can probably think of other solutions.

DON'T USE: The real estate agent told Mr. Quigley that he does not understand the building code problem. (*Who* does not understand the building code problem—the real estate agent or Mr. Quigley?)

USE: In his conversation with Mr. Quigley, the real estate agent admitted that he does not understand the building code problem.

DON'T USE: Dr. Johnson and Mr. Polson met with contractors, but he refused to discuss the appointment. (Who does the *he* refer to—Dr. Johnson, Mr. Polson, or both?)

USE: When he and Mr. Polson met with contractors, Dr. Johnson refused to discuss the appointment.

In Chapter 8, you learned how to use the case forms of personal pronouns. Recall that nominative case personal pronouns are used for subjects and complements of verbs. Objective case personal pronouns are used for objects of verbs and prepositions. Remembering these uses of the nominative and objective cases will help you understand this learning unit.

Before you begin to study antecedent agreement in number and gender, note that three personal pronouns have no antecedents. The first-person pronoun *I* requires no antecedent because it stands for the speaker. The second-person pronoun *you* usually does not have an antecedent. When *it* is used in a general sense such as *It rained last night*, the pronoun *it* does not have an antecedent. Be aware, however, that the use of *it* can make a sentence vague.

NUMBER AGREEMENT

A personal pronoun and its antecedent *must* agree in number. If the antecedent is singular, the pronoun must be singular. If the antecedent is plural, the pronoun must be plural. Note the following examples of singular and plural pronoun agreement:

Singular: The new insurance agent could do more business if *she* sold insurance during the evening hours. (*She* is the subject of the verb *sold*. Subjects of verbs must be in the nominative case.)

Plural: The *retail store owners* are concerned about decreasing sales, but *they* do not know what steps should be taken. (*They* is the subject of the verb phrase *do not know*. Subjects of verbs must be in the nominative case.)

Making antecedents and pronouns agree is not always as easy as in the preceding sentences. Situations do occur that can give problems. Five of these situations are given in the subheadings that follow. Under each subhead is the solution to the problem situation.

In studying the following rules, you can often use your "ear." When the correct pronoun form is also the one that sounds correct to you, move on to the next rule. Concentrate only on the rules that are unfamiliar to you. You will note that several examples use the adjective pronouns discussed in Chapter 8.

Singular Antecedents Connected by *and*

1. **When two or more singular antecedents connected by *and* refer to only one person or thing, the pronoun must be singular.** Confusion can result when the reader does not know if the writer is referring to one person or two people. If one person is being discussed, then a modifier, such as *my* or *the*, is used only once—before the first antecedent.

   My confidante and business associate is always here when I need *her*. (If two people were being discussed, *my* would be repeated before the words *business associate*.)

   The president and district superintendent budgets *his* time carefully. (If two people were being discussed, *the* would be repeated before the words *district superintendent*.)

   If a subject is obviously plural, the modifier (*my*, *the*, etc.) need not be repeated.

   My mother and father own their own business.

2. **When two or more singular antecedents connected by *and* refer to different persons or things, the pronoun must be plural.**

   Mr. Odom and *Mr. Grazzo* decided that *they* would combine *their* resources.

The assistant *cashier* and the chief *appraiser* have just submitted their resignations. (The word *the* is repeated to show clearly that two people are involved.)

Antecedents Connected by *or* or *nor*

1. **Singular antecedents connected by *or* or *nor* require singular pronouns.**

   Ms. Barbetti or Ms. Shimoto must put *her* signature on this bill of lading.

   Neither Ron nor Carl could find *his* copy of the bill of sale.

2. **If the antecedents consist of a singular noun and a plural noun connected by *or* or *nor*, write the singular noun first. Then, make the pronoun plural to agree with the second, or nearer, antecedent.**

   Neither Mr. Foster nor his *salespeople* would admit that *they* had overheard the conversation.

   The president or his *advisers* should devote part of *their* time to the proposed merger.

Antecedents Preceded by *each, every, neither,* or *many a*

When antecedents are preceded by *each, every, neither,* or *many a*, pronouns are always singular. A singular pronoun should be used even when two such antecedents are joined by *and.*

Every girl at the baseball tryouts hoped that *she* would be chosen.

Every man and boy in the barber shop expressed *his* opinion.

Many a young man has placed *his* trust in this product.

Collective Noun as Antecedent

1. **When the antecedent is a collective noun referring to a group as a single unit, use the singular pronoun *its*.** A noun is collective if in its singular form it names a group of people or objects such as *jury, audience, faculty, committee, number,* or *family.* Company names are usually considered to be collective nouns.

   The audience expressed *its* pleasure with resounding applause.

   The jury has reached *its* decision.

   Stein & Grayson, Inc., has sold *its* Chicago properties.

2. **When the antecedent is a collective noun referring to individuals acting separately or independently of one another, use a plural pronoun.**

   The audience are going *their* separate ways.

   The crew continued to argue among *themselves.*

Although these sentences are grammatically correct, they sound awkward. In such situations, you can eliminate any possibility of awkwardness by using a subject that is obviously plural, as in the following sentences:

Members of the audience are going *their* separate ways.

Crew members continued to argue among *themselves.*

When using the collective noun *number* as a subject, remember that *the number* is always singular and *a number* is always plural.

The number of people in attendance was small.

A number of the workers have changed *their* minds.

Parenthetical Expressions

A **parenthetical expression** is a word or group of words that has been placed in a sentence as an aside and tends to interrupt the natural flow. **Parenthetical expressions or words that appear between an antecedent and a pronoun do not influence the form of the pronoun that should be used.** Do not be misled by parenthetical expressions beginning with *rather than, as well as, in addition to, together with,* or similar words. Such expressions are usually set off with commas. (See Chapter 17 for a discussion of parenthetical elements.)

The accountant, rather than any of the other officers, will be asked for *his* opinion of this venture.

Mr. Norman Brett, in addition to several other members of the organization, has offered *his* unqualified support.

GENDER AGREEMENT

Pronouns must agree with their antecedents in gender. When antecedents are definitely male or female, this rule is easy to follow. Note the following sentences:

Roger Weiskof plans to open *his* own office in New Orleans. (The masculine gender is correct here.)

Louise Handley will be given considerable authority if *she* accepts the new position. (The feminine gender is correct here.)

Our new product has many fine features, but *it* has not been well accepted by the public. (The neuter gender is correct here.)

The traditional use of a male pronoun for antecedents that could refer to female and male is no longer accepted. Sometimes, however, writers want to emphasize the two sexes.

1. **When writers want to be precise in their gender references, they may use both masculine and feminine pronouns.**

Every employee will be expected to punch *his* or *her* timecard.

Each person must present *his* or *her* request in writing.

2. If two singular antecedents of different genders are used, two pronouns of different genders must be used.

> No man or woman can afford to gamble with *his* or *her* future by regarding English lightly.

> Although the preceding sentence is grammatically acceptable, it is awkward and, therefore, should be avoided. A smoother wording would be *College students cannot afford to gamble with their futures by regarding English lightly.* Remember that grammatical correctness is only one test of good writing or speaking. The goal of writing is to use the English language effectively so that readers understand the message writers want to express.

YOU TRY IT

Circle the correct pronoun of the two or more pronouns shown in parentheses.

1. The manager and business officer has a car for (her, their) own use.

2. Dr. Gray and Dr. Macaulay were convinced that (they, she, he) could be more effective by working together.

3. Neither William nor his investigation committee members were ready to report (their, his) findings.

4. Every manager in the company has been given (their, his or her) evaluation.

5. The board of trustees decided to limit (their, its) meeting to four hours.

6. The committee disagreed among (itself, themselves) about the recommendations.

7. The staff members, as well as the director, offered (their, his) congratulations.

8. Every manager must submit (their, his or her) evaluation.

Answers: 1. her 2. they 3. their 4. his or her 5. its 6. themselves 7. their 8. his or her

---

LEARNING UNIT 9-2
COMPANY NAMES AND INDEFINITE PRONOUNS AS ANTECEDENTS

Learning Unit 9-1 discussed the agreement of number and gender of personal pronouns. This learning unit discusses when to consider a company as a group of individuals and how to use indefinite pronouns as antecedents.

COMPANY NAMES AS ANTECEDENTS

Since a company is made up of individuals, a writer might have those individuals in mind when using the company's name, as in the following example:

I wrote to the Kingston Company and told *them* to expect an early delivery. (A collective noun, *Kingston Company*, is used here with a plural pronoun, *them*.)

The example above is acceptable as long as it is not awkward. This sentence would have sounded strange if the singular pronoun *it* had been used instead of *them*. Remember, however, that company names are *usually* considered singular. As a general rule, use singular verbs and pronouns when referring to the names of companies.

INDEFINITE PRONOUNS AS ANTECEDENTS

The English language has many words that may function as noun substitutes even though they do not possess the properties of personal pronouns. Indefinite pronouns belong to this group. **Indefinite pronouns**, as their name indicates, point out persons, places, and things that are unidentified or not immediately identifiable.

The indefinite pronouns include the following:

all	each	few	nobody	some
any	either	many	none	somebody
anybody	every	many a	nothing	someone
anyone	everybody	more	one	something
anything	everyone	most	others	
both	everything	neither	several	

Like nouns, indefinite pronouns have the same form in the nominative and objective cases and they change forms in the possessive case. Indefinite pronouns can function as adjectives, subjects of verbs, and antecedents of personal pronouns. As the subject of a verb, the indefinite pronoun must agree with the verb in gender and number. This is also true when indefinite pronouns are antecedents of personal pronouns. (Chapter 12 discusses indefinite pronoun agreement with verbs.)

Some indefinite pronouns are always singular, some are always plural, and some are singular or plural. These three groups of indefinite pronouns are discussed in Chapter 12.

Indefinite Pronouns That Are Always Singular

The indefinite pronouns in the following list are always singular. They often function as singular antecedents for personal pronouns.

anybody	every	neither	someone
anyone	everybody	nobody	something
anything	everyone	nothing	
each	everything	one	
either	many a	somebody	

Note the incorrect and correct examples on the following page.

INCORRECT: *Everyone* should know *their* shortcomings. (*Everyone*, the singular indefinite pronoun, does not agree with *their*, the plural personal pronoun.)

CORRECT: *Everyone* should know *his or her* shortcomings. (*Everyone*, the singular indefinite pronoun, does agree with *his or her*, the singular personal pronouns.)

INCORRECT: *Anyone* in this position must be careful about *their* appearance. (*Anyone*, the singular indefinite pronoun, does not agree with *their*, the plural personal pronoun.)

CORRECT: *Anyone* in this position must be careful about *his or her* appearance. (*Anyone*, the singular indefinite pronoun, does agree with *his or her*, the singular personal pronouns.)

Gender bias must be avoided in business writing. Do *not* use a masculine pronoun to refer to a singular indefinite pronoun. The following sentence is no longer acceptable:

Anyone in this position must be careful about *his* appearance.

Using *his or her* can become awkward. Avoid using this pair more than once in a sentence. Often the sentence can be rewritten in the plural and the single indefinite pronoun avoided.

People in this position must be careful about *their* appearance.

Sometimes you can recast the sentence as follows:

In this position, *you* must be careful about *your* appearance.

WRITING TIP

Write *some one, any one,* or *every one* as two words only when an *of*-phrase (prepositional phrase beginning with *of*) follows.

Any one of these floppy disks should prove satisfactory.
Every one of our engineers has a bachelor's degree.

Indefinite Pronouns That Are Always Plural

The indefinite pronouns that follow are always plural. When used as antecedents of personal pronouns, the personal pronouns must also be plural.

both	few	many	others	several

Several of the jobbers expressed *their* opinions about excessive discounts. (The plural indefinite pronoun *several* agrees with the plural personal pronoun *their*.)

Many of the workers objected to *their* new chairs. (The plural indefinite pronoun *many* agrees with the plural personal pronoun *their*.)

Note the prepositional phrases that follow the indefinite pronouns in the sentences on the previous page. These phrases indicate that the writer needs a plural indefinite pronoun and a plural personal pronoun.

Pronouns That May Be Singular or Plural

The following indefinite pronouns may be either singular or plural:

all	any	more	most	none	some

The word to which the pronoun refers is the determining factor. Again, the object of the prepositional phrase following these pronouns determines whether the pronoun should be considered singular or plural. This is illustrated in the two sentences that follow.

Most of the representatives found *their* correct seats.

All of the employees received *their* bonuses.

Note the following exceptions:

None of the applicants brought *his or her* pen. (The pronoun *none* is singular whenever its meaning is clearly *not one.*)

Any of the younger children can reach *his or her* bookshelf. (*Any* is singular whenever its meaning is clearly *any one.*)

YOU TRY IT

Circle the correct pronoun of the two shown in parentheses.

1. Although the Star Travel Agency was small, (it, they) served a large number of clients.

2. Everyone should write (their, his or her) congressional representative about this crucial environmental issue.

3. Anyone who agreed to the terms was given (his or her, their) new contract.

4. (Every one, Everyone) who was (any one, anyone) attended the benefit banquet.

5. (Anyone, Any one) of these will do.

6. Both of them turned in (his or her, their) reports.

Answers: 1. it 2. his or her 3. his or her 4. Everyone, anyone 5. Any one 6. their

---

LEARNING UNIT 9-3
USAGE PROBLEMS OF MISCELLANEOUS PRONOUNS

This learning unit discusses several miscellaneous pronouns. The first topic is the correct use of *who* and *whom.* A discussion of the compounds *whoever* and *whomever* follows. The correct use of *whose* is the third topic. Then, you will learn how to use *whose, who, which,* and *that.*

CORRECT USE OF *WHO* AND *WHOM*

Who and *whom* are relative pronouns and interrogative pronouns that often cause difficulty for writers. The correct use of these pronouns depends on whether they are used in the nominative or objective case. *Who* is a nominative case pronoun and functions as the subject or subject complement of a verb. Subject complements always follow intransitive linking verbs. *Whom* is an objective case pronoun and functions as the object of a verb or a preposition and as the subject or object of an infinitive. In this learning unit, we are only concerned with *whom* as the object of a verb and preposition.

Note the following uses of *who* and *whom.* In most of these examples, you will see immediately that the nominative case calls for the pronoun *who* and the objective case calls for the pronoun *whom.*

Who requested the information? (*Who* is the subject of the transitive verb *requested.*)

Who placed this order for 50 bags of cement? (*Who* is the subject of the transitive verb *placed.*)

Whom does Joan like? (Reverse this sentence and you have *Joan likes whom? Whom* is the direct object of the transitive verb *likes.*)

To *whom* should this letter be addressed? (*Whom* is the object of the preposition *to.*)

Whom were you with on Thursday afternoon? (*Whom* is the object of the preposition *with.* Try this wording: *You were with whom on Thursday afternoon?*)

The following method for making the choice between *who* and *whom* is similar to the cross-out method used in Chapter 8 for determining the correct case for compound sentence elements. Remember that *he, she, they, who,* and *whoever* are subjects and must be in the nominative case. The *m* forms (*him, them, whom, whomever*) and *her* are objects and must be in the objective case. We will use the following sentence to illustrate the cross-out method:

Mr. Maxwell is the salesperson (*who/whom*) spoke with you on the telephone.

1. Temporarily cross out all the words in the sentence *except* the clause that begins with the pronoun that you question.

   ~~Mr. Maxwell is the salesperson~~ (*who/whom*) spoke with you on the telephone.

   You are now working with the adjective clause in the sentence.

2. Invert the clause if necessary so it is in the subject-verb-object order; also cross out parenthetical expressions.

3. Find the verb of the clause. *Spoke* is the verb. Subjects of verbs are in the nominative case. *Who* is the correct choice, since it is the subject of the verb. *Whom* is in the objective case.

The sentence should read as follows:

> Mr. Maxwell is the salesperson *who* spoke with you on the telephone.

In the following sentences, the *who* or *whom* is used correctly. Check the sentences with the cross-out method. Hints are given in parentheses following the sentences.

> *Who* would you guess has seniority in this office? (Did you recognize *would you guess* as a parenthetical expression and cross it out? Expressions of this type tend to interrupt the natural flow of a sentence.)

> I have no way of knowing *who* it is. (In the noun clause *who it is, it* is the subject and *who* is a subject complement.)

> The job was given to the applicant *whom* you recommended. (The adjective clause *whom you recommended* modifies *applicant.* The subject of *recommended* is *you*; the direct object is *whom.*)

> We were surprised to hear about Dr. Lonberg, *whom* Mr. Cassady trusted completely. (*Whom* is the direct object of *trusted.*)

> *Whom* did you have in mind? (You might try this wording: *You did have whom in mind.*)

> *Whom* were you with on Thursday afternoon? (*Whom* is the object of the preposition *with.* Try this wording: *You were with whom on Thursday afternoon.*)

Sometimes you may have difficulty when a sentence has an inverted word order and you are not sure whether the pronoun functions as a subject or object. Remember that you must look at how the pronoun functions in its own clause. Reverse the word order so that you have the subject-verb-object order. If the verb has no other subject, the pronoun must be the subject, and you will use the nominative case *who.* If the verb has a subject, the pronoun must be the objective case *whom.*

CORRECT USE OF *WHOEVER* AND *WHOMEVER*

If you know how to use *who* and *whom*, you will know how to use their compound forms *whoever* and *whomever. Whoever*, like *who*, is in the nominative case; *whomever*, like *whom*, is in the objective case. The difficulty you may have is to remember that you must look *within* the clauses that themselves function as objects and subjects. Recall from Chapter 6 that noun clauses can function as major sentence elements. *Whoever* and *whomever* are either the subject or object of the noun clause and *not* the object of the main clause.

The cross-out method also works for *whoever* and *whomever.* Observe their roles in the following sentence:

> Prices of the concert will be set by the company or by (*whoever/ whomever*) hired the group. (The noun clause is [*whoever/whomever*] *hired the group,* and it functions as the object of the preposition *by.*

You must look at how *whoever/whomever* is used *in* the noun clause and not look at it as the object of the preposition.)

Cross out everything in the sentence except the noun clause.

~~Prices of the concert will be set by the company or by~~ (*whoever/whomever*) hired the group. (*Hired* is the verb. Subjects of verbs are in the nominative case. *Whoever* is the subject.)

The sentence should read as follows:

Prices of the concert will be set by the company or by *whoever* hired the group.

Here is another example:

We will need the name of (*whoever/whomever*) you select.

~~We will need the name of~~ (*whoever/whomever*) you select. (*Select* is the verb of the noun clause. The subject of *select* is *you*. Invert the clause to its subject-verb-object order. *You select* (*whoever/whomever*). The object of a verb is in the objective case. *Whomever* is in the objective case.)

The sentence should read as follows:

We will need the name of *whomever* you select.

CORRECT USE OF *WHOSE*

The word *whose* is a possessive case form. Do not confuse it with the contraction *who's*, meaning *who is*.

Whose typewriter is in need of repair?

We need more people like Dodson, *whose* work is always neat and well organized.

Who's the new advertising executive for the Acme Agency?

CORRECT USE OF *WHO*, *WHICH*, AND *THAT*

The word *who* (or *whom*) is used to refer to persons. *That* may be used to refer to animals or inanimate objects; it may also refer to people spoken of as a class or type, as shown in the third sentence below. *Which* may be used to refer to animals or inanimate objects.

Amelia Maznicki is the insurance agent *who* conducted the investigation.

We have not yet received the machine *that* we ordered.

She is not the kind of supervisor *that* the trucking division needs.

Please work on the Agee account, *which* must be ready by tomorrow.

You should know the following rule for *that* and *which*: Use *that* to introduce an idea that is essential (restrictive) to the meaning of the sentence; use *which* to introduce an idea that is nonessential (nonrestrictive) to the meaning of the sentence. Nonessential ideas are usually set off with commas.

Essential: Jacob prefers that all the information in the report be kept confidential.

Nonessential: All the information, which is confidential, can be found in this report.

WRITING TIP

Be careful of sentences that contain expressions such as *one of the men who*. Study the following sentence:
She is one of the employees who seem happy with their work. (*Who seem happy with their work* is an adjective clause modifying *employees*. Its subject, *who*, is plural because the antecedent of *who*, *employees*, is plural; therefore, a plural verb, *seem*, has been used. *Their* agrees in number with its antecedent, *employees*.)

YOU TRY IT

Circle the correct pronoun of the two shown in parentheses.

1. With (who, whom) will you travel?

2. Mr. Tsu, (whose, who's) clients are large companies, exceeded the sales goals for July.

3. Ms. Machida is the agent (who, whom) handled those reservations.

4. She is the candidate (who, whom) the polls show will receive the most votes.

5. (Who, Whom) did you recommend for the position?

6. Star Travel is the company (that, who) booked the most rooms for the cruise.

7. I told Star Travel that (they, it) should answer the phone more quickly.

8. (Whoever, Whomever) at Star Travel conducted the research on this tour did a great job!

Answers: 1. whom 2. whose 3. who 4. who 5. Whom 6. that 7. they 8. Whoever

---

CHAPTER SUMMARY: A QUICK REFERENCE GUIDE

PAGE	TOPIC	KEY POINTS	EXAMPLES
173	**Number agreement**	**1. Singular antecedents connected by *and*.** **a.** Two or more antecedents referring to one person or thing require singular pronouns.	My manager and friend received *his* promotion yesterday.

PAGE	TOPIC	KEY POINTS	EXAMPLE
		b. Two or more antecedents referring to different persons or things require plural pronouns.	Lynn and Kurt thought *they* should sign the petition.
		2. Antecedents connected by *or* or *nor*.	
		a. Singular antecedents require singular pronouns.	Either June or Beth should give *her* consent to the new proposal.
		b. Singular and plural nouns together: Write singular noun first, then plural noun. Then make pronoun plural to agree with plural noun.	Neither the owner nor the workers found *their* missing tools.
		3. Antecedents preceded by *each*, *every*, *neither*, or *many* **a.** A singular pronoun is always used.	Every man in the company received *his* new tie.
		4. Collective noun as antecedent.	
		a. When referring to a group as a single unit, use the singular pronoun *its*.	The committee filed *its* report.
		b. When referring to individuals acting separately, use a plural pronoun.	A number of workers looked for *their* missing keys.
		5. Parenthetical expressions. When these appear between an antecedent and a pronoun, they do not influence the pronoun form.	June, as well as her coworkers, showed *her* disapproval for the intrusion.
175	**Gender agreement**	**1. Do *not* use male pronouns when referring to antecedents that could also be female.** **2. Sometimes it is necessary to be precise in gender references.** **3. For two singular antecedents of different genders, use two different pronouns.**	*No*: Everyone left *his* coat in the hall. *Yes*: Everyone left *his* or *her* coat in the hall. Every employee received *his* or *her* bonus. No girl or boy has received *her* or *his* new uniform.
176	**Company names as antecedents**	**Company names are *usually* considered singular.** If this is awkward, use a plural pronoun.	I told the Haring and Sons Company that *its* main office closed too early.
177	**Indefinite pronouns as antecedents**	**1. Always singular:** *anybody, anyone, anything, each, either, every, everybody, everyone, everything, many a, neither, nobody, nothing, one, somebody, someone, something.*	*Something* is wrong with the new computer.

PAGE	TOPIC	KEY POINTS	EXAMPLE
		2. **Always plural:** *both, few, many, others, several.* 3. **Singular or plural:** *all, any, more, most, none, some.*	*Several* of the new supervisors went to the convention. *All* of the new salespersons are in the store.
180	**Correct use of *who* and *whom***	1. ***Who* is the nominative case, so use it as the subject of verb or as a subject complement.** 2. ***Whom* is in the objective case, so use it as the object of verb or a preposition and as the subject or object of an infinitive.** 3. ***Cross-out method:*** **a.** Cross out all words except the clause that begins with the pronoun. **b.** Invert clause if necessary; cross out parenthetical expressions. **c.** Find the verb of the clause. For verb subjects, use nominative case pronouns. For verb objects, use objective case pronouns.	*Who* made this telephone call? To *whom* are you speaking?
181	**Correct use of *whoever* and *whomever***	Use the same rule as for *who* and *whom.*	*Whoever* wants to come is welcome. *Whomever* did you choose?
182	**Correct use of *whose***	Do not confuse *whose* with *who's* (*who is*).	*Who is* the new president?
182	**Correct use of *who*, *which*, and *that***	*Who* (or *whom*) is used for persons; *that* is used for animals or inanimate objects or people spoken of as a class; *which* is used for animals or inanimate objects. Use *that* for essential ideas. Use *which* for nonessential ideas.	

QUALITY EDITING

Star Travel has prepared the following draft of an advertising flyer. Edit the draft.

Whose ready for a trip? Whomever has the yearning to

see exotic places and learn about the worlds' cultures

should take themselfs derectly to Star Travel Star Travel is the company who takes care of each and every-one of their cleints like family. Ms. Arenas, the owner and office manager, made travvel their life's work. If who you conduckt bissiness with matters, you will find Star Travel a delight. Call today and ask for the Star agent whom will make your next trip the most exciting never!

APPLIED LANGUAGE

1. Write a sentence that begins with each of the following words:

   a. Who _____

   b. Whom _____

   c. Whose _____

   d. Who's _____

   e. Whoever _____

   f. Whomever _____

2. When a company name is the antecedent, the pronoun used is usually *it*. Sometimes, however, the pronoun *they* can be used, especially when referring to people in the company. Write a sentence with *Star Travel* as the antecedent and the pronoun *it*. Then, write one with *Star Travel* and the pronoun *they*.

   a. _____

   b. _____

3. Write a short letter to Star Travel describing your experiences on a cruise ship. Use the words *who* and *whom* in both clauses and phrases. (Suggestions: *the steward who*, *whomever I played tennis with*, and *whose sport clothes were.*)

WORKSHEET 9

Name _____

Date _____

PRONOUN USAGE

Part A. Agreement of Antecedents and Personal Pronouns

Each of the sentences below contains an italicized pronoun. Write the antecedent of that pronoun in the space provided.

EXAMPLE:

This paragraph is clear, but *it* lacks emphasis. _____**paragraph**_____

1. This tour would be popular with businesspeople and *their* families.

   _____

2. Stephanie may lose her seat if *she* does not return her reservation card.

   _____

3. The ship does not have a tennis court, but *it* does have a shuffleboard court.

   _____

4. Chet's shyness should have been a handicap; instead, *it* endeared him to his clients.

   _____

5. Neither the marketing director nor the trainees lost *their* patience.

   _____

6. Captain Dewitt, as well as the cruise line, expressed *his* willingness to help.

   _____

7. The travel information booklets were sent from the travel agency to *its* clients.

   _____

8. The loan officer, after talking with the committee members, gave *her* approval.

   _____

In each of the sentences below, two pronouns appear in parentheses. Circle the correct pronoun.

9. If Ms. Wong or Mrs. Andrews decides to go, (she, they) must make prior arrangements.

10. The firm has offices in several large cities, but (its, their) headquarters are in Boston.

11. Every woman on the panel has (her, their) own ideas about the project.

12. Mr. Jay and Mr. Arrizán said (he, they) would answer the client's inquiry.

13. Each agent contacted (their, his or her) own clients.

14. Neither airlines nor railways reported (its, their) passenger counts for the year.

15. Neither the airline nor the railway reported (its, their) passenger counts for the year.

16. Each director and each manager expressed (his or her, their) appreciation to the chairperson.

17. Neither Lavonne nor the other designers finished (her, their) work before midnight.

18. Ms. Barrett, rather than the others, corrected the error, but (she, they) said nothing.

19. If Mark or Luke has a solution, (he, they) should speak up.

20. If Mark and Luke have a solution, (he, they) should speak up.

21. The art director and exercise enthusiast used (his, their) lunch hour to work out.

22. Neither Christopher nor his colleagues told us that (he, they) had been to the conference.

23. Elizabeth, as well as the other editors, gave (her, their) full support.

24. Four trainees, together with two of the instructors, will give (his or her, their) presentation on Friday.

25. Our committee, as you know, has done (its, their) best.

26. The designer and the programmer asked how (she, they) could help.

27. Two of the programmers, together with Ms. Strong, will test it (themselves, herself).

28. Every person in this company has (his or her, their) own specific duties.

Part B. Company Names and Indefinite Pronouns as Antecedents
Circle the correct pronoun of the two pronouns in parentheses.

1. Sanchez & Sanchez, Inc., has opened (its, their) new office in Springfield.

2. Everyone in Julia's group has (her, their) own manual.

3. The Fitness Health Club created (its, their) own workout fashion line.

4. (Anyone, Any one) with experience in tourism will be considered.

5. If any of the sulfur has spilled, clean (it, them) up immediately.

6. If any of the tickets are missing, find (it, them) immediately.

7. Several of the passengers complained, but I doubt that (he or she, they) will sue the bus line.

8. Either route will be satisfactory if (it, they) goes through Memphis.

9. Both routes will be satisfactory if (it, they) go through Memphis.

10. Each woman member voted to increase (her, their) volunteer time.

Part C. Using Miscellaneous Pronouns Correctly
Circle the correct pronoun of the two or more pronouns in parentheses.

1. The clerk with (who, whom) you spoke is Jason.

2. Call me immediately if you discover (who, whom) has the brochures.

3. We need the name of a company (who, whom, that) makes travel videos.

4. (Whoever, Whomever, Whom ever) accepts this work will supervise the new agents.

5. Davidson sent a cryptic response to the person (who, whom) arranged the tour.

6. She addressed the letter "To (Whom, Who) It May Concern."

7. I would like to know (who, whom) will be the next bursar.

8. The laser printer (that, who) prints the boarding passes is out of order.

9. Please work on the Babcocks' itinerary, (which, that) must be ready tomorrow.

10. (Whose, Who's) computer monitor is broken?

SECTION 3

WORDS THAT ASSERT

CHAPTER 10

VERBS: TYPES, PRINCIPAL PARTS, REGULAR AND IRREGULAR

Nouns may be the most loaded words, but verbs are the most dramatic.

—John Fairfax and John Moat, *The Way to Write*

LEARNING UNIT GOALS

LEARNING UNIT 10-1: IDENTIFYING VERB TYPES

- Define and identify transitive and intransitive verbs. (pp. 193–96)
- Define and identify main and helping verbs. (p. 196)
- Use verb types correctly in your writing. (pp. 193–96)

LEARNING UNIT 10-2: PRINCIPAL PARTS OF VERBS; REGULAR AND IRREGULAR VERBS

- Define the four principal parts of verbs. (pp. 197–99)
- Explain the effect of regular and irregular verbs on the past verb form and past participle verb form. (pp. 199–201)

LEARNING UNIT 10-3: THREE TROUBLESOME VERB PAIRS

- Know how to use *lay/lie, set/sit,* and *raise/rise.* (pp. 202–4)

From your study of nouns, you know that nouns usually bring pictures to your mind—a screwdriver, a scissors, a friend. However, without verbs, nouns are lifeless, like "still" pictures. Verbs provide the action to make things happen; verbs are the *heart* of a sentence.

Learning Unit 10-1 reviews the verb types that you have been using in your sentence analysis. Learning Unit 10-2 takes you to the next logical step in your verb study—the principal verb parts. Knowing the principal verb parts makes it possible to understand regular and irregular verbs—the second topic in Learning Unit 10-2. Learning Unit 10-3 discusses three pairs of troublesome verbs. As you progress through the three verb chapters, you will see how the principal verb parts and helping verbs combine to express verb voice, mood, and tense (Chapters 11 and 12).

LEARNING UNIT 10-1
IDENTIFYING VERB TYPES

Verbs can be classified as (1) transitive and intransitive, (2) main and helping, and (3) regular and irregular. This learning unit reviews the first two classifications. Learning Unit 10-2 discusses regular and irregular verbs.

TRANSITIVE AND INTRANSITIVE VERBS

Chapters 3 and 4 introduced you to the transitive/intransitive verb classification. In Chapter 5, you used this classification to analyze sentences. Now you are ready to add the active voice and passive voice to the transitive/intransitive verb classification.

Transitive Verbs

Test yourself on your ability to identify transitive verbs by deciding which of the following six verbs are not complete in themselves and require direct objects:

The custodian *locked*	I *listened*
John *hesitated*	She *stated*
The clerk *spilled*	The meeting *ended*

This, of course, was easy. The verbs *locked, spilled,* and *stated* are transitive. They become more meaningful with the following direct objects: *The custodian locked the door. The clerk spilled the ink. She stated the facts.* Each direct object completes the meaning of the verb.

Remember that words, phrases, and clauses may all function as direct objects. Study the following sentences (the direct objects are in italics):

Mr. Husaki prepares our tax *returns.* (The direct object, *returns,* receives the action of the transitive verb *prepares.*)

Why did you leave your previous *position?* (The direct object, *position,* receives the action of the transitive verb phrase *did leave.* Note how the subject is between the two parts of the verb phrase. This often happens in questions.)

The superintendent reprimanded *Steven* and *me*. (The noun *Steven* and pronoun *me* function as a compound direct object.)

Everyone knows *that you will succeed*. (The direct object is a noun clause. Noun clauses, as you may recall, can function as major sentence elements.)

A transitive verb, then, must *transfer* its action to a receiver (a noun or pronoun) to complete its meaning. The receiver is usually a direct object (active voice), but it can be a subject (passive voice).

In Chapter 1, you learned how the active voice changes to the passive voice. The following example illustrates this:

Active Voice: The sales manager reported a loss in computer sales. (*Reported* is a transitive verb. *Who* reported? Manager. *Manager* is the subject. Manager reported *what*? Loss. *Loss* is the direct object.)

Passive Voice: A loss was reported by the sales manager. (*Was reported* is a transitive verb phrase. *What* was reported? Loss. *Loss* is the subject. Loss was reported by *whom*? The prepositional phrase *by the sales manager* functions as a direct object. The direct object, *loss*, in the active-voice sentence becomes the subject and receives the action of the verb *was reported* ; the prepositional phrase becomes the direct object.)

Chapter 11 explains the active voice and passive voice in detail. Since only the transitive verb can change from the active voice to passive voice, we add the active voice and passive voice to the transitive verbs section of the transitive/intransitive classification outline given in Chapters 3 and 4.

1. Transitive verbs (must have objects to complete their meaning)
    a. Active voice (direct object receives the action of the verb)
    b. Passive voice (subject receives the action of the verb)
2. Intransitive verbs (do not have objects to complete their meaning)
    a. Complete verbs (action stops with the verb)
    b. Linking verbs (link nouns or pronouns in the subject to a subject complement in the predicate)

Intransitive Verbs

Intransitive verbs, as you know, require no additional word or direct object to complete their meaning. Since they do not have a direct object, intransitive verbs cannot have a passive voice. Words may follow intransitive verbs, but they will be used as modifiers or subject complements—not as direct objects. Intransitive verbs are either *complete verbs* or *linking verbs*.

Complete Verbs.   From previous chapters, you know that complete verbs are intransitive verbs that stop with the action of the verb and need no other words to make a complete statement. They cause few problems. The words that follow complete verbs are modifiers. Complete verbs *do not* have complements. Chapter 5 gave an example of how to analyze a sentence with a complete verb. Other examples of sentences with complete verbs follow.

I *listened* to her problems.

Mr. Green *talked* for an hour.

The agent *walked* slowly to the door.

As you can see, none of these sentences have *direct objects*. Also, you could stop each sentence after the verb and have a complete thought. The prepositional phrases—*to her problems, for an hour,* and *to the door*—function as adverbs.

Linking Verbs.   Linking verbs act as equal signs connecting the sentence subject to subject complements, which may be nouns, pronouns, or adjectives. From your experience with sentence analysis, you know that a subject complement that renames the same person or thing referred to by the subject is a predicate noun or pronoun. A subject complement that describes the subject is a predicate adjective.

As discussed in Chapters 3 and 4, the most common linking verb is a form of the verb *to be*. It is always a linking verb when used alone. You know most of the forms of *to be* (*am, is, are, was, were,* and verb phrases that end in *be, been,* or *being*). The sense verbs (*feel, smell, taste, sound, look*) and the verbs *appear, become,* and *seem* have a meaning similar to the verb *to be* and are often used as linking verbs. The sense verbs can be either transitive or intransitive, so check the dictionary if you are not sure you are using them correctly. *Appear* and *seem* are always intransitive verbs and must have subject complements. *Become* is intransitive when it means to come into existence and transitive when it means to suit or be suitable to.

In the following sentences, note how each subject complement (in italics) renames or modifies the subject:

Lorenzo is our new service *manager.* (The subject complement, *manager,* is a predicate noun that renames the subject *Lorenzo.*)

He appears *happy* in his work. (The subject complement, *happy,* is a predicate adjective that modifies the subject *he.*)

The trip that you have described sounds *exciting.* (The subject complement, *exciting,* is a predicate adjective that modifies the subject *trip.*)

Recall that Chapter 3 gave a study tip on substituting a form of the verb *to be,* such as *is* and *are,* for the word that you think may be a linking verb. If the sentence makes sense, you probably have a linking verb. This method does have exceptions, however, so check your dictionary to see if it lists the word you question as intransitive (*vi.*).

STUDY TIP

The linking verb is the only verb that can stand between a noun or pronoun and an adjective. For example, He *appears* happy or The manager *looks* tired.

The ability to distinguish transitive verbs and intransitive (complete and linking) verbs is important when you want to use an unfamiliar verb. If the dictionary says the verb is transitive, you know that in the active voice, you must use an object to complete its meaning. Also, you can use the verb in the passive voice. For example, the verb *encroach* is intransitive; therefore, it needs no other word to complete its meaning. *Encroach* cannot be used in the passive voice or with an object in the active voice. A construction such as *the thief encroached our balcony* would show that you do not understand the correct usage of *encroach*.

MAIN AND HELPING VERBS

In Chapter 3, you learned that the final verb, or main verb, in a verb phrase determines whether the verb phrase is transitive or intransitive. Used alone, main verbs are transitive when they express an action that needs a direct object. They are intransitive when they express (1) complete action or (2) link the subject to a subject complement. Main verbs can change in form for tense, person, and number (see Chapters 11 and 12).

From Chapter 3, you also know that the words that precede main verbs in a verb phrase are helping verbs. Table 10-1 groups the helping verbs so you can easily remember them. Of these 23 helping verbs, the nine helping verbs in the last column are always helping verbs. The verbs in the other columns function either as helping verbs or main verbs.

Knowing the following common combinations of helping verbs will help you recognize verb phrases:

could be	has been	must have been	should have been
could have been	have been	shall be	will be
had been	must have	should have	would have

Following are some examples of verb phrases used in sentences:

The new building *will be constructed* in June.

Jim *should be asked* to return from his vacation on Wednesday.

The president *must have* the report tomorrow.

In a verb phrase, other words may come between helping verbs and a main verb (see Chapter 4).

Table 10-1 Helping Verbs

Forms of *to Be*	Forms of *to Have*	Forms of *to Do*	Other Helping Verbs*
am, are, is was, were be, being, been	have, has had	do, does did	can, will, shall could, would, should may, might, must

* The verbs in this column are always helping verbs.

YOU TRY IT

A. Write the letter of the description in Column B next to the verb type in Column A. Use each letter only once.

COLUMN A   COLUMN B

_____ **1.** transitive verb   **a.** is used when the subject is the receiver of an action

_____ **2.** intransitive verb **b.** may be transitive or intransitive

_____ **3.** linking verb   **c.** can be active or passive

_____ **4.** passive voice   **d.** connects a subject to a subject complement

_____ **5.** complete verb   **e.** does not require a direct object

_____ **6.** modern verb   **f.** consists of a main verb and its helping verbs

_____ **7.** helping verb   **g.** is not a part of standard grammar classification

_____ **8.** verb phrase   **h.** precedes a main verb in a verb phrase

_____ **9.** main verb   **i.** needs no other words to make a complete statement

B. Complete the following sentences by inserting a helping verb. Refer to Table 10-1 as necessary.

1. Mr. Rath _____ given Grace an excellent review.

2. The committee _____ meeting tomorrow.

3. You _____ learned several new procedures at the seminar.

4. I _____ writing a new personnel manual after Susan becomes vice president.

5. We _____ decided to offer the job to Ms. Politzer, but she

_____ accepted a previous offer.

Answers: **A.** 1. c 2. e 3. d 4. a 5. i 6. g 7. h 8. f 9. b Possible Answers: **B.** 1. has/had 2. will be 3. have 4. shall be 5. had, had

LEARNING UNIT 10-2
PRINCIPAL PARTS OF VERBS; REGULAR AND IRREGULAR VERBS

As you study the principal parts of verbs in this learning unit, you will realize why you must learn about regular and irregular verbs, the second topic in this learning unit. Understanding the principal verb parts and the regular and irregular verbs prepares you for studying Chapters 11 and 12.

PRINCIPAL PARTS OF VERBS

In Chapter 3, you learned that if you put *to* in front of a word and it makes sense, you probably have a verb. This is the infinitive form of the verb.

Although infinitives are not true verbs but *verbals* (see Chapter 12), the verb of the infinitive is the basic, or root, form of the verb. This verb is normally identical with the first-person singular form that we call the *present* form—the first principal part of a verb.

The four principal parts of verbs are the first-person singular of the present form (*walk*); the first-person singular of the past form (*walked*); the past participle (*walked*); and the present participle (*walking*). All verb forms come from the first three principal parts of the verb—present, past, and past participle. The present participle comes from the present verb form.

STUDY TIP

You can remember the three principal verb parts by this formula: I *walk* today (he *walks* today), I *walked* yesterday, and I *have walked* some time in the past. Later in this unit you will see that *walk* is a regular verb. If you want to use an irregular verb in the formula, you could say: I *write* today (he *writes* today), I *wrote* yesterday, and I *have written* some time in the past.

The **present verb form** is the action or state-of-being verb alone—without any helping verbs. When the present verb form stands alone, it expresses a present or recurring action. The verb form given in the dictionary is the present verb form. It is sometimes used with helping verbs to form other verb forms.

The **past verb form** comes from the *present verb form* and shows the action occurring in the past; it functions as a single-word verb. Past verb forms can be *regular* or *irregular* verbs.

STUDY TIP

In this chapter, we discuss the principal parts of verbs in a general sense. When you study tense in Chapter 12, the present verb form and past verb form are called the present tense and past tense. Verbs are challenging. To help you, we are avoiding using terms that you have not studied.

To understand the *past participle* and *present participle* verb forms, you must know something about participles. A **participle** is a hybrid verb form that functions both as a verb and an adjective. Participles can *only* function as verbs when used with the helping verb forms *to be* and *to have*. In Chapter 12, you will see that participles belong to a group called *verbals* and can function alone as adjectives. Remember that a participle alone *cannot* function as the predicate verb of a sentence.

The form of the **past participle** depends on whether the verb is *regular* or *irregular*. The past participle always needs a helping verb. When the past participle is used with tense, the helping verb is a form of *to have*; when it is used with the passive voice, the helping verb is a form of *to be*. You will learn more about this in Chapter 11.

Present participles by themselves modify nouns or pronouns. The present participle is a present verb form with *ing* added. As a verb, the present participle is always used with a form of the helping verb *to be*. Chapter 11 also discusses present participles.

Now let's distinguish between regular and irregular verbs.

REGULAR AND IRREGULAR VERBS

Table 10-2 shows the principal parts of regular verbs. **Regular verbs** add a *d* or *ed* for the past verb form and the past participle. To form the present participle, the *ing* is always added to the present verb form and used with a form of the helping verb *to be*.

The *past* and *past participle* verb forms of an **irregular verb** are *not* made by the addition of *d* or *ed*. Often, a vowel change is involved. Some irregular verbs like *cost*, *hit*, and *put* do not change their spelling for the past and past participle forms. Since these verbs are irregular in form, you usually must memorize them. When in doubt, however, you can always go to the dictionary, which shows all the principal parts of irregular verbs. The dictionary does not show the principal parts of regular verbs.

Table 10-3 gives some common irregular verbs. You probably already know many of these irregular verbs. Concentrate on learning the verbs that you do not know.

WRITING TIP

Knowing the irregular verb forms will avoid two common errors: (1) using the past participle for the past tense and (2) using the past tense instead of the past participle with a helping verb. Two common errors you may have heard or seen are the following:

INCORRECT: I *seen* him last week.
CORRECT: I *saw* him last week.

INCORRECT: *Have* the dogs *drank* their water?
CORRECT: *Have* the dogs *drunk* their water?

Table 10-2 Principal Parts of Regular Verbs

Present	Past	Past Participle (used with forms of *to have* and *to be*)	Present Participle (used with forms of *to be*)
hope	hoped	hoped	hoping
listen	listened	listened	listening
look	looked	looked	looking
refuse	refused	refused	refusing

Table 10-3 Principal Parts of Common Irregular Verbs

Present	Past	Past Participle (used with forms of *to have* and *to be*)	Present Participle (used with forms of *to be*)
be (am, are, is)	was, were	been	being
bear	bore	borne	bearing
beat	beat	beaten	beating
begin	began	begun	beginning
bid (offer a price)	bid	bid	bidding
bite	bit	bitten	biting
blow	blew	blown	blowing
bring	brought	brought	bringing
build	built	built	building
burst	burst	burst	bursting
choose	chose	chosen	choosing
cling	clung	clung	clinging
come	came	come	coming
cost	cost	cost	costing
deal	dealt	dealt	dealing
dig	dug	dug	digging
do	did	done	doing
draw	drew	drawn	drawing
drink	drank	drunk	drinking
drive	drove	driven	driving
eat	ate	eaten	eating
fall	fell	fallen	falling
fight	fought	fought	fighting
fly	flew	flown	flying
forecast	forecast	forecast	forecasting
forget	forgot	forgotten	forgetting
freeze	froze	frozen	freezing
get	got	got, gotten	getting
give	gave	given	giving
go	went	gone	going
grind	ground	ground	grinding
grow	grew	grown	growing
hang (an object)	hung	hung	hanging
hide	hid	hidden	hiding
hit	hit	hit	hitting
hurt	hurt	hurt	hurting
know	knew	known	knowing
lay	laid	laid	laying
lead	led	led	leading
lie	lay	lain	lying
mean	meant	meant	meaning
meet	met	met	meeting
pay	paid	paid	paying
put	put	put	putting
quit	quit	quit	quitting
ring	rang	rung	ringing
rise	rose	risen	rising
run	ran	run	running
see	saw	seen	seeing
seek	sought	sought	seeking
set	set	set	set
shake	shook	shaken	shaking
shine	shone, shined	shone, shined	shining
shrink	shrank (shrunk)	shrunk	shrinking
sing	sang	sung	singing
sink	sank	sunk	sinking
sit	sat	sat	sitting
speak	spoke	spoken	speaking
spend	spent	spent	spending
spring	sprang	sprung	springing
sting	stung	stung	stinging
strike	struck	struck, stricken	striking
swear	swore	sworn	swearing
swim	swam	swum	swimming
swing	swung	swung	swinging
take	took	taken	taking
teach	taught	taught	teaching

Table 10-3 Principal Parts of Common Irregular Verbs (Continued)

Present	Past	Past Participle (used with forms of *to have* and *to be*)	Present Participle (used with forms of *to be*)
tear	tore	torn	tearing
tell	told	told	telling
throw	threw	thrown	throwing
wear	wore	worn	wearing
win	won	won	winning
wring	wrung	wrung	wringing
write	wrote	written	writing

Many verbs are both regular and irregular, depending on the meaning of the sentence. In the following sentences, the past forms of the verbs *fly, shine,* and *hang* are used correctly:

The pilot *flew* the new plane.   The batter *flied* out to the left fielder.
The sun *shone* brilliantly.   The workers *shined* their shoes.
I *hung* the picture in my office.   In the story, the judge *hanged* the
                                          outlaw for starting the fire.

YOU TRY IT

A. Write a sentence using each of the verbs or verb phrases listed. Then, in the space provided to the left, write P1 (present), P2 (past), P3 (past participle), P4 (present participle) to indicate which principle part of the verb is used in the phrase. Finally, underline the principal part of the verb.

EXAMPLE:

___P3___   have told

I have <u>told</u> them to avoid the busy highway.

_____ **1.** wanted

_____ **2.** was building

_____ **3.** had written

_____ **4.** would have emphasized

_____ **5.** think

B. Review Table 10-3. Select three verbs that you either find difficult or think that other people find difficult. Write three sentences, each one using the past form of each of the verbs you selected. Then write another set of three sentences using the past participle form of each verb.

Past:

1. _____

2. _____

3. _____

Past Participle:

1a. _____

2a. _____

3a. _____

Answers: 1. P2 2. P4 3. P3 4. P3 5. P1

LEARNING UNIT 10-3
THREE TROUBLESOME VERB PAIRS

Three troublesome verb pairs are *lay/lie*, *set/sit*, and *raise/rise*. Table 10-4 gives their principal parts. (Which verb in Table 10-4 is a regular verb?)

LAY/LIE

When you study Table 10-4, you can see that the past tense of *lie* is the same as the present tense of *lay*. This makes for the confusion in the usage of the two words.

Lay is a transitive verb meaning to put or place something down. Since you understand transitive and intransitive verbs, you know that *lay* needs an object for its action. So when you use *lay* in the active voice (the direct object receives the action), it must have a direct object.

He *lays* bricks expertly. (Present form; *bricks* is the direct object.)

He *laid* the specifications on Mr. Ware's desk. (Past form; *specifications* is the direct object.)

He *was laying* bricks at 8:00 A.M. (Present participle; *bricks* is the direct object.)

Table 10-4 Principal Parts of *Lay/Lie*, *Set/Sit*, and *Raise/Rise*

Present	Past	Past Participle (used with forms of *to have* and *to be*)	Present Participle (used with forms of *to be*)
lay	laid	laid	laying
lie	lay	lain	lying
set	set	set	setting
sit	sat	sat	sitting
raise	raised	raised	raising
rise	rose	risen	rising

Lie is an intransitive verb meaning to recline or rest in a horizontal position. Intransitive verbs cannot take an object or be used in the passive voice (the subject receives the action). To use *lie* correctly, you would say:

He seldom *lies* down during the day. (Present form.)

He *lay* down for several minutes this morning. (Past form.)

He *was lying* on the couch when the telephone rang. (Present participle.)

The report *is lying* on my desk. (Present participle.)

SET/SIT

Set is a transitive verb meaning to cause to sit or to place. As a transitive verb, *set* needs an object.

Set the large lamp on that table. (Present form; *lamp* is the direct object.)

The workers *are setting* buoys in the lake to mark the underwater rocks. (Present participle; *buoys* is the direct object.)

The owner *set* the price too high. (Present form; *price* is the direct object.)

Sit is an intransitive verb meaning to seat oneself. You would *set* something down, but you can never *sit* anything down.

I *sat* here yesterday. (Past form.)

The eagle *has been sitting* in the oak tree all day. (Present participle.)

I always *sit* there. (Present form.)

RAISE/RISE

The transitive verb *raise* means to cause or help to rise to a standing position. It also has other meanings such as to build something and to grow something. As a transitive verb, it takes an object.

Can you *raise* the chair to a higher level? (Present form; *chair* is the direct object.)

Do you know how to *raise* squash? (Present form; *squash* is the direct object.)

The owner *raised* the rent. (Past form; *rent* is the direct object.)

The intransitive verb *rise* means to assume an upright position, especially from a lying, kneeling, or sitting position or to move upward. Since *rise* is an intransitive verb, it does not have an object.

Rent prices *are rising* again. (Present participle.)

The broker said interest rates *will rise*. (Present form.)

The flood waters *are rising*. (Present participle.)

STUDY TIP

You may have noticed that the second letter of the three intransitive verbs *lie*, *sit*, and *rise* is *i*. This will help you remember which of the troublesome verb pairs is intransitive.

You can see that the important factor in these three troublesome verb pairs is whether the verb is transitive or intransitive. Since you have been deciding whether verbs are transitive or intransitive in your sentence analysis, you should have no problem using the troublesome verbs *lay/lie*, *set/sit*, and *raise/rise*.

YOU TRY IT

For each of the sentences below, circle the verb in parentheses that is appropriate for the sentence.

1. Apex Agriculture (raises/rises) corn and soy beans.

2. The workers (set/sat) the seedlings in frames in the greenhouse.

3. They (laid/lay) the hay over the new plants to keep them warm.

4. The field had (laid/lain) dormant for two years.

5. The dairy hands (rose/raised) early to start the milking.

6. The broken tractor (set/sat) in the middle of the field.

Answers: 1. raises 2. set 3. laid 4. lain 5. rose 6. sat

CHAPTER SUMMARY: A QUICK REFERENCE GUIDE

PAGE	TOPIC	KEY POINTS	EXAMPLES
193	**Transitive verbs**	The action of a transitive verb must be transferred to a receiver (a noun or pronoun) to complete its meaning; the receiver is usually a direct object (active voice) but can be a subject (passive voice).	June sold the green *rocker*.
194	**Intransitive verbs**	Intransitive verbs do not have direct objects to complete their meaning; they use only the active voice. The two types of intransitive verbs are (1) complete verbs and (2) linking verbs.	*Complete verb*: During the meeting, the manager *slept*. *Linking verb*: Her supervisor *seems* pleased about the situation.

PAGE	TOPIC	KEY POINTS	EXAMPLES
196	**Main verbs**	Used alone, a main verb can express action, occurrence, or state of being. Main verbs are transitive or intransitive; they change form for tense, person, and number.	Every worker on the loading dock *looked* for the missing crate. The air in the loading dock *is* damp.
196	**Helping verbs**	Helping verbs (see Table 10-1) precede main verbs and form a verb phrase with the main verb.	The key *has been lost* for a week.
197	**Principal parts of verbs**	**1.** The *present verb form* is the action or state-of-being verb alone, without helping verbs; it expresses present or recurring action and is listed in the dictionary. Sometimes the present verb form is used with helping verbs to form other verb forms.	I *learn* new skills from my manager every day. He *learns* new skills from his manager every day.
		2. The *past verb form* comes from the present verb form. It shows past action and functions as a single-word verb. The form of the verb depends on whether the verb is a regular or irregular verb.	I *learned* all these skills from my manager.
		3. The form of the *past participle* depends on whether the verb is a regular or irregular verb; it needs helping verbs.	I *have learned* many skills from my manager in the past.
		4. *Present participles* are the present verb form with *ing* added. They modify nouns or pronouns and are used with a form of the helping verb *to be*.	I *am learning* many skills from my manager today.
199	**Regular verbs**	Regular verbs add *d* or *ed* for the past verb form and past participle.	*learn, learned, learned, learning*
199	**Irregular verbs**	Irregular verbs are not formed by adding *d* or *ed*. Often a vowel change occurs. Some irregular verbs do not change their spelling. Table 10-3 gives the common irregular verbs. Use your dictionary when in doubt.	*come, came, come, coming*
202	**Lay/lie**	*Lay* is a transitive verb meaning to put down. *Lie* is an intransitive verb meaning to recline or rest in a horizontal position.	*Lay* your clothes on the chair. Jim is *lying* on the floor.

PAGE	TOPIC	KEY POINTS	EXAMPLES
203	**Set/sit**	*Set* is a transitive verb meaning to cause to sit or to place. *Sit* is an intransitive verb meaning to seat oneself.	*Set* the lamp in the corner. I *sat* in the green chair yesterday.
204	**Raise/rise**	*Raise* is a transitive verb meaning to cause or help to rise to a standing position. *Rise* is an intransitive verb meaning to move upward.	He *raised* all the windows. Clothing prices *are rising* every year.

QUALITY EDITING

You work for the Apex Agricultural Company. Your coworker has asked you to check a letter to be sent to a customer about how to care for some young plants the customer ordered. Help your coworker by editing the letter. Check the verbs carefully! Also, check for any errors in spelling, end-of-sentence punctuation, plurals, and any other grammatical problems.

Dere Ms. Thatcher:

Thanking you for your order of one hundred dozen tomatoes plants? These smal plants is quiet frajil. You must handel them with grate care. Sit them in frames in your greenhouse until any danger of frosting has passed. Do not allow them plants to lay in a drafty area. When the whether is wharm, the litel plants can be sat in rows three feets apart. This distanse will allowed the plants pelenty of roam to grow.

In order for the plants to be helthy, you must rise them according the to instructions on the card enclosed with this litter. Do not foreget to watter and fertilitize regular

If you want additional asistance, call Alice Moore or I at Apex at (201) 555-6794.

Sincerely,

David Summers

APPLIED LANGUAGE

A. Column A lists helping verbs. Column B lists main verbs in their first person present form. Combine the helping verb in Column A with the main verb in Column B to create a verb phrase. Enter the verb phrase in Column C. **Note**: You may change the main verb if necessary, but do not change the helping verb. There may be more than one correct answer.

EXAMPLES: has + program = *has programmed*

are + consider = *are considering*

COLUMN A	COLUMN B	COLUMN C
Helping Verb:	Main Verb:	Verb Phrase:
1. will have	write	_____
2. are	think	_____
3. should have been	ask	_____
4. will be	seek	_____
5. had been	give	_____

B. Complete the following sentences by adding a main verb and an object(s) as necessary. Then, underline the main verb.

1. Jason would have _____.

2. Ms. Silver must have been _____.

3. Send the documents to the company that has been _____

_____.

4. Some people may _____.

5. The manager could have been _____.

C. Insert the correct form of the verb given in parentheses into the blank in the following sentences.

EXAMPLE: (drive) By noon George had _____ 60 miles.

1. (fight) The firm had long _____ to market its products in Europe.

2. (choose) Which of the three was _____ to serve on the committee?

3. (come) How many of the executives have _____ to the conference?

4. (build) The office tower was _____ in nine months.

5. (lead) Howard had _____ the previous meetings.

Name _____

Date _____

VERBS: TYPES, PRINCIPAL PARTS, REGULAR AND IRREGULAR

Part A. Transitive Verbs: Active and Passive Voice

Transitive verbs can be in the active or the passive voice. In each of the sentences below, the transitive verbs are in the passive voice. Rewrite the sentences so that they are in the active voice.

EXAMPLE: The letter was given to me by Larry.

      Larry gave me the letter. _____

1. The farm cooperative was managed by Ms. Symes.

_____

2. The equipment invoices were paid by the accounts payable manager.

_____

3. The customers were contacted by the sales representative.

_____

4. The mail list will be updated by Susan.

_____

5. The milk processing is controlled by the new computer.

_____

6. The feed is delivered by Acme.

_____

7. The revenue reports will be prepared by Carlos.

_____

8. The truckloads of oranges will be sold by the Dinar Auction Company.

_____

9. The entire crop was ruined by the flood.

_____

10. The unhappy customer was not satisfied with Fruithaven's response.

_____

Part B. Helping Verbs

Select a helping verb from the list below and write a sentence that combines a helping verb with the main verb shown in parentheses.

Helping Verbs:

could have been	must have	has been
might	would have	could have
will be	have been	might be
are	did	
could be	must have been	
is	will have been	
should	had been	

Regular Main Verbs:

EXAMPLE: must have + prepare = _____**The manager must have prepared the report.**_____

1. (want) _____.

2. (prepare) _____.

3. (talk) _____.

4. (discuss) _____.

5. (report) _____.

6. (inquire) _____.

7. (display) _____.

8. (complete) _____.

9. (deliver) _____.

10. (order) _____.

Irregular Main Verbs:

EXAMPLE: should have + write = _____**The manager should have written the report earlier.**_____

11. (write) _____.

12. (pay) _____.

13. (buy) _____.

14. (sell) _____.

15. (think) _____.

16. (drive) _____.

17. (say) _____.

18. (go) _____.

19. (do) _____.

20. (start) _____.

Part C. Irregular Verbs

For each of the following sentences, circle the correct verb in parentheses.

1. The farmer had been (gave/given) notice of foreclosure.

2. The automated dairy equipment will (milk/milking) the cows at scheduled intervals.

3. By September, the executive will have (flown/flew) a million miles.

4. Maria Elias was (drove/driven) to the airport.

5. The water (cost/costed) more than the feed.

6. The silo will be (built/builded) in June.

7. Sal Ferraro might have (brung/brought) the report to the meeting.

8. Have you (began/begun) the installation yet?

9. Do you know who was (chose/chosen) for the personnel review committee?

10. Has the accounting office (payed/paid) the invoice?

Part D. Lay/Lie, Set/Sit, Raise/Rise

For each of the following sentences, circle the correct verb from the pair in parentheses.

1. (Set/Sit) the reports on the desk.

2. Ms. Honesto (raised/rose) to the occasion.

3. (Lay/Lie) these materials out for everyone to see.

4. Has Irwin (raised/risen) the question?

5. Golden Poultry Farms (raises/rises) chickens.

6. (Set/Sit) at the end of the table.

7. Karen did not (set/sit) any books on the exhibit shelf.

8. Jerome had (laid/lain) the cash on the credenza.

9. (Lay/Lie) down.

10. The reports were (laying/lying) in plain sight.

CHAPTER 11

VERBS: VOICE, MOOD, TENSE

> While it is quite true that a verb expresses action or state of being . . . , it is even more important to remember that a verb is a word we use to show changes in the *time* of an action.
>
> —Carlin Kindalen, *Basic Writing*

LEARNING UNIT GOALS

You know that to express your thoughts, you must call things by their right names. The English language also provides words that you can use to *assert* something about things. These words are verbs—words that give power to a sentence. The strength of your written communication depends on your knowledge of the verb.

The expression of voice, mood, and tense use the principal verb parts you learned about in Chapter 10. Learning Units 11-1, 11-2, and 11-3 illustrate how the three elements of voice, mood, and tense make it possible to write sentences that accurately portray what you want to say. Learning Unit 11-4 discusses the progressive and emphatic forms of verbs.

LEARNING UNIT 11-1
VERB VOICE

You should know from using the active voice in your writing exercises how quickly it improves your writing. All writers should prefer the active voice. This unit reinforces the value of the active voice by explaining how transitive verbs change voice.

Voice shows whether a speaker or writer wants to emphasize the subject (active voice) or the object (passive voice). If the main verb is in the active voice, the subject is the doer of the action. If the main verb is in the passive voice, the subject is acted upon. In the passive voice, the doer often appears as the object of the preposition *by*. As stated in Chapter 10, the passive voice is formed from a past participle and a form of the helping verb *to be*. Note the following sentences:

Active Voice: We *received* your order on January 16.

Passive Voice: Your order *was received* by us on January 16.

Active Voice: Mr. Paulson *has made* five excellent sales this morning.

Passive Voice: Five excellent sales *have been made* by Mr. Paulson this morning.

You know that a transitive verb must transfer its action to a receiver (a noun or pronoun) to complete its meaning. Most of the transitive verb sentences we have been studying were in the active voice and have transferred the action to direct objects. Look again at the passive voice sentences above. What happened to the direct objects that were in the predicate of the active voice sentences?

In the passive voice, the writer moves the direct object from the natural position of doer-action-receiver to the unnatural position of receiver-action-doer. Now the subject receives the action of the verb. The direct object in the active voice becomes the subject in the passive voice and completes the meaning of the verb.

Let's look at two more sentences written in the active and the passive voice. Note again how the passive voice uses a past participle and a form of the verb *to be*.

Active Voice: Selina *receives* the supplies for the assembly department.

Passive Voice: The supplies for the assembly department *are received* by Selina.

Active Voice: Brent *read* the sales report.

Passive Voice: The sales report *was read* by Brent.

The verb in the first two sentences—*receive*—is a regular verb because it adds *d* to form the past verb form. What about the verb *read*? Since Table 10-3 in Chapter 10 does not list *read* as an irregular verb, let's check the dictionary. From the dictionary, we learn that *read* is an irregular verb that does not change its spelling. The past verb form and the past participle are also *read*.

The advantages of the active voice are:

1. The action of the active voice moves naturally from left to right. Readers can follow this action at a greater speed than the action of the passive voice because they must reverse the direction from right to left. This directs a reader's mind backward.

2. Verbs in the active voice are stronger than verbs in the passive voice.

3. The active voice makes a sentence more direct and less wordy.

Remember, however, that the passive voice has a purpose. Writers use the passive voice (1) when they want to give the object in the sentence more emphasis than the subject and (2) when they do not know the subject of a verb or do not want to show the subject of a verb. For example, of the two sentences that follow, the second (passive voice verb) is more diplomatic.

Active Voice: You made two errors in this report.

Passive Voice: Two errors were made in this report.

WRITING TIP

Often, writers use the passive voice automatically. After you have written a paragraph, make a special effort to look for sentences in the passive voice that you can change to the active voice. This will tighten your writing and sharpen your reader's mental pictures of the images you present.

Also, avoid mixing the active and passive voice in a sentence. For example, instead of writing *The check you sent December 10 has been received by me,* write *I received the check you sent on December 10.*

YOU TRY IT

The following sentences are written in the passive voice. Rewrite the sentences so that they are in the active voice.

EXAMPLE:

Passive: The building was designed by A. J. Babcock & Associates.

Active: **A. J. Babcock & Associates designed the building.**

1. A new accounting system was installed by Benton Business Systems.

2. The freight was delivered to the dock by AirFreight's cartage agent.

_____

3. The customers complained that they had not been given good service by the sales representative.

_____

_____

4. The machine parts were ordered by Sonja in the operations department.

_____

5. Kelly was given a promotion by the president.

_____

6. Three of the managers were criticized by the committee.

_____

7. The hearings were conducted by Senator March.

_____

8. The reference manual was written by Professor Moreno.

_____

9. Parker Dodge was relocated to Denver by his company.

_____

10. The direct mail campaign was designed and implemented by our marketing consultant, Dr. Tajitsu.

_____

_____

LEARNING UNIT 11-2
VERB MOOD

The meaning of _mood_ in grammar is completely different from the emotional _mood_ of real life. In grammar, the **mood** of a verb refers to the _way_ a listener or reader should react to the action or state of the verb.

English grammar has three moods. Sentences in the **indicative mood** make a positive statement of fact or ask a question. Sentences in the **imperative mood** issue a command or make a request. Imperative mood sentences are in the present verb form, second person; and the subject (you) is often omitted. Note the following examples of the indicative mood and imperative mood:

The shipping department has a new manager. (Indicative mood.)

Do you know the new manager of the shipping department? (Indicative mood.)

Keep the storage room locked. (Imperative mood.)

Please answer all the questions. (Imperative mood.)

The indicative mood and the imperative mood do not present problems. However, you should understand the third mood—the *subjunctive mood*—to avoid some common grammatical errors.

Writers use the **subjunctive mood** primarily to make a statement contrary to fact (something that is not true) or to express a wish. Words often used with the subjunctive mood are *if, though, unless,* and *that.* Frequently, you will find the subjunctive in a dependent clause introduced by *if* or *that.*

The present subjunctive, singular or plural, uses the verb *to be*; the past subjunctive, singular or plural, uses the word *were.* All other verbs with third-person singular subjects use the subjunctive present, singular only, without the *s* ending. Helping verbs such as *might* and *should* can also express subjunctive meanings.

> I demanded that the new employee *be* (not *is*) on time. (The present subjunctive uses I *be,* you *be,* he *be,* we *be,* you *be,* and they *be* instead of the indicative I *am,* you *are,* he *is,* we *are,* you *are,* and they *are.*)

> If I *were* (not *was*) you, I would leave. (The past subjunctive uses I *were* and you *were* instead of the indicative I *was* and he *was.*)

> I would suggest that he *come* (not *comes*) immediately. (Verbs with third-person singular subjects use the subjunctive present, singular only, without the *s* ending.)

> If Elaine has nothing to say, she *should leave.* (Helping verbs are often used to express the subjunctive.)

Note the following two uses of the subjunctive mood:

1. **Use the subjunctive (*were*) to express an idea that is clearly contrary to fact.** In the past verb form, which actually expresses present time, the sign of the subjunctive is *were.* Use this verb with *all* subjects, as in the following sentences:

   > If Mr. Wiley *were* here, he would suggest a solution.

   > If the facts *were* known, Professor Johnson would not have completed the contract.

   > I wish it *were* possible for me to go with you.

   Although the last sentence above is expressed as a wish, the idea conveyed by the subjunctive is clearly contrary to fact.

2. **Use the subjunctive (*to be*) in *that* clauses after verbs expressing a desire, recommendation, demand, motion, suggestion, request, necessity, or resolution.**

   > It is my wish that Everett Larson *be* promoted. (Desire expressed.)

   > The supervisor recommended that all overtime work *be* eliminated. (Recommendation expressed.)

   > I demand that the machine *be* repaired immediately. (Demand expressed.)

   > The president moved that the meeting *be* adjourned. (Motion expressed.)

I suggest that Heather and Jeff *be* given more responsibility. (Suggestion expressed.)

Marta requested that the information *be* stored on a hard disk. (Request expressed.)

It is necessary that he *be* allowed to leave immediately. (Necessity expressed.)

The resolution has been made that he *be* so notified by telegram. (Resolution expressed.)

Writers usually have no difficulty with subjunctive constructions when they use verbs other than *to be*. Note the following sentences:

My wish is that she *go* to the convention.

The owner demands that Barbara *finish* the project by Friday.

Today, most writers have replaced the subjunctive with the indicative. The following sentences use the indicative mood:

If Mr. Hodges *was* ill on Tuesday, the disbursements may not have been made.

If Craig *was* on television, he probably appeared in a commercial.

Note that these sentences do not express an idea clearly *contrary to fact*. Mr. Hodges *may* have been ill on Tuesday, and Craig *may* have appeared on television.

YOU TRY IT

Complete the following sentences using the subjunctive mood.

EXAMPLE: I could fly if **I were a bird.** _____

1. Emma Miller wished that she _____.

2. Mr. Tsu requested that the data _____.

3. I could design wonderful reports if _____.

4. Nancy asked that she _____.

5. His supervisor expected that he _____.

Possible Answers: 1. were in London. 2. be deleted. 3. I were an artist. 4. be excused. 5. attend the convention.

LEARNING UNIT 11-3
VERB TENSE

What does time mean to you? Your first answer probably is that your *present* time always seems to "run out." You never have enough time to do what you would like to do. You may add that in the *past*, you seemed to have more time for yourself. Then, looking toward your *future* time, you probably decide you should *make* time for yourself. In your little excursion through time, you have described the three divisions of time—present, past, and future.

These three divisions of time are called the **simple** or **primary tenses**. A verb's **tense** tells the *time* when an action occurs or the condition exists. Simple tenses give actions as taking place today, yesterday, and tomorrow. The tenses come from the present, past, and past participle verb forms that you learned about in Chapter 10.

Tense can imply more than the *time* of the action. The use of tense also can tell the *stage* of the action. Is the action continued or completed (perfected)? To tell you this, the English language has three more tenses: *present perfect*, *past perfect*, and *future perfect*.

We will begin by studying the simple tenses.

SIMPLE TENSES

Recall from Chapter 10 that although infinitives are not true verbs but *verbals* (which will be explained in Chapter 12), the verb of the infinitive is also a first-person singular verb form, or a **present tense** verb. The present tense of the verb states a general truth; tells that the action is in progress, continued, and habitual; and can express future action.

Jim's accountant *helps* me with my taxes. (Sentence states a general truth.)

The electricians *pay* for all their equipment and supplies. (Sentence states a general truth.)

I *am enjoying* my new work. (Sentence expresses action in progress.)

You *are working* too fast. (Sentence expresses action in progress.)

I hope Jeff *arrives* before the meeting ends. (Sentence expresses future action.)

Past tense verbs describe an action that happened or existed at some time in the past. You will recall from Chapter 10 that regular verbs add *d* or *ed* to the present verb, or *present tense*, to form the past verb, or *past tense*. Irregular verbs have irregular ways of forming the past tense.

Jim's accountant *helped* me with my taxes.

The electricians *paid* for all their equipment and supplies. (This irregular verb is listed in Table 10-3 in Chapter 10.)

Remember the following two rules:

1. **Unlike nouns, the verb form ending in *s* is likely to be *singular*.** Note carefully the endings of verbs that have third-person subjects.

Singular	**Plural**
He knows	They know
She hesitates	They hesitate
Benson writes	They write
Jacob comes	They come
She owes	They owe

Table 11-1 Present, Past, and Future Tenses of *Drive*

Present Tense		Past Tense		Future Tense	
Singular	*Plural*	*Singular*	*Plural*	*Singular*	*Plural*
I drive (am driven)	We drive (are driven)	I drove (was driven)	We drove (were driven)	I will drive (will be driven)	We will drive (will be driven)
You drive (are driven)	You drive (are driven)	You drove (were driven)	You drove (were driven)	You will drive (will be driven)	You will drive (will be driven)
He, she, *or* it drives (is driven)	They drive (are driven)	He, she, *or* it drove (was driven)	They drove (were driven)	He, she, *or* it will drive (will be driven)	They will drive (will be driven)

2. **Do not put a verb in the past tense simply because another verb in the same sentence happens to be in the past tense.** Use the present tense to express an idea that continues to be true.

> DON'T USE: What did you say your name *was?*
>
> USE: What did you say your name *is?* (The person being spoken to still has the same name.)
>
> DON'T USE: He told me that he *didn't* like this brand.
>
> USE: He told me that he *doesn't* like this brand. (The person being spoken about still doesn't like a particular brand.)

The **future tense** expresses an action or condition that has not yet taken place. Usually, the future tense is formed by using *will* as a helping verb. Although the helping verb *shall* is sometimes used, most speakers and writers find it more natural to use *will. Shall* is sometimes used to express determination, emphasis, or a promise.

The plumber *will install* the new water heater tomorrow.

The brick layer *shall finish* the front of the house next Monday.

You also can form the future tense with the present participle (a present verb with *ing* added) and *will be.*

The entire road crew *will be working* through the night.

Table 11-1 gives the simple tenses of the first-person, second-person, and third-person forms of the irregular verb *drive.* The table first shows the verbs in the active voice and then shows the passive voice in parentheses.

WRITING TIP

When you are deciding on the correct tense to use, remember to look at the time implied by the other words in a sentence.

PERFECT TENSES

The simple tenses show the natural division of time into present, past, and future. The perfect tenses show actions that have been completed or will be completed before a given time. They combine forms of the verb *to have* with the past participle.

Present Perfect Tense

The **present perfect tense** is formed with the helping verbs *have* and *has* and a past participle. It shows action that is completed at the time of writing, is continuing into the future, and may occur again.

Actions that have just been completed:

I *have* just *finished* reading the instructions.

The computer in my office *has been repaired.*

Actions beginning in the past and continuing into the future:

We *have worked* for this company for seven years.

He *has been helping* with the delivery of supplies.

Past actions that may occur again:

I *have helped* Harry on three occasions.

She *has had* three promotions since the beginning of the year.

Past Perfect Tense

The **past perfect tense** is formed with the helping verb *had* and a past participle. It shows an action completed before some point of time in the past, which may be specified or simply implied.

I tried to telephone Wayne, but he *had* already *left* the store. (*Left* is the past participle of the irregular verb *leave*. Wayne's leaving occurred *before* the writer tried to call.)

Although it was only 3:00 P.M., we *had sold* every camera in stock.

Jim told us that he *had signed* the contract.

Future Perfect Tense

The **future perfect tense** is formed with the helping verbs *will have* and a past participle. It shows an action that will be completed before another future action or by a specified future time.

The employees *will have left* by the time the bell rings. (Action completed before another future action.)

By 9:30 P.M. the meeting *will have ended.* (Action completed by a specific future time.)

The auditor *will have examined* the books by the time you return from your vacation. (Action completed before another future action.)

You *will have been working* for three hours by the time the store closes. (Action completed by a specific future time.)

Table 11-2 Present Perfect, Past Perfect, and Future Perfect Tenses of *Drive*

Present Perfect Tense		Past Perfect Tense		Future Perfect Tense	
Singular	*Plural*	*Singular*	*Plural*	*Singular*	*Plural*
I have driven (have been driven)	We have driven (have been driven)	I had driven (had been driven)	We had driven (had been driven)	I will have driven (will have been driven)	We will have driven (will have been driven)
You have driven (have been driven)	You have driven (have been driven)	You had driven (had been driven)	You had driven (had been driven)	You will have driven (will have been driven)	You will have driven (will have been driven)
He, she, *or* it has driven (has been driven)	They have driven (have been driven)	He, she, *or* it had driven (had been driven)	They had driven (had been driven)	He, she, *or* it will have been driven (will have been driven)	They will have driven (will have been driven)

Table 11-2 gives the perfect tenses of the first-person, second-person, and third-person forms of the irregular verb *drive*. The table first shows the verb in the active voice and then shows the passive voice form in parentheses. Note that the active and passive voice use the helping verbs *have, have been, has,* and *has been* to form the present perfect tense; *had* and *had been* to form the past perfect tense; and *will have* and *will have been* to form the future perfect tense.

Now that you have studied the six verb tenses, let's relate these tenses to the principle verb parts you learned about in Chapter 10. Table 11-3 shows that the basic forms of the six tenses only use three principal parts: present, past, and past participle.

You should be aware at this point in your verb study that verbs, like nouns, have *properties*, which are characteristics belonging only to verbs. The properties of verbs are voice, mood, tense, person, and number. In this chapter, you learned about voice, mood, and tense. Chapter 12 discusses subject-verb agreement of person and number.

Table 11-3 Basic Forms of the Six Tenses of *Drive*

Tense	Basic Form	Principal Part Used
Present	I drive	Present
Past	I drove	Past
Future	I will drive	Present
Present perfect	I have driven	Past participle
Past perfect	I had driven	Past participle
Future perfect	I will have driven	Past participle

YOU TRY IT

A. Circle the verb in parentheses that is in the correct tense.

1. Returning the envelope to me, he said that he (doesn't/didn't) accept junk mail.

2. He (doesn't/didn't) accept junk mail until he received a coupon that saved him $100. Now he does accept it.

3. The staff (met/will have met) by the time the mail is delivered.

4. The staff (had met/will have met) by the time the mail was delivered.

B. Look again at sentence number 3.

5. When the writer created this sentence, had the staff met yet? _____

C. Look again at sentence number 4.

6. Did the staff meet before or after the mail came? _____

Answers: 1. doesn't 2. didn't 3. will have met 4. had met 5. no 6. after

LEARNING UNIT 11-4
PROGRESSIVE AND EMPHATIC VERB FORMS

Two special verb forms can be used with the three simple tenses and the three future tenses—the progressive and emphatic verb forms. This learning unit explains how to use these verb forms.

PROGRESSIVE VERB FORMS

The six **progressive verb forms** express an action or event in progress at the time indicated. They combine the correct form of the helping verb *to be* with the present participle (*ing* added to the present tense).

Table 11-4 gives the progressive forms for the six tenses of the irregular verb *drive*. The passive voice is given for the present, past, and future progressive forms.

Note the following uses of the progressive forms:

1. The *present progressive form* shows something happening at the time the sentence is written.

   The president *is studying* the financial report.

2. The *past progressive form* shows a past action continuing, often within certain expressed or implied limits.

   The receiving clerk *was getting* many deliveries this month.

3. The *future progressive form* shows that an action in the future will continue for some time. This action can depend on something else stated in the sentence.

   If the interest rates continue to increase, the company *will be looking* for ways to cut costs.

4. The *present perfect progressive form* describes ongoing action likely to continue in the future.

   I *have been singing* for many years.

Table 11-4 Six Progressive Forms of *Drive*

Progressive Form	Singular	Plural
Present Progressive	I am driving (am being driven) You are driving (are being driven) He, she, *or* it is driving (is being driven)	We are driving (are being driven) You are driving (are being driven) They are driving (are being driven)
Past Progressive	I was driving (was being driven) You were driving (were being driven) He, she, *or* it was driving (was being driven)	We were driving (were being driven) You were driving (were being driven) They were driving (were being driven)
Future Progressive	I will be driving (will have been driven) You will be driving (will have been driven) He, she, *or* it will be driving (will have been driven)	We will be driving (will have been driven) You will be driving (will have been driven) They will be driving (will have been driven)
Present Perfect Progressive	I have been driving You have been driving He, she, *or* it has been driving	We have been driving You have been driving They have been driving
Past Perfect Progressive	I had been driving You had been driving He, she, *or* it had been driving	We had been driving You had been driving They had been driving
Future Perfect Progressive	I shall have been driving You will have been driving He, she, *or* it will have been driving	We shall have been driving You will have been driving They will have been driving

5. The *past perfect progressive form* describes an ongoing past condition that has been ended by something said in the sentence.

   James *had been selling* well until the store closed.

6. The *future perfect progressive form* expresses an action or condition ongoing until some specific time in the future.

   I *shall have been working* for hours by the time you arrive.

   In January, our furnace *will have been operating* for 20 years.

WRITING TIP

When the action in your sentence is unfinished, use a progressive form.

EMPHATIC VERB FORMS

The helping verb *to do* (*does, did*) followed by a present tense verb is used for the **emphatic verb forms**. As you might expect, this form is used to

Table 11-5 Present Emphatic and Past Emphatic Forms of *Drive*

Present Emphatic Forms		Past Emphatic Forms	
Singular	*Plural*	*Singular*	*Plural*
I do drive	We do drive	I did drive	We did drive
You do drive	You do drive	You did drive	You did drive
He, she, *or* it does drive	They do drive	He, she, *or* it did drive	They did drive

emphasize something. Table 11-5 shows that the emphatic forms are used only in the present and past tenses of the active voice.

Here are some examples of the emphatic forms used in sentences.

He *does drive* to work every day. (Present emphatic tense, third-person singular.)

They *do drive* to every stockholders' meeting. (Present emphatic tense, third-person plural.)

I *did walk* to the store during my lunch hour. (Past emphatic form, first-person singular.)

We *did sing* at the company picnic. (Past emphatic form, first-person plural.)

YOU TRY IT

Write a sentence using each of the progressive verb phrases below. In each sentence, be sure to tell *the time of* the action.

EXAMPLE: will have been selling

Our competitors will have been selling their product for several months before our product is available.

EXAMPLE: had been printing

The document had been printing without problems until the paper jammed.

1. was sending

2. have been writing

3. will have been working

4. is reading

5. will be expanding

6. had been progressing

CHAPTER SUMMARY: A QUICK REFERENCE GUIDE

PAGE	TOPIC	KEY POINTS	EXAMPLES
214	**Verb voice**	Voice shows whether a speaker or writer wants to emphasize the subject (*active voice*) or the object (*passive voice*). Active voice advantages are as follows: **1.** Readers follow action at greater speed. **2.** Stronger verbs. **3.** Direct and less wordy.	The broker *advertised* the bond issue. The bond issue *was advertised* by the broker.
216	**Verb mood**	Mood refers to the *way* a listener or reader should react to an action or state of a verb. *Indicative mood* gives a positive statement of fact or asks a question. *Imperative mood* gives a command or request. Imperative mood sentences are in the present verb form, second person; and the subject (you) is often omitted. *Subjunctive mood* is used to make statements contrary to fact or to express a wish. Subjunctive mood is often used with *if, though, unless,* and *that.* Subjunctive mood does the following: **1.** Uses *were* to express an idea contrary to fact. **2.** Uses *be* in *that* clauses to express a desire, recommendation, demand, motion, suggestion, request, necessity, or resolution.	The hardware store sold its entire stock of snow shovels. Be sure to go to the bank today. If Jim *were* vice president, the company would succeed. The owner asked that she *be* included in the discussion.
219	**Simple tenses**	Tense tells the *time* when an action occurs or the condition exists. Tense also tells the stage of action—continued or completed. **1.** The *present tense* states a general truth; tells action in progress, continued, and habitual; and can express future action.	Lynn *sells* beauty salon equipment for the Ace Salon Company.

PAGE	TOPIC	KEY POINTS	EXAMPLES
		2. To form the *past tense*, regular verbs add *d* or *ed*, and irregular verbs form the past tense in irregular ways. Two rules: (1) Verbs ending in *s* are usually singular. (2) Do not use the past tense because another verb in the same sentence is in the past tense.	The salespeople *counted* every item for the year-end inventory.
		3. The *future tense* expresses an action or condition that has not yet taken place. Usually, it uses the helping verb *will*.	The store buyer *will find* a clothing company that meets the needs of our customers.
221	**Perfect tenses**	Perfect tenses show actions that have been completed or will be completed before a given time. They combine the verb *to have* with the past participle.	
		1. The *present perfect tense* uses *have* and *has* with the past participle to show actions (a) just completed, (b) beginning in past and continuing to future, and (c) past actions that may occur again.	Every employee *has looked* for the missing order. She *has typed* for two hours.
		2. The *past perfect tense* uses *had* with the past participle to show action completed before some point in time in the past, specified or implied.	The dentist *had kept* the office open late for three nights. Although the day was long, Julian *had completed* every job given him.
		3. The *future perfect tense* uses *will have* with the past participle to show action completed before another future action or by a specified future time.	The inventory *will have been completed* when you return.
223	**Progressive verb forms**	The six progressive verb forms express an action or event in progress at the time indicated. They are formed with the helping verb *to be* and the present participle. These six progressive verb forms are the present progressive, past progressive, future progressive, present perfect progressive, past perfect progressive, and future perfect progressive.	The store *will be selling* all its leather jackets.
224	**Emphatic verb forms**	The emphatic verb forms are formed with the helping verb *to do* (*does, did*) followed by a present tense verb.	Diane *did look* in that file for the missing sales slip.

QUALITY EDITING

Sharon Anderson wrote the following draft of a memo to Jose Macias in Human Resources about an accident that happened at work. Edit Sharon's memo before it is sent to Jose.

DATE: June 22, 19XX

TO: Jose Macias, Human Resources

FROM: Sharon Anderson

SUBJECT: Accident at Work

Yesterday Andrea Myers has been working at her desck when she was being called to the lobby. When she has been going down the stares, she had fallen down. Their had been a tare in the carpet and she was catching her toe in the lose carpet at that moment. Andrea would not be falling if she was not in such a hurry.

As a result of her fall, Andrea is spraining her ankel. She has being told to avoid putting her wait on the ankel for severel Days and will be stay home until next Thursday. Andrea's absence will have been four work days before she returns. Since her assistance, Chan and margaret, are already been working on important projects, I sugest that a manager from the sales department be assigned to cover Andrea's duties.

Andrea will have bene contacting you concerning benefits. The company attorny will contacted Andrea by next Tuesday. The carpet will have being fixed before the end of the day.

APPLIED LANGUAGE

A. Write a sentence using each of the verbs below.

EXAMPLE: will have been waiting

By the time we return to the office, our client will have been waiting for an hour.

EXAMPLE: if . . . were

If we were the managers, we would complete the evaluations in May.

1. had been traveling

_____

2. had been written

_____

3. if . . . were

_____

4. will be broken

_____

5. writes

_____

6. went

_____

7. designs

_____

8. avoided

_____

B. Assume that you are a word processing operator and you use a computer every day in your work. Your supervisor found you typing a letter on a typewriter today and inquired about it. Write a short memo to your supervisor with the following information: Explain what happened to your computer, how you are getting your work done now, and how the problem will be corrected.

W ORKSHEET 11

VERBS: VOICE, MOOD, TENSE

Part A. Voice

The following sentences are written in the passive voice. Rewrite the sentences so that they are in the active voice.

EXAMPLE:

Passive: The windows had been cleaned by the maintenance crew.

Active: **The maintenance crew had cleaned the windows.** _____

1. The work has been given to the staff by the supervisor.

_____

2. An accountant was hired by the new vice president.

_____

3. An order for a dozen monitors had been received by the sales department.

_____

4. A safety check of the premises will be conducted by the OSHA representative.

_____

5. The staff was given an extra vacation day by the executive committee in recognition of their excellent work on the project.

_____

_____

6. Four of the employees had been moved to the new offices by Acme Movers.

_____

7. The computers' drives were examined by the MIS director.

_____

8. The inks will have been tested by American Product Testing, Inc., before Friday.

_____

9. The boxes were sent to the warehouse by Yolanda.

_____

10. The documents were not received by the conference center in time for the meeting.

_____

Part B. Mood

Complete the sentences below. Note that each one contains a verb in the subjunctive mood. (Some sentences have two parts. Complete both parts.) Use your imagination! Then, underline the verb that is in the subjunctive mood.

EXAMPLE: If the executive were <u>more communicative</u>, she would <u>have more cooperation from her staff</u>.

1. Georgia demanded that Ryan be _____.

2. Ms. Lopez suggested that the company be _____.

3. If I were _____, I would _____.

4. If it were possible for me to _____, I would.

5. The committee recommended that the company finish _____.

6. If Mr. Yee were to _____, he would _____.

Part C. Tense

A verb appears in parentheses in each of the sentences below. Write the correct tense of the verb in the blank. (Some sentences contain two verbs.) **Note**: Read the entire sentence before you respond.

EXAMPLE: Yesterday, the firm (sell) _____ **sold** _____ 400 cartons of disks.

1. Last week, the firm (sell) _____ 900 cartons of disks, but because of the snowstorm, it only (sell) _____ 400 cartons.

2. We expect that the order (prepare) _____ by next week.

3. The project must (complete) _____ last month because the client has already paid the invoice.

4. The company (compensate) _____ you for your mileage.

5. The FineTime Printing Company (print) _____ the program materials.

6. Mr. Faraday (write) _____ to the president before he (hear) _____ of the decision.

7. I don't know if the committee (meet) _____ yet.

8. Joanna (help) _____ with the inventory before the firm promoted her to manager.

9. We (employ) _____ by this firm for four years.

10. Yesterday, the sales representative (tell) _____ us that the

freight (deliver) _____ next Tuesday.

11. The items behind the cabinet (hide) _____ from view.

12. No one would believe that Charlotte Holcomb (make) _____ the sale.

13. We reached the building at 9:00 A.M., only to discover that the

auditor (arrive) _____ an hour earlier.

14. By noon tomorrow, the claims processor (make) _____ her decision.

15. If you prepare the letter now, I (sign) _____ it before I leave.

16. By the time we reach Minneapolis, the talks (end) _____.

17. If the budget allows, we (purchase) _____ a notebook computer next month.

18. For months I (tell) _____ him to repair the equipment.

19. Dr. Mason (be) _____ here only an hour when she was called away.

20. Next year, the company (hire) _____ someone to handle new orders.

Part D. Progressive and Emphatic Verb Forms
Write the progressive and emphatic forms of each of the underlined verbs in the space provided.

	was using	did use
EXAMPLE: I <u>used</u> the computer.		
1. You <u>walk</u> to work.		
2. He <u>spoke</u> to charitable groups.		
3. I <u>work</u> long hours.		
4. They <u>update</u> their equipment.		
5. We <u>manage</u> our own printing.		
6. You <u>help</u> Mr. Lazio.		
7. She <u>listened</u> carefully.		
8. Ms. Fortune <u>writes</u> the contracts.		
9. I <u>spoke</u> Italian with the visitor.		
10. Howard and Avery <u>sought</u> the best available price.		

CHAPTER 12

SUBJECT-VERB AGREEMENT AND VERBALS

If the structure of a sentence is needlessly indirect, try simplifying it. Look for opportunities to strengthen the verb.

—Diana Hacker, *Rules for Writers*

LEARNING UNIT GOALS

The term *agreement* is not new to you. Chapter 9 discussed the agreement of antecedents and personal pronouns. Sentence agreement occurs when one sentence structure changes its form to agree with another sentence structure. Personal pronoun agreement occurs when the form (number and gender) of the personal pronoun agrees with its antecedent. Verb agreement occurs when the form (person and number) of the verb agrees with its subject noun or pronoun.

Learning Unit 12-1 gives the basic subject-verb agreement rules. Learning Unit 12-2 gives special subject-verb number agreement rules. The last chapter learning unit, Learning Unit 12-3, explains verbals.

LEARNING UNIT 12-1
BASIC SUBJECT-VERB AGREEMENT RULES

Most of you know that a *verb must agree with its subject in person and number.* So far our study of verb properties (voice, mood, and tense) has focused on the verb. The person and number agreement of subjects and verbs, however, focuses on the subject.

To make a verb agree with its subject, you must locate the subject. To locate the subject, you must first find the verb. Since you have been locating the subject in your sentence analysis, this learning unit briefly reviews the steps for locating a subject. Then, you will learn the basic rules for subject-verb agreement.

Do you wonder why you must pay special attention to the subject in subject-verb agreement? The person (first, second, or third) and number (singular or plural) of the subject *control* the verb form. When you determine the person and number of the subject, you go back to the verb and make it agree with the subject. The subject *directs* the choice of person and number verb forms.

LOCATING THE SUBJECT

In Chapter 5, you will recall that before you began the sentence analysis flowchart steps, you crossed out all the prepositional phrases to avoid the error of using a modifier as a major sentence part. Now you also cross out parenthetical phrases that make the subject appear plural (*as well as, together with, in addition to, rather than, along with, including,* or *accompanied by*). The following two steps for locating the subject were explained in Chapters 4 and 5:

Step 1. Find the verb or verb phrase. To find the verb or verb phrase, add *yesterday* or *tomorrow* before the sentence you are studying. The verb is the word (or words if the verb is a verb phrase) that changes with the change in time.

Step 2. Find the subject. To find the subject, insert *who?* or *what?* before the verb. The answer is the subject, which is always a noun or pronoun. Remember that the subject may be after the verb in some sentences (especially sentences beginning with the adverbs *there* or *here*). This is also true of questions. You can usually reverse these sentences and put them in the subject-verb order.

The sentences that follow illustrate these two steps. In this learning unit and most of Learning Unit 12-2, the subject is underlined once, and the verb or verb phrase is underlined twice.

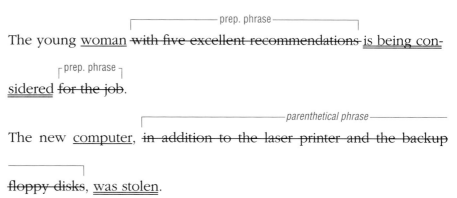

The correct <u>combination</u> <u>is</u> the first and last items mentioned. (Note that in this sentence the linking verb connects the singular subject *combination* with the plural subject complement *items.* Following the sentence analysis steps will avoid the error of thinking the subject complement is the subject.)

The first and last <u>items</u> mentioned <u>are</u> the correct combination. (This sentence is the opposite of the sentence above. The linking verb connects the plural subject *items* with the singular subject complement *combination.*)

Here <u>is</u> the <u>invoice</u> I need. (The <u>invoice</u> I need <u>is</u> here.)

<u>It</u> <u>is</u> the answer you need. (When the singular pronoun *it* replaces the subject of the sentence, the pronoun controls the verb even if the subject, which is not named in the sentence, is plural.)

SUBJECT-VERB PERSON AGREEMENT

Recall from Chapter 8 that the personal pronoun shows person by a change in form. First-person subjects are the person(s) speaking (*I* for singular and *we* for plural). Second-person subjects are the person(s) spoken to (*you* for both singular and plural). Third-person subjects are the person(s) or thing(s) spoken of (*he/she/it* for singular and *they* for plural; various singular and plural nouns are also third-person subjects). How do we relate persons to verbs?

Verbs express action. This action may be the action of the person(s) speaking, spoken to, or spoken of. So we say that verbs have three persons. The sentence subject determines the person of the verb, and verbs change form to agree with the subject person.

The verb *to be* is the only verb that changes form with the first, second, and third persons: *I am, you are, he/she/it is, I was,* and *you were.* Other verbs only change their third-person singular form in the present tense. For example, when we speak of the action of the first person, we say, *I walk*; when we speak of the action of the third person, we say, *he (she, it) walks*. The subject in the third person can vary from the singular and plural nouns of people, such as *manager(s)*, to the singular and plural nouns of things, such as *computer(s)*, and the singular and plural pronouns, such as *he, she, it*, and *they*.

Correct subject-verb person agreement usually comes naturally. Subject-verb number agreement, however, can be challenging.

BASIC SUBJECT-VERB NUMBER AGREEMENT RULES

Like many things, knowing something *must* happen is often not enough to *make* it happen. The rule that subject and verb must agree is easily stated. To make this happen, however, writers must overcome the difficulties that are responsible for the errors they make. Subject-verb number agreement can be separated into agreement with singular and plural subjects and agreement with compound subjects.

Agreement with Singular and Plural Subjects

Since number agreement is about *singular* and *plural*, this is your first decision about the subject. Is the subject singular or plural?

When you decide if your subject is singular or plural, you follow this rule: *Singular subjects need singular verbs; plural subjects need plural verbs.*

The <u>accountant</u> <u>knows</u> the balance sheet contains errors. (The subject and verb are singular.)

The junior <u>accountants</u> <u>were</u> responsible for the errors. (The subject and verb are plural.)

Agreement with Compound Subjects

Compound Subjects Connected by *and*.   When a compound subject is connected by *and*, it is usually plural and requires a plural verb. (*Exception*: When a compound subject refers to a single item, use a singular verb.)

Mortgage <u>bankers</u> and mortgage <u>brokers</u> <u>provide</u> the same basic service. (*Mortgage* is a noun functioning as an adjective answering the question *what kind?* about *bankers* and *brokers*.)

The <u>horse and buggy</u> <u>was</u> popular in 1890. (*Horse and buggy* is a compound subject referring to a single unit.)

Two Singular Subjects Joined by *or, nor, either . . . or*, or *neither . . . nor*.   When two singular subjects are joined by *or, nor, either. . .or*, or *neither . . . nor*, use a singular verb. If one subject is singular and one is plural, place the plural form closer to the verb and use a plural verb.

Neither <u>Doris</u> nor her <u>brother</u> <u>is willing</u> to cooperate.

Neither <u>Doris</u> nor her <u>brothers</u> <u>are willing</u> to cooperate.

YOU TRY IT

A. Complete the following sentences. For items 1–4, check to make sure that your verb *agrees* with the subject provided. For items 5–8 select the appropriate verb from the two in parentheses. For each sentence circle the correct word to indicate whether the subject is plural or singular.

1. The supervisor, who oversees the work of four assemblers,_____

_____. (Plural/Singular)

2. Janet Gaynes and her staff members _____

_____. (Plural/Singular)

3. Delayed by the late train, the travelers _____

_____. (Plural/Singular)

4. Neither Mr. Myers nor Ms. No _____

_____. (Plural/Singular)

5. The owner and manager, Mrs. Kelly, (send/sends) _____

_____. (Plural/Singular)

6. Pen and ink (is/are) _____

_____. (Plural/Singular)

7. The owner and the manager (is/are) speaking about _____

_____. (Plural/Singular)

8. Bread and butter (is/are) _____

_____. (Plural/Singular)

B. Can you think of any other noun/noun combinations that might make a singular subject? Here are a few to get you started: bottle and stopper, potatoes and gravy, and pencil and paper. List five additional combinations.

1. _____

2. _____

3. _____

4. _____

5. _____

Answers: **A.** 1. singular 2. plural 3. plural 4. singular 5. sends, singular 6. is, singular 7. are, plural 8. is, singular **B.** 1. peanut butter and jelly 2. pen and ink 3. cat and dog 4. men and women 5. ham and eggs

LEARNING UNIT 12-2
SPECIAL SUBJECT-VERB NUMBER AGREEMENT RULES

In special situations, writers sometimes have difficulty determining whether a subject is singular or plural. This learning unit presents the special subject-verb number agreement rules that will solve most of your number difficulties.

AGREEMENT WITH NOUN SUBJECTS THAT CAN BE SINGULAR OR PLURAL

Some nouns that end in *s* can be singular in meaning (take a singular verb) or plural in meaning (take a plural verb). Note the following:

1. **Nouns such as *billiards*, *civics*, *mathematics*, *news*, *measles*, *mumps*, and *series* are always singular in meaning.**

    Mathematics <u>is</u> an intriguing subject.

    The <u>news</u> about the new airplane <u>is</u> exciting.

2. **Nouns such as *politics*, *statistics*, *athletics*, and *aerobics* are singular when they refer to fields of knowledge or an activity but plural in other situations.**

    <u>Politics</u> <u>has played</u> an important role in her life. (Singular.)

    The <u>politics</u> of the organization <u>are</u> difficult to understand. (Plural)

3. **Nouns such as *wages*, *tongs*, *pliers*, *trousers*, *scissors*, and *odds* are always plural in meaning.**

    The <u>pliers</u> used by the mechanic <u>are</u> new.

    The <u>odds</u> <u>are</u> against finding the lost ledger.

AGREEMENT WITH COLLECTIVE NOUNS

Collective nouns may be singular or plural. If the noun refers to a group as a single unit (*army*, *audience*, *class*, *committee*, *crowd*, *family*, *majority*, or *jury*), use a singular verb. If the sentence suggests that members of the group are acting independently of one another, use a plural verb.

The <u>audience</u> <u>is</u> quiet and attentive.

The <u>audience</u> (or <u>Members</u> of the audience) <u>are</u> going their separate ways.

Recall from Chapter 9 that it was suggested to reword a sentence such as the previous sentence by using the words in parentheses. This avoids any confusion.

AGREEMENT WITH INDEFINITE PRONOUNS

Table 12-1 reviews the singular, plural, and singular or plural indefinite pronouns discussed in Chapter 9. Refer to this table as you study the following rules.

Indefinite Pronouns Used as Subjects

The following rules apply when indefinite pronouns are used as subjects:

1. **The *always singular* indefinite pronouns take singular verbs.**

Table 12-1 Indefinite Pronouns: Singular, Plural and Singular or Plural

Always Singular	Always Plural	Either Singular or Plural
anybody/anyone/anything	both	all
each/either	few	any
every/everybody/everyone/everything	many	more
many a/neither	several	most
nobody/nothing/one	others	none
somebody/someone/something		some

Everybody <u>seems</u> pleased with the results.

Each of our products <u>has been accepted</u> by the public.

Neither of his comments <u>was overheard</u>.

2. **The *always plural* indefinite pronouns take plural verbs.**

Both of the tall ladders <u>are</u> broken.

Several dental assistants <u>have found</u> new jobs.

3. **The *singular or plural* indefinite pronouns take singular or plural verbs.** The prepositional phrase that follows these pronouns indicates whether the pronouns are singular or plural. (*Exceptions*: As stated in Chapter 9, *none* is singular whenever its meaning is clearly *not one*; *any* is singular whenever its meaning is clearly *any one*.)

Some of the floppy disks <u>have been lost</u>.

Most of the building <u>is made</u> of old brick.

None of the workers <u>takes</u> his or her tools to the job. (The pronoun *none* means *not one* in this sentence, so the verb is singular.)

Any of the accountants <u>has</u> difficulty with his or her financial records. (The pronoun *any* means *any one* in this sentence, so the verb is singular.)

Each, Every, or *Many a* Used as Adjectives Modifying Subjects

If *each, every,* or *many a* is used as an adjective modifying a subject, that subject is singular even though it may have more than one part.

Every <u>man</u> and <u>woman</u> <u>has</u> the opportunity to give.

Many a <u>person</u> <u>is</u> likely to be disappointed.

AGREEMENT WITH TITLES AND COMPANY NAMES

The title of one book, song, magazine, article, or the name of a company is singular even though the title or name itself may be compound or plural in form.

Three Angry Men <u>is</u> a fine novel.

Three angry <u>men</u> <u>are waiting</u> to see you.

<u>Tustin, Fox & Benson, Inc.,</u> <u>has submitted</u> a bid.

<u>Tustin and Fox</u> <u>are</u> excellent accountants. (*Tustin and Fox* is a compound subject and must have a plural verb.)

AGREEMENT WITH NOUNS EXPRESSING AMOUNTS

Nouns that express amounts (distance, fraction, money, percentage, time, etc.) are singular when they refer to people or things thought of as a unit and plural when they are thought of as individuals or individual items.

Forty <u>feet</u> of cord <u>is</u> still on the shelf. (The subject refers to *cord*, which is singular.)

A <u>third</u> of the shipment <u>has been stolen</u>. (The subject refers to *shipment,* which is singular.)

A <u>third</u> of the boxes <u>have been stolen</u>. (The subject refers to *boxes,* which is plural.)

Four hundred <u>dollars</u> <u>is</u> a fair price.

Four hundred <u>chairs</u> <u>have been ordered</u>.

Thirty <u>minutes</u> <u>is</u> more time than you need.

AGREEMENT WITH RELATIVE PRONOUN CLAUSES

Recall from Chapter 6 that dependent noun clauses are usually introduced with the relative pronouns. In Chapter 13, you will learn that dependent adjective clauses also are introduced with relative pronouns. Frequently, dependent noun and adjective clauses use the relative pronouns *who, which,* and *that* as subjects. When the antecedents of these relative pronouns are singular, they must be followed with a singular verb; when the antecedents are plural, they must be followed with a plural verb.

Here is the *supervisor* who *wins* the award. (*Who* refers to the singular noun *supervisor,* so the verb *wins* is singular.)

The committee *members* that *were* late could not find seats. (*That* refers to the plural noun *members,* so the verb *were* is plural.)

The words *one of the* and *the only one of* can cause problems. *One of the* usually refers to a plural antecedent, so the verb following *who* must be plural. *The only one of* usually refers to a singular antecedent, so the verb following *who* must be singular.

Lila is *one of the* managers who *see* the meaning of the change in policy. (*Who* refers to the plural *managers,* so the verb following *who* must be plural.)

Diane is *the only one of* those managers who *has paid* her association dues. (*Who* refers to the singular *one* because only one manager paid her dues. You could omit the prepositional phrase *of those managers,* and the sentence would still mean the same.)

STUDY TIP

Remember that to determine whether a subject is singular or plural, you should first consider the meaning of the subject, not the form of the subject. Also note that subjects are usually made *plural* by adding *s* (or *es*) and verbs are usually made *singular* by adding *s*. So you can follow the general guideline that in subject-verb agreement you have only one added *s*.

YOU TRY IT

Complete the following sentences. Make sure that your verb *agrees* with the subject provided. Then, circle the correct word in parentheses to indicate whether the subject is plural or singular.

1. The series of reports _____

_____. (Plural/Singular)

2. Statistics on the survival of new businesses _____

_____. (Plural/Singular)

3. The wages paid by Oliver and Associates _____

_____. (Plural/Singular)

4. The committee on the building's environment _____

_____. (Plural/Singular)

5. Each of the company's new products _____

_____. (Plural/Singular)

6. Most of the papers_____

_____. (Plural/Singular)

7. Most of the effort _____

_____. (Plural/Singular)

8. Sonja is one of those people who _____

_____. (Plural/Singular)

Answers: 1. singular 2. plural 3. plural 4. singular 5. singular or plural depending on meaning 6. plural 7. singular 8. plural

---

LEARNING UNIT 12-3
VERBALS

Verbals are special verb forms that can function as nouns, adjectives, and adverbs. By themselves, verbals cannot function as predicates. Understanding verbals will be helpful when you study punctuation. The use of verbals will also increase variety in your sentence structure.

The English language has three verbals: the infinitive, the participle, and the gerund. The infinitive and participle have been discussed briefly in earlier chapters. The term *gerund* may be a new term for you. Let's begin with infinitives.

INFINITIVES

The **infinitive**, you may recall, is the present form of the verb preceded by *to*, such as, *to be, to study, to give,* or *to understand.* Although the word *to* is usually present, it may be omitted in certain constructions such as *He helped me choose a secretary* or *Mr. Lomax let me use his drill.* Infinitives can function as nouns, adjectives, or adverbs.

1. **As nouns, infinitives can function as subjects, direct objects, predicate nouns, appositives, and objects of prepositions.**

   To fail is not a calamity. (Infinitive functioning as a subject.)

   Mary was asked *to help.* (Infinitive functioning as a direct object.)

   Jane's goal has always been *to sell.* (Infinitive functioning as a predicate noun.)

   Jane's goal, *to sell,* was realized. (Infinitive functioning as an appositive.)

   Bruce was so tired that he had no choice but *to sleep.* (Infinitive functioning as an object of a preposition.)

2. **As adjectives, infinitives modify nouns and pronouns.**

   Now is the time *to leave.* (*To leave* modifies the noun *time.*)

3. **As adverbs, infinitives modify verbs, adjectives, or other adverbs.**

   Ms. Jenkins is difficult *to please.* (*To please* modifies the adjective *difficult.*)

Infinitives can be combined with objects and/or modifiers to form **infinitive phrases**. Like other phrases, the infinitive phrase functions as a single part of speech (nouns, adjectives, and adverbs).

Most speakers and writers do not have difficulty with the infinitive. The following two suggestions, however, may prove helpful:

1. **Avoid awkward split infinitives.** Placing one or more words between *to* and the verb form often results in an awkward construction.

DON'T USE:	I want you *to, as quickly as possible, finish* this project.
USE:	I want you *to finish* this project *as quickly as possible.*
DON'T USE:	No one in the room was able *to perfectly complete* the test.
USE:	No one in the room was able *to complete* the test *perfectly.*

Experienced writers occasionally split an infinitive if the resulting construction is sufficiently smooth. Usually, however, they avoid the use of split infinitives.

2. **Use the present infinitive (*to do*, *to say*, *to go*, etc.) except when the action expressed by the infinitive occurs before the action of the governing verb.** In this situation, use the perfect infinitive (*to have done, to have said, to have gone,* etc.). Both the present and the perfect infinitive may be used in the passive voice.

> *Present Tense:* to tell, to be told
>
> *Perfect Tense:* to have told, to have been told

> Some people like *to write* letters. (Present infinitive, active voice.)

> The merchandise is known *to be damaged.* (Present infinitive, passive voice.)

> Brad Proxmire is said *to have stolen* from the company. (The perfect infinitive is used because the stealing preceded the action of the main verb.)

> The merchandise is known *to have been damaged* in transit. (Note the use of the perfect infinitive in the passive voice. The merchandise is known now to have been damaged earlier.)

PARTICIPLES

Recall that in Chapter 10 you learned that participles can function as verbs when used with forms of the helping verbs *to be, to have,* and *to do.* In this chapter, the term **participle** refers to a verb form that functions alone as an adjective.

In the sentence *The president described the coming events, coming* is a participle. It is a form of the verb *to come* and is used here as an adjective to modify the noun *events.* The participle may appear in any of several forms, as shown below.

	Present Participle	**Past Participle**	**Perfect Participle**
Active voice:	finding	—	having found
Passive voice:	being found	found	having been found

You can expand a participle into a **participial phrase** by adding objects, complements, and modifiers. Like a participle, a participial phrase functions as an adjective and modifies a noun or pronoun.

Dangling participial (verbal) phrases will be discussed in the last section of this learning unit. Before you consider such phrases, you should master these two basic rules:

1. **Use the present participle, past participle, or perfect participle correctly.** If the action expressed occurs at the same time as the action of the main verb, use the present participle. If the action

expressed occurs before the action of the main verb, use the past participle or the perfect participle, whichever is suitable.

> *Walking* rapidly, Ms. Haynes made excellent progress.

> *Hoping* for a promotion, Mr. Colburne accepted every difficult assignment.

> The report *read* by Ms. Hemandez has 50 pages. (Note the use of the *past participle* to indicate a completed action.)

> *Having walked* rapidly for an hour, Ms. Haynes looked exhausted. (Perfect participle is used to express action before that of the main verb.)

> *Having failed* to get the promotion, Mr. Colburne was bitter. (First, he failed; then, he was bitter.)

2. **Avoid the awkward use of absolute phrases.** An absolute phrase consists of a noun or pronoun and a participle. Absolute phrases function as modifiers of the entire sentence to which they are attached.

> DON'T USE: *Wednesday being a holiday*, we were not required to work.
>
> USE: *Because Wednesday was a holiday*, we were not required to work.

GERUNDS

The **gerund** is a verb form that ends in *ing* and functions as a noun. Because it is a verbal *noun*, a gerund may be used as a subject, an object, or a complement. Unlike other nouns, however, gerunds may take an object. Gerunds cause no difficulty in sentences such as the following:

> *Writing* can become tiresome. (Gerund is used as subject.)

> The new bookkeeper is slow in his *thinking*. (Gerund is used as object of a preposition.)

> Edmond's favorite pastime is *hiking*. (Gerund is used as subject complement.)

Gerund phrases contain a gerund with modifiers or a complement functioning as a noun. They are frequently misplaced in sentences and will be discussed later in this learning unit. In our day-to-day use of gerunds, we should be especially careful to observe this rule: *Make certain that a noun or pronoun used to modify a gerund is in the possessive case.* The possessive case form, of course, does the work of an adjective.

> DON'T USE: I heard about *Graham* getting the order.
>
> USE: I heard about *Graham's* getting the order. (*Graham's* modifies the gerund *getting* and is the possessive case. *Getting* is a gerund because it is the *ing* form of the verb *to get* and serves as the object of the preposition *about*; thus it functions as a *noun*. The noun *order*, you will note, serves as the object of the gerund.)

DON'T USE: We were not aware of *him* leaving the firm.

USE: We were not aware of *his* leaving the firm. (*His* modifies the gerund *leaving* and is in the possessive case.)

DON'T USE: Did *Clyde* winning the contest surprise you?

USE: Did *Clyde's* winning the contest surprise you? (The surprise was not *Clyde*; it was *his winning*.)

As you study these sentences, observe that in each of the preferred versions a noun or pronoun in the *possessive case* has been used to modify a gerund. Remember that a gerund, which itself may take an object, will always serve as a subject, an object, or a complement.

DANGLING VERBAL PHRASES

Verbals (infinitives, participles, and gerunds) appear frequently in phrases that function as modifiers, and such phrases are often used to introduce sentences. **Verbal phrases** should be properly placed in a sentence. If a verbal phrase relates to another word in the sentence, it should be placed so that its relation to that word is perfectly clear. If a verbal phrase is not so placed, it is said to dangle. A verbal phrase may be said to be *misplaced* if it is too far removed from the word(s) to which it relates and to be *dangling* if it relates to no other word in the sentence.

If a sentence begins with *Having left Seattle in March*, the next word(s) should tell *who* left Seattle in March. If a sentence begins with *In correcting the mistake*, the next word(s) should tell who corrected the mistake. Following are five pairs of sentences that begin with verbal phrases. Note that in each logical version, the introductory phrase is followed immediately by the word to which it is related (in italics).

CONFUSING: Reading very rapidly, the manuscript took only 45 minutes.

LOGICAL: Reading very rapidly, *she* finished the manuscript in only 45 minutes.

CONFUSING: Having retired early, the years ahead looked bright.

LOGICAL: Having retired early, *Sam* looked forward to the years ahead.

CONFUSING: By using this new technique, ten minutes will be saved.

LOGICAL: By using this new technique, *you* will save ten minutes.

CONFUSING: While writing the letter, her telephone rang four times.

LOGICAL: While writing the letter, *Lois* received four telephone calls.

CONFUSING: To determine its value, the property was appraised.

LOGICAL: To determine its value, *we* appraised the property. (Although the infinitive phrase modifies *appraised*, it relates to the pronoun we, which some grammarians refer to as its subject.)

A verbal phrase can also be misplaced at the end of a sentence. Note the following example:

CONFUSING: The witness hesitated before answering the attorney,
looking somewhat bewildered.

LOGICAL: The witness, looking somewhat bewildered, hesitated
before answering the attorney.

Because *you* (understood) is the subject of a command or request, verbal phrases are easily controlled in grammatical constructions such as these: *To save time, use the new computer. In addressing Mr. Pulaski, don't mispronounce his name.*

ANALYZING SENTENCES WITH VERBAL PHRASES

Verbals and verbal phrases were not included in the sentence analysis explained in Chapter 5 because you had not yet studied them. From this learning unit, you know that (1) infinitives can function as nouns and (2) gerunds always function as nouns. This means that as nouns, infinitives and gerunds can function as major sentence parts—subjects, direct objects, and predicate nouns. When you see an infinitive and gerund in a sentence, you now can consider them as part of your analysis.

To analyze sentences with infinitive phrases and gerund phrases, first identify the phrase. Think of the phrase as one word. Then follow with the sentence analysis flowchart steps to find the verb, subject, direct object, and predicate noun. If the phrase does not fit in the sentence analysis flowchart, you have a modifier. Let's analyze the following sentence. If the sentence has a prepositional phrase, remember to cross it out.

The new accountant hoped *to find the old records.*

Step 1. Find the phrase. Infinitive phrases usually begin with the infinitive *to* followed by a verb. In this sentence, *to find the old records* is an infinitive phrase (infinitive and any modifiers or complements).

Step 2. Think of the phrase as one word: tofindtheoldrecords.

Step 3. Follow the sentence analysis flowchart steps to determine if the infinitive phrase functions as a subject, direct object, or predicate noun (subject complement).

  a. Find the verb or verb phrase. The word that changes when *tomorrow* is added before the sentence is *hoped. Hoped* is the verb.

  b. Find the subject. *Who* hoped? Accountant. *Accountant* is the subject.

  c. Does the sentence have a direct object? Accountant hoped *what?* Accountant hoped *tofindtheoldrecords.* This sentence has a direct object. *To find the old records* is the direct object.

$$vt \quad \overline{\qquad\qquad DO \qquad\qquad}$$
The new <u>accountant</u> <u>hoped</u> *to find the old records.*

When a phrase does not fit the sentence analysis flowchart steps, you know it functions as a modifier.

In the following sentence, we have only stated the result of the sentence analysis steps:

Todd traveled ~~in Wisconsin~~ *to get his information*. (The prepositional phrase is crossed out. The infinitive phrase is *to get his information*. *Traveled* is an intransitive complete verb. *Todd* is the subject. Complete verbs are followed by modifiers. The infinitive phrase does not fit in the sentence analysis flowchart steps so it is a modifier of *traveled*.)

YOU TRY IT

Each of the following sentences contains a problem related to a verbal. Rewrite each sentence to correct the problem. You may need to add some information to create a correct sentence. **Note**: A sentence may have more than one problem!

EXAMPLE: Before making a decision, the data were analyzed.

Joanne analyzed the data before making a decision.

1. Writing in haste, important information was forgotten by the supervisor.

2. Working quickly, the paperwork was completed in a day.

3. The committee made every effort to efficiently complete the report.

4. To help calmly discuss the problem, the opinions were written out on the easel paper.

5. Did the firm getting the product onto the market before the competition surprise you?

6. Everyone was shocked at the news about the product being faulty.

Possible Answers: 1. Writing in haste, the supervisor forgot important information. 2. Working quickly, Nathan completed the paperwork in a day. 3. The committee made every effort to complete the report efficiently. 4. To help discuss the problem calmly, they wrote their opinions on the easel paper. 5. Did the firm's getting the product onto the market before the competition surprise you? 6. Everyone was shocked at the news about the product's being faulty.

CHAPTER SUMMARY: A QUICK REFERENCE GUIDE

PAGE	TOPIC	KEY POINTS	EXAMPLES
236	**Locating subject**	**1.** Cross out prepositional phrases and parenthetical phrases.	This *painting* ~~of old cars~~ *is* a classic.

PAGE	TOPIC	KEY POINTS	EXAMPLES
		2. Using the sentence analysis steps, find the verb or verb phrase; then find the subject. 3. Some subjects appear after the verbs, especially sentences beginning with *there* or *here* and questions.	The *warehouse is* damp and cold. Here *is* the *report* I promised.
237	**Basic subject-verb person agreement rules**	The subject determines the person of the verb. Verbs change to agree with the subject person. *To be* is the only verb that changes form with the first, second, and third persons. Other verbs change in the third-person singular form of the present tense.	*You were* correct in assuming the door was locked.
238	**Basic subject-verb number agreement rules**	**Singular and plural subjects:** Singular subjects need singular verbs; plural subjects need plural verbs. **Compound subjects:** Compound subjects connected with *and* usually require a plural verb. When two singular subjects are joined by *or, nor, either . . . or,* or *neither . . . nor,* use a singular verb; when one subject is singular and one is plural, place the plural close to the verb and use a plural verb.	The *president is* in New York. All *stock clerks are* in the lunch room. The office *desks* and *chairs are* new. Either *John* or *Beth is* available.
239	**Special subject-verb number agreement rules**	**Noun subjects either singular or plural:** 1. Nouns always singular in meaning: *billiards, civics, mathematics, news,* etc. 2. Nouns singular when fields of knowledge or activity, otherwise plural: *politics, statistics, athletics, aerobics.* 3. Nouns always plural in meaning: *wages, tongs, pliers, trousers,* etc. **Collective nouns:** They may be singular or plural. If they refer to a group as a single unit, they are singular; if they refer to members acting independently, they are plural. **Indefinite pronouns:** Use Table 12-1. When *each, every,* or *many a* are used as adjectives, the subject is singular. **Titles and company name:** Titles of one book, song, magazine, article, and company name are singular.	The game of *billiards is* fascinating. *Statistics is* a difficult course. *Tongs are* frequently used for this job. The *committee has* not *made* a decision. *Each* worker *is* receiving a pay increase. *McCall's* is a magazine many women enjoy.

PAGE	TOPIC	KEY POINTS	EXAMPLES
		Nouns expressing amounts. Distance, fractions, money, etc., are singular when they refer to people or things as a unit; plural when they are thought of as individuals or individual items.	Four *houses* were visited.
		Relative pronoun clauses: When antecedents are singular, they are followed by a singular verb; when plural, they are followed by a plural verb.	This is the *salesperson* who *is* famous.
244	**Infinitives**	**Verbals** are special verb forms that can function as nouns (subjects, direct objects, predicate nouns, appositives, objects of prepositions), adjectives, and adverbs but not as predicates by themselves.	Lynn likes *to shop.*
		Infinitives are the present form of the verb preceded by *to*; they function as nouns, adjectives, and adverbs.	*To find* a new location is difficult.
		Infinitive phrases are infinitives combined with an object and/or modifiers; they function as nouns, adjectives, and adverbs.	
		Avoid split infinitives. Use the present infinitive except when action occurs before verb, then use the perfect infinitive.	
245	**Participles**	Participles are used as adjectives. 1. Use the *present participle* if the action is the same time as the main verb. Use the *past participle* or *perfect participle* if action is expressed before the main verb action. 2. Avoid awkward use of absolute phrases. **Participial phrases** are participles, objects, complements, and/or modifiers; they function as adjectives.	*Laughing* loudly, the manager left the room.
246	**Gerunds**	Gerunds end in *ing* and function as nouns. As a verbal noun, they are used as subjects, objects, or complements. They may take an object. Make sure that the noun or pronoun used to modify a gerund is in the possessive case. **Gerund phrases** are gerunds with modifiers or a complement functioning as a noun.	*Walking* is an exercise enjoyed by many.

PAGE	TOPIC	KEY POINTS	EXAMPLES
247	**Dangling verbal phrases**	A verbal phrase must be placed so the word it relates to is clear; otherwise it is called dangling. Verbal phrases can be misplaced both at the beginning and the end of a sentence. Introductory verbal phrases are followed by a comma.	*Confusing:* By using this new technique, ten minutes will be saved. *Logical:* By using this new technique, *you* will save ten minutes.

QUALITY EDITING

Your colleague in the marketing department has asked you to check the draft of a memo. Edit the memo. Watch carefully for agreement between subjects and verbs and for problems with verbals. Also, watch for other errors including spelling and end-of-sentence punctuation.

MEMO

DATE: April 2, 19XX

TO: The Marketing Staff

FROM: Timothy Lyons

SUBJECT: New Orders

Company statistics indicates that 80 percent of our product sales comes from 20 percent of our cutsomers. This situation has dire consequenses for the company future. If that small number of customers was to switch to another companies product, we would have difficulty selling enough to, for very long, stay in business.

We need to quickly identify the customers who is most likely to change products and contacting them. Our personel communication with byers will halp to strongly motivate him to only buy from us.

APPLIED LANGUAGE

Below are four sets of phrases. Write three sentences for each set. **Note**: Do not change the wording of the phrase.

EXAMPLE:

Participle: *Using the new computer program*, the team finished the report three days early. (*Using the new computer program* is used as a modifier.)

Infinitive: The team decided *to use* the new computer program. (The infinitive phrase *to use* is being used as a noun, the object of the verb, decided.)

Gerund: *Using the new computer program* allowed the team to finish early. (The phrase *using the new computer program* is used as a noun, the subject of the sentence.)

Participle:	Infinitive:	Gerund:
1. entering the data carefully	to enter the data carefully	entering the data carefully
2. reporting on the sales of the product	to report on the sales of the product	reporting on the sales of the product
3. writing effectively	to write effectively	writing effectively
4. leaving the meeting early	to leave the meeting early	leaving the meeting early

1. a. Participle _____

_____

b. Infinitive _____

_____

c. Gerund _____

_____

2. a. Participle _____

_____

b. Infinitive _____

_____

c. Gerund _____

_____

3. a. Participle _____

_____

b. Infinitive _____

_____

c. Gerund _____

_____

4. a. Participle _____

_____

b. Infinitive _____

_____

c. Gerund _____

_____

WORKSHEET 12

Name _____

Date _____

VERBS: SUBJECT-VERB AGREEMENT AND VERBALS

Part A. Basic Subject-Verb Agreement Rules

In the space provided on the left, write an S if the subject is singular or a P if it is plural. Then, complete each sentence. Be sure that your verb agrees in person and number with the subject.

_____ **1.** The committee _____.

_____ **2.** The first and last items _____.

_____ **3.** Joanne Shepherd and her staff _____.

_____ **4.** Neither the senior accountant nor the vice president _____.

_____ **5.** The chief executive officer and president _____.

_____ **6.** He and his assistant _____.

_____ **7.** Prestige Services, Inc., _____.

_____ **8.** The four committees _____.

_____ **9.** Harriet Jones _____.

_____ **10.** Everyone in accounting _____.

Part B. Special Subject-Verb Number Agreement Rules

Circle the appropriate verb of the two in parentheses.

1. The series of mistakes (were, was) detrimental to the financial health of the company.

2. Statistics (indicates, indicate) that 80 percent of our sales come from 20 percent of our customers.

3. Rosario's wages (is, are) being increased.

4. Each of the committee members (has, have) deliberated on the problem.

5. Both of the new employees (was, were) assigned to the marketing department.

6. Some of the computers (has, have) been moved.

7. Some of the space on that computer disk (is, are) filled with the old database.

8. None of the employees (receive, receives) bonus pay.

9. One of the warehouse doors (is, are) damaged.

10. *Swimming with the Sharks* (is, are) highly recommended reading.

Part C. Verbals

Circle the preferred verbal of the two in parentheses.

1. (Being, Having been) judged guilty, the defendant was sent to jail.

2. Even after 200 years, she is known to (be, have been) a great leader.

3. Mr. Flores was pleased with (Lou, Lou's) handling of the problem.

4. (Speaking, Having spoken) much too rapidly, Hans was not understood by his audience.

5. We cannot understand (him, his) buying the older model.

6. (Standing, Having stood) at the front of the room, she made her presentation brilliantly.

7. Are you aware of (Irving, Irving's) big sale?

8. Preparing the report in haste, (George had made errors, it contained errors).

9. To do the job properly, (use a computer, a computer should be used).

10. While giving his demonstration, (Mr. Ziegler used visual aids, I am glad Mr. Ziegler used visual aids).

Part D. Verbals

Indicate whether each of the following sentences is correct or incorrect by placing a C before the correct sentences and an I before those that contain errors. Then, rewrite the incorrect sentences so that they are correct.

_____   **1.** In selecting a plan to fill her needs, both cost and benefits should be seriously considered by Olivia.

_____

_____

_____   **2.** Pauline decided to review the contracts carefully.

_____

_____

_____   **3.** To assiduously avoid any appearance of wrongdoing was Robert's first concern.

_____

_____

_____   **4.** Having two warehouses and four trucks, the firm does not need to deliver goods by a common carrier.

_____

_____

_____   **5.** While talking with me last week, Lynn mentioned the need for additional support staff.

_____

_____

_____   **6.** Please be sure to thoroughly read the directions before you install the program.

_____

_____

_____   **7.** Realizing that I was not ready to make a decision, the agent reluctantly left my office.

_____

_____

_____   **8.** The executives asked Quentin to report on the financial status of the firm.

_____

_____

_____   **9.** The benefits manager tried to carefully explain the health insurance program to each new employee.

_____

_____

_____   **10.** Working late all week, the project should be finished by Friday.

_____

_____

SECTION 4

WORDS THAT MODIFY AND CONNECT

CHAPTER 13

ADJECTIVES

Used by a fine writer, the adjective is a word of unrivaled delight. It is adornment, the jewelry, the leaves on the tree. It brings the picture into exact focus; or in other words it focuses the reader's eye exactly as the writer wishes.

—John Fairfax and John Moat, *The Way to Write*

LEARNING UNIT GOALS

LEARNING UNIT 13-1: RECOGNIZING ADJECTIVES; COMPARISON OF ADJECTIVES

- Explain the two main types of adjectives. (pp. 261–62)
- Use the adjective types correctly in your writing. (pp. 261–64)

LEARNING UNIT 13-2: COMPOUND ADJECTIVES; INDEPENDENT ADJECTIVES

- Know the correct use of compound adjectives. (pp. 265–67)
- Recognize independent adjectives that must be separated by commas. (pp. 267–68)

LEARNING UNIT 13-3: ADJECTIVE USAGE

- Understand the placement of adjectives and the choice of strong adjectives. (pp. 268–69)
- Recognize how phrases and clauses function as adjectives. (pp. 269–71)

Have you noticed that when you look at a painting, you often see small patches of bright colors? For example, when Norman Rockwell painted the *Doctor and Doll*, he painted a book, the stethoscope, and the little girl's hat and mittens bright red. Strong colors give artistic emphasis to a painting. Strong descriptive adjectives give artistic emphasis to the writer's word pictures.

Now that you can recognize the main parts, or basic elements, of a sentence, you are ready to learn more about the modifiers, or secondary elements, of a sentence. The modifiers—adjectives and adverbs—describe or limit the basic sentence elements.

This chapter discusses the adjective—the modifier of nouns and pronouns. Learning Unit 13-1 will help you recognize the different types of adjectives, and it explains the comparison of adjectives. In Learning Unit 13-2, you learn about compound adjectives and independent adjectives. Learning Unit 13-3 discusses adjective usage.

LEARNING UNIT 13-1
RECOGNIZING ADJECTIVES; COMPARISON OF ADJECTIVES

The English language has many types of adjectives. This learning unit organizes these adjectives into groups, some of which you may already know. Then, you will learn the forms that adjectives assume when they make comparisons.

RECOGNIZING ADJECTIVES

In Chapter 3, adjectives were defined as *descriptive* or *limiting* words that modify a noun or a pronoun. They can also be word groups (phrases and clauses) that function as single adjectives.

Adjectives answer the questions *how many?*, *what kind?*, and *which one?* Sometimes you will find that you can use some additional questions to identify adjectives. To these three questions, you can add the related questions *how much?* and *whose?* Note that adjectives do not answer the question *what?* Subjects and direct objects answer this question.

Since adjectives usually precede the nouns and pronouns they modify, look for the nouns and pronouns in a sentence. Adjectives may, of course, also appear after the nouns and pronouns they modify. The adjectives in the following sentences are in italics.

The *financial* report shows the *close* relationship between the *small* companies. (These adjectives *describe* the nouns that follow, answering the question *what kind?* As **descriptive adjectives**, they name a quality or condition of the noun [or pronoun] they modify).

This report shows the relationship between the *two* companies. (These adjectives *limit* the nouns that follow, answering the questions *which one?* and *how many?* As **limiting adjectives**, they point out or indicate a number or quantity of the noun [or pronoun] they modify).

Classifying adjectives makes them easier to recognize. Table 13-1 gives an overview of the various types of adjectives. Many of these adjectives have been mentioned in previous chapters. These chapters and their learning units (LU) are given in parentheses.

Table 13-1 Overview of Types of Adjectives

Type of Adjective	Adjectives Belonging to Group	Examples
DESCRIPTIVE ADJECTIVES:	Adjectives that name a quality or condition of the noun or pronoun they modify; answer *what kind?*	*old* house, *torn* coat, *heavy* box
Proper Adjectives	(1) Proper nouns used as adjectives and (2) adjectives formed from proper nouns.	*Florida* oranges; *Texas* grapefruit; *Victorian* art; *Hawaiian* culture
Predicate Adjectives	Adjectives describing subjects (subject complements) and direct objects (object complements). Subject complements follow linking verbs; object complements follow direct objects. (Chapter 4, LU 4-1; Chapter 5, LU 5-3; Chapter 10, LU 10-1)	The shipping clerk was *absent.* The president found the report *inaccurate.*
Nouns Used as Adjectives	Nouns are frequently used as adjectives modifying other nouns. (Chapter 6, LU 6-2)	*table* lamp, *credit* balance, *college* credit
Participle Adjectives	Present, past, and perfect participles used alone function as adjectives. (Chapter 12, LU 12-3)	*growing* company, *singing* waiter, *used* table
LIMITING ADJECTIVES:	Adjectives that point out or indicate a number or quantity; answer *how many?, how much?, which one?,* and *whose?*	*four* chairs, *no* time, *each* employee
Articles	*Definite: the*–refers to a specific person, place, or thing. *Indefinite: a* and *an*–refer to any one of a class of people, places, or things. Use *a* before words beginning with a consonant sound. Use *an* before words beginning with a vowel sound. Exceptions: Before a word beginning with (1) a silent *h*, use *an;* (2) a *o* sounding like a vowel, use *an;* (3) long *u* (or *eu*), use *a;* (4) and a word such as *one,* use *a.* 1. When two adjectives modify the same noun, use the article only once. 2. When the same article identifies two different nouns, repeat the article. 3. When the same article modifies two nouns naming one person, do not repeat the article. 4. If one noun is modified by *a* and another by *an,* repeat each article. (Chapter 3, LU 3-2)	*a* factory, *a* manager, *a* storeroom *a* hotel, *a* historical society, *a* union, *a* university; but *an* heir, *an* honor, *an* umbrella *an* egg, *an* upper berth, *an* envelope *The* old and worn gloves were lost. *The* vice president and *the* treasurer were absent. (Two people.) *The* secretary and treasurer is an appointed office. (One person.) We hired *a* painter and *an* electrician. (Two people.)
Demonstrative Adjectives	Pronouns that point out and function as adjectives: *this* and *that* (singular) and *these* and *those* (plural). Never use *them* for *those.* (Chapter 8, LU 8-3)	*this* kind, *these* kinds *this* company, *these* companies *that* type, *those* types *that* office, *those* offices
Indefinite Adjectives	Pronouns functioning as adjectives: *any, each, every, few, many, most, several, some,* and *others.* (Chapter 8, LU 8-3; Chapter 9, LU 9-2)	*each* suggestion, *few* employees, *many* letters, *most* stockholders
Interrogative Adjectives	Pronouns functioning as adjectives: *what, which,* and *whose.* (Chapter 8, LU 8-3)	*Which* report did you write? (*Which* modifies report. Without the noun *report, which* would be an interrogative pronoun.)
Numerical Adjectives	*one, first, two, second,* and *others*	*three* days, *fourth* class
Possessive Adjectives	Pronouns functioning as adjectives: *my, your, his, her, our,* and *their.* Possessive nouns can also be used as adjectives. (Chapter 8, LU 8-1)	*my* book, *your* car, *our* school *supervisor's* desk, *Jack's* house
Relative Adjectives	Pronouns functioning as adjectives: *what, which, whose, whatever, whichever,* and *whoever.*	This is the supervisor *whose* brother worked in the warehouse. (*Whose* modifies *brother* and connects clause that follows to *supervisor.*)

WRITING TIP

Since the article *a* serves no purpose in expressions such as *sort of (a) person* or *kind of (a) product,* do not use it after *sort of* and *kind of.* For example,

> **AVOID: What kind of a position do you want?**
>
> **USE: What kind of position do you want?**
>
> **AVOID: He is the type of a leader that this organization needs.**
>
> **USE: He is the type of leader that this organization needs.**

COMPARISON OF ADJECTIVES

An abridged dictionary, you will recall, does not contain every word in the English language. The dictionary you use may not list words like *fewer, fewest, taller,* or *tallest,* although these are common words. These words are not included because they are *regular* variations of the comparison forms of descriptive adjectives.

The average person makes frequent use of descriptive adjectives in three different forms, or *degrees.* Most of us naturally add *er* or *est* to one-syllable adjectives or to two-syllable adjectives that end in *y.* In expressing variations of other adjectives, we also automatically use *more, most, less,* or *least* appropriately.

Most adjectives (and adverbs) have three degrees of comparison: *positive degree, comparative degree,* and *superlative degree.*

The **positive degree** is the simple form of the adjective that names a quality and is listed in the dictionary; it makes no comparison.

The **comparative degree** compares two people, things, or ideas by adding *er* or prefixing *more* (or *less*) to the positive degree.

> Interest rates on personal loans are *higher* than those on business loans.

> The company is interested in the *less expensive* of the two properties.

> Tomorrow may be even *colder* than it is today.

The comparative degree is also used when comparing a person or an object with others in the same class. The words *other* or *else* are often used when comparing persons or objects in the same class.

> The Bank of Malibu has *greater* assets than any *other* California bank.

> Ms. Thurston is *more* conservative than anyone *else* on the management team.

> This new photocopier is *faster* than any *other* photocopier in the office.

The **superlative degree** indicates an extreme when referring to more than two (people, objects, etc.) by adding *est* or *most* (or *least*) to the positive degree.

Ms. Jackson is the *wealthiest* small store owner in the city.

Of all the novelty dealers, John Kindel had the *highest* sales.

The Acme representative was the *most* persuasive of the four.

Note that in the comparative and superlative degrees, most adjectives (and few adverbs) of one syllable usually add *er* (comparative) and *est* (superlative). Adjectives of two syllables can add either *er* or *more* to form the comparative degree, or they can add *est* or *most* to form the superlative degree. Adjectives (and adverbs) of three or more syllables always add *more* for the comparative degree and *most* for the superlative degree. When you have a choice to make, use what sounds the best.

The following list includes the three comparison degrees of several adjectives.

Positive	**Comparative**	**Superlative**
fine	finer	finest
cold	colder	coldest
funny	funnier	funniest
careful	more (or *less*) careful	most (or *least*) careful
difficult	more (or *less*) difficult	most (or *least*) difficult

Some adjectives, such as *circular, empty, perfect, unique,* and *universal,* do not lend themselves to comparison. Since words of this type suggest extremes, you would not say *more perfect* or *more unique.* These adjectives are called **absolute adjectives**.

YOU TRY IT

A. Write the comparative and superlative forms of each adjective given below.

Positive: Comparative: Superlative:

1. strong   _____   _____

2. informative   _____   _____

3. clear   _____   _____

4. talented   _____   _____

5. easy   _____   _____

B. Review Table 13-1. For each italicized adjective in Column A, write the letter of the type of adjective given in Column B. The first item is completed as an example. **Note:** Use each letter only once.

COLUMN A   COLUMN B

Adjectives in Context: Type of Adjective:

_____ **1.** The *alert* clerk found the error.   **a.** participle adjective

_____ **2.** The timecards are ready for *data* entry.   **b.** predicate adjective

_____ **3.** The *printing* mechanism malfunctioned.   **c.** noun used as adjective

_____ **4.** *Her* work is exceptional.   **d.** descriptive adjective

_____ **5.** Her work is *exceptional.*   **e.** numerical adjective

_____ **6.** The *fourth* file contains the contracts.   **f.** possessive adjective

---

LEARNING UNIT 13-2
COMPOUND ADJECTIVES; INDEPENDENT
ADJECTIVES

The two topics in this learning unit, compound adjectives and independent adjectives, cause problems for many writers. The guidelines and rules that follow will help you avoid these problems.

COMPOUND ADJECTIVES

Compound adjectives are two or more words used as a single adjective *before the noun it modifies.* The general rule is to use a hyphen between the compound adjective when it is necessary to show a close association and avoid misreading. By combining words in this way, writers can coin their own modifiers and, even more important, clarify their intended meaning. Hyphens are usually not needed when the compound adjective appears after the noun.

Use Hyphens	**Hyphens Not Needed**
work-connected injury	injury is work connected
much-discussed theories	theories are much discussed
often-quoted book	book is often quoted
coin-operated machine	machine is coin operated
decision-making conference	conference is on decision making

An important guideline for hyphenating compound adjectives is to first consult the dictionary when you are not sure about hyphenating a compound adjective. You can often save time if you immediately look up the word in the dictionary (be sure it is listed as an adjective). You will be surprised how many times the word you question is in the dictionary.

Many guidelines and rules exist for the hyphenating of compound adjectives. Handbooks such as *The Chicago Manual of Style* often give complete guidelines for using compound words and words with prefixes and suffixes.[1] Some of the more common guidelines follow.

---

[1] *The Chicago Manual of Style*, 14 ed. (Chicago: The University of Chicago Press, 1993).

Compound Adjectives of Quantity or Degree

Use a hyphen with a compound adjective when one of the words expresses quantity or degree such as *high-*, *low-*, *large-*, *small-*, *part-*, *full-*, *single-*, *multiple-*, *upper-*, *lower-*, *middle-*, and so on. (Hyphens are usually not used when a compound adjective follows a noun.)

high-income rate	small-time thief	single-premium policy
low-income housing	part-time worker	multiple-unit housing
large-scale model	full-time employee	upper-class group

You will find some of these hyphenated compounds listed in the dictionary.

Number Combinations of Compound Adjectives

Use a hyphen for number combinations of compound adjectives that precede (*not* follow) the word modified.

With Hyphen	**Without Hyphen**
ten-year lease (not ten-years lease)	lease for ten years
20-year contract (not 20-years contract)	contract for 20 years

Some number combinations are written with hyphens regardless of where they are placed in a sentence. Compound numbers from twenty-one to ninety-nine are always written with hyphens.

forty-five   sixty-one   eighty-two

Most fractions that must be spelled out are written with hyphens. (If either the numerator or the denominator is hyphenated, do not place a hyphen between them.)

one-third	eleven forty-fifths	BUT: a half
three-fourths	thirty-one thirty-seconds	a third

Note: Some writers do not use a hyphen in a fraction that functions as a noun. (*Two thirds* of our members were present. BUT: A *two-thirds* vote was required.)

When you use a series of partially expressed compound number adjectives that require hyphens, insert the hyphens as in the following example:

All four-, five-, and six-story buildings will be inspected.

This rule also pertains to a series of compound adjectives that are not number adjectives.

Compound Adjectives with Prefixes and Suffixes

Many compound adjectives that include the prefixes *self, all, ex,* or *great* and the suffix *elect* use a hyphen.

governor-elect	ex-treasurer
self-conscious	all-powerful
self-respect	great-uncle

Exceptions to this rule include *selfish, selfsame, selfhood,* and *selfless.*

Use a hyphen with words that would be difficult to understand if a hyphen did not separate a prefix or suffix from a root word.

> He *re-signed* the contract. (Hyphen avoids confusion with *resigned* as in the sentence *He resigned from his job.*)

> The figure in the center of the drawing is *bell-like.*

> Our supervisor is interested in *pre-Columbian* artifacts. (Use a hyphen between a prefix and a proper name.)

When Not to Use a Hyphen

Do not use a hyphen to connect an adverb ending with *ly* (*carefully, nearly,* etc.) with a following adjective.

> a carefully prepared statement   a fairly reliable estimate

Do not use a hyphen if the compound adjective consists of words that are easily recognized as compound nouns requiring no hyphen.

> New England states   post office box
> Rocky Mountain scenery   real estate broker
> high school administrator   ice cream cone

INDEPENDENT ADJECTIVES

Independent adjectives are two or more consecutive adjectives that appear before a noun and modify the noun equally. These adjectives must be separated with a comma. The expression *long, dull report* refers to a report that is long and dull. However, the expression *long typewriter carriage* obviously refers to a typewriter carriage that is long. The two adjectives, *long* and *typewriter*, do not function as a unit and do not require a hyphen. Also, these two adjectives do not modify *carriage* independently and do not require commas.

Adjectives are independent when you can (1) add the word *and* between the adjectives and (2) reverse the order of the adjectives without changing the meaning of the sentence. Note the following sentence:

> He is a *capable, efficient, hard-working* bookkeeper.

In this sentence, you could say that the bookkeeper is a *capable and efficient and hard-working bookkeeper.* You could also say that the bookkeeper is a *hard-working, efficient, capable bookkeeper.* The adjectives in this sentence are independent adjectives and must be separated by commas.

The last adjective in a series of independent adjectives is never separated by a comma from the noun it modifies. Often a series of such adjectives will be arranged with the shorter words first.

When consecutive adjectives are *cumulative,* that is, any adjective in the series modifies the total idea that follows, commas are not used between the adjectives. You cannot add *and* between cumulative adjectives and make sense. Also, you cannot reverse cumulative adjectives and make sense.

Several small walnut end tables are still in stock. (You could not add *and* between the adjectives [*small walnut end*] and you could not reverse them and make sense. Chapter 17 also discusses adjectives in a series.)

YOU TRY IT

A. Fill in the two missing words in the statement below.

The basic rule for hyphenating compound adjectives is to use a hyphen when the compound

adjective comes (1) _____ the noun it modifies but not when it comes

(2) _____ the noun.

B. Underline the compound adjectives in the following sentences. Then, add any necessary hyphens.

1. The work schedule was a middle management responsibility.

2. The five hour meeting exhausted the participants.

3. The report was almost complete.

4. His North Dakota business was prospering.

5. The three contracts must be signed within forty eight hours.

Answers: **A.** 1. before 2. after **B.** 1. middle-management 2. five-hour 3. almost complete 4. North Dakota 5. forty-eight.

LEARNING UNIT 13-3
ADJECTIVE USAGE

This learning unit discusses the placement of adjectives, choosing strong adjectives, adjective phrases and clauses, and frequently misused adjectives. Understanding each of these topics will help you use adjectives effectively.

ADJECTIVE PLACEMENT

Adjectives usually precede the nouns they modify. To achieve greater emphasis or variation, however, a writer may occasionally place adjectives after the noun. The adjectives in the following sentences illustrate this:

Current records, *routine* or *special*, are easily located in these files.

The salesperson, *young* and *inexperienced*, sold little merchandise.

Adjectives can also appear in the predicate. When you analyze intransitive linking verb sentences, the subject complement is frequently a descriptive *predicate adjective*. Note the following examples:

The new format will be *satisfactory.* (It is a *satisfactory format.*)

The information is *confidential.* (It is *confidential information.*)

CHOOSE STRONG ADJECTIVES

Choose adjectives that add exactness or color to a description or an explanation. The expression *salty ocean* tells a reader little, since all oceans are salty. General adjectives such as *good, beautiful, nice,* and *fine* usually add little to an idea and may even detract from the strength of a statement. *I recently read a good book* is a comment that says little about the book. A book may be *inspiring* or *thought provoking* or *suspenseful* or *exciting* or *humorous* or *educational* or *revealing* or *refreshing* or *wholesome* or *unique*; the possibilities are almost limitless.

Adjectives that have been used too often with certain nouns become trite. Avoid these overworked adjectives. Remember the words of Mark Twain: "As to the adjective: When in doubt, strike it out." Skillful writers choose varied adjectives that give exact pictures, contribute an accent, or make appeal to the imagination or emotions of the reader.

WRITING TIP

When you find yourself using a common adjective, check your thesaurus or synonym book to see if you can find a more exacting adjective.

ADJECTIVE PHRASES AND CLAUSES

You will recall from your study of sentence analysis that phrases can function as adjectives. Prepositional phrases are often used as adjectives. From Chapter 12, you know that infinitive phrases and participial phrases can also function as adjectives. When used as adjectives, these phrases can be called **adjective** (or **adjectival**) **phrases**. Note the following sentence:

The person *in the black suit* did not attend the meeting. (The prepositional phrase functions as an adjective modifying *person* by answering the question *which one?*)

Dependent clauses also can function as adjectives; they are called **adjective clauses**. Relative pronouns (*who, whom, whose, which,* and *that*) or relative adverbs (*where, when,* and *why*) frequently introduce adjective clauses. (Relative adverbs are discussed in Chapter 14.)

The manager *who answered your inquiry* is also a vice president. (The adjective clause *who answered your inquiry* modifies *manager* by answering the question *which one?*)

You should be aware that both noun clauses and adjective clauses can begin with the words *that, which, who, whom,* and *whose.* (Note that noun clauses can begin with *what;* adjective clauses do not.) To determine if you have a noun clause or an adjective clause, you must consider how the clause functions in the sentence. Remember that in the sentence analysis flowchart, noun clauses are important because they can function as major sentence parts just like nouns. Adjective clauses function as modifiers.

If you are not sure if you have an adjective or noun clause, begin by locating the clause as explained in Chapter 6. Think of the clause as one word. Then, follow the steps of the sentence analysis flowchart. The noun clause will fit one of the steps (unless it is an appositive). If you have finished the flowchart steps and find the clause you are concerned about does not fit anywhere, you have a modifier, or adjective clause.

Let's first analyze a sentence with a noun clause. This will provide a review of the noun clause sentence analysis illustrated in Chapter 6.

The new manager forgot that he was a clerk.

Step 1. Find the clause. Look for the clue word, a subject, and a verb. In this sentence, the clause is *that he was a clerk.*

Step 2. Think of the clause as a one-word noun: The new manager forgot thathewasaclerk.

Step 3. Use the sentence analysis flowchart steps:
   a. Find the verb or verb phrase. The word that changes with time when you add the word *tomorrow* before this sentence is *forgot,* the verb.
   b. Find the subject. *Who* forgot? Manager. *Manager* is the subject.
   c. Does the sentence have a direct object? Repeat the subject and the verb and ask *what?* or *whom?* Manager forgot *what?* Manager forgot thathewasaclerk. This sentence has a direct object. In this sentence, *forgot* is transitive, meaning to lose the remembrance of. (*Forgot* can also be an intransitive verb meaning to cease remembering at a given time, such as forgot about paying a bill.)
   d. What is the direct object? The noun clause *that he was a clerk* functions as a direct object.

The new <u>manager</u> <u>forgot</u> that he was a clerk.

STUDY TIP

Be aware that sometimes a dependent adjective clause does not have a word that joins it to the rest of the sentence. For example, the sentence *Jim knows that he will be early* could be written *Jim knows he will be early.*

The following sentence has an adjective clause (adjective clauses often begin with the clue word *that*):

The office furniture that the president ordered is late.

Step 1. Find the clause. Look for the clue word, a subject, and a verb. In this sentence, the dependent clause is *that the president ordered.*

Step 2. Think of the clause as a one-word noun: The office furniture thatthepresidentordered is late.

Step 3. Use the sentence analysis flowchart steps:

    a. Find the verb or verb phrase. When you add *yesterday* before this sentence, *is* changed to *was*, a past form of the verb *to be.*

    b. Find the subject. *What* is? Furniture. *Furniture* is the subject.

    c. Does the sentence have a direct object? No. We know that when the verb *to be* (*is, are, were*, etc.) stands alone, it never has a direct object. So the answer here is no. *The verb is intransitive.*

    d. Does the sentence have a complete or linking verb? The sentence does not have a complete verb. You cannot stop the sentence after furniture is. You know that *is* used alone is always a linking verb, so this sentence has a linking verb.

    e. Give the subject complement. *Late* is a predicate adjective that describes the subject *furniture.*

                      —Dependent adjective clause —⌐vi.l. SC PA

The office <u>furniture</u> that the president ordered <u>is</u> late.

In this sentence, *that* in the dependent adjective clause refers to *furniture* in the independent clause. The clause functions as an adjective because it answers *which one?* about the noun *furniture.* Since adjective clauses are modifiers, they are not included in the sentence analysis flowchart.

STUDY TIP

The important distinction between an adjective clause and a noun clause is that the adjective clause *modifies* a subject, object, or complement. A noun clause *is* a subject, object, or complement. Also, the first word in an adjective clause has an antecedent elsewhere in the sentence; the first word in a noun clause does not.

Words, phrases, and clauses that function as adjectives can add interest, precision, and color to the nouns and pronouns they modify. Business writers can often avoid misunderstandings by using adjectives skillfully to achieve maximum clarity.

FREQUENTLY MISUSED ADJECTIVES

1. **Later, latter.** The adjective *later*, which is the comparative form of *late*, generally refers to time. *Latter*, in contrast to *former*, refers to the second of two mentioned. It may also be used to mean near to the end.

> The debate will be continued *later*.

> He praised the *latter* of the two speakers mentioned.

> Most of these changes took place during the *latter* part of the century.

2. **First, last.** When using either of these words with a number (*first four* or *last five*), write the number last.

> Read the *first ten* pages of the manual.

> His accounts covered the *last 45* days of the operation.

3. **Farther, further.** The first dictionary definition of the adjective *farther* is "the more distant or remote." The first definition of the adjective *further* is "additional; more." Then, the dictionary indicates that *farther* and *further* are interchangeable. Careful writers, however, make a distinction between the two words.

> How much *farther* is it to the museum? (The adjective *farther* refers to the distance to the museum.)

> The president wants *further* information. (The adjective *further* refers to additional information that the president desires.)

4. **Less, fewer.** These are the comparative forms of *little* and of *few*. *Less* usually is used to modify singular nouns; *fewer* is used to modify plural nouns. *Less*, however, may be used with plural nouns that name single units of time, distance, or money.

> The new photocopying machine uses *less* paper than the old machine.

> The strike lasted for *less* than three weeks. (*Weeks* are single units of time.)

> You will find *fewer* employees on the sixth floor.

> There have been *fewer* complaints since the pay raise was granted.

YOU TRY IT

A. Replace the italicized adjectives in the following sentences with stronger adjectives. Write the new adjective in the space provided. Be creative!

EXAMPLE: Leo is a *nice* person. _____**thoughtful**_____

1. The report was *good*. _____

2. The president's message was *fine*. _____

3. The *pretty* picture was hung in the lobby. _____

4. No one responded to the *bad* advertising. _____

5. The *bad* weather affected our attitudes. _____

B. Complete each of the following sentences by adding an adjective phrase or clause in the blank space.

1. The company that _____ went bankrupt.

2. The woman _____ ran down the hall.

3. The clerk who _____ was given a reward.

4. The sales of the product _____ increased dramatically.

5. Everyone who _____ was delighted with the architect's design.

C. From the two choices given in parentheses, circle the one that best completes the sentence.

1. Jennifer writes very (good/well).

2. Those data appear in the (later/latter) half of the report.

3. The (first ten pages/ten first pages) summarize the document.

4. The project was completed in (fewer/less) days than expected.

Possible Answers: **A.** 1. decisive 2. inspirational 3. intriguing 4. misleading 5. stormy. **B.** 1. gave too much credit to customers 2. with the telegram in her hand 3. found the accounting error 4. designed to find the bugs in the software program, 5. saw the new building. **C.** 1. well 2. latter 3. first ten pages 4. fewer

CHAPTER SUMMARY: A QUICK REFERENCE GUIDE

PAGE	TOPIC	KEY POINTS	EXAMPLES
261	**Recognizing adjectives**	**Descriptive adjectives** name a quality or condition of the noun (or pronoun) they modify. **Limiting adjectives** point out or indicate a number or quantity of the noun (or pronoun) they modify. Table 13-1 gives an overview of the types of pronouns.	The company bought everyone *blue* desk chairs. Look for *two* workers wearing *their* jackets.
263	**Comparison of adjectives**	The *positive degree* is the basic adjective found in the dictionary. The *comparative degree* compares two people, things, or ideas by adding *er* or prefixing *more* (or *less*) to the positive degree. It is also used for comparing a person or object with others in the same class (also used with either *other* or *else*).	The shipping manager was carrying a *heavy* workload. The parts manager carried a *heavier* workload than the shipping manager.

PAGE	TOPIC	KEY POINTS	EXAMPLES
		The *superlative degree* indicates an extreme when referring to more than two (people, objects, etc.) by adding *est* or *most* (or *least*) to the positive degree.	The accounting manager carried the *heaviest* workload.
265	**Compound adjectives**	**1.** Use a hyphen when one word expresses quantity or degree.	*high-, low-, large-, small-, part-, full-, single-, multiple-, upper-, lower-, middle-*
		2. Use a hyphen for number combinations of compound adjectives preceding (not following) the word modified.	*ten-year* lease, *20-day* trip
		3. Some words always use hyphens:	
		a. Compound numbers from twenty-one to ninety-nine.	after *thirty-one* years
		b. Most fractions are spelled out and written with hyphens.	*one-half, one-fourth*
		c. Use hyphens for series of partially expressed compound adjectives.	*12-, 14-, and 16-year* contracts
		4. Many compound adjectives that include the prefixes *self, all, ex,* or *great* and the suffix *elect* use a hyphen. Use a hyphen for words that would be difficult to understand if the hyphen did not separate a prefix or suffix from the root word.	*president-elect* left the office *re-create* (to create again)
		5. Do not use a hyphen to connect adverbs ending in *ly* with a following adjective.	*fairly long* contract
		6. Do not use a hyphen if the words in the compound are easily recognized.	high school reunion
267	**Independent adjectives**	Separate by commas two or more consecutive adjectives that appear before a noun and modify the noun equally. Adjectives are independent when you can (1) add the word *and* between the adjectives and (2) reverse the order of the adjectives. The last adjective in a series is never separated from the noun by a comma. If consecutive adjectives are cumulative, commas are not used between the adjectives.	*long, cold* winter *new oak* table
268	**Adjective placement**	Usually precede nouns they modify; for variation, two adjectives may follow a noun; predicate adjectives follow linking verbs.	The security guard, *cold* and *wet,* left the building. The employee became *ill.*

PAGE	TOPIC	KEY POINTS	EXAMPLES
269	**Choose strong adjectives**	Avoid overworked adjectives; "when in doubt, take it out."	*Avoid*: The supervisor is a *fine* person.
269	**Adjective phrases and clauses**	**Adjective phrases**: prepositional, infinitive, and participial phrases. **Adjective clauses**: Often introduced by relative pronouns (*who, whom, whose, which, that*) or relative adverbs (*where, when, why*).	Lynn is the lady *with the red glasses*. James liked the story *that the supervisor told*.
272	**Frequently misused adjectives**	1. **Later**, **latter**. *Later* generally refers to time. *Latter* refers to the second of the two mentioned. 2. **First**, **last**. When used with a number, write the number last. 3. **Farther**, **further**. *Farther* means "more distant or remote." *Further* means "additional, more." These can be used interchangeably; writers often make the distinction. 4. **Less**, **fewer**. *Less* usually modifies singular nouns; it may be used with plural nouns naming single units. *Fewer* modifies plural nouns.	The meeting is *later* than you think. Use the *latter* of your two ideas. Buy the *first* two complete tool sets you can find. How much *farther* can you go? Corbin said she wanted no *further* difficulties. This car uses *less* gas. *Fewer* employees went to the convention this year than last year.

QUALITY EDITING

You work for a software publisher that has recently produced a new program. Your boss wants you to correct the following product description before it is printed in the next catalog. Check for errors in grammar, spelling, end-of-sentence punctuation, redundancies, or any other problems. Replace weak adjectives.

```
Write Your Business Plan in Four Hours with PlanWrite

Software


To begin starting a new businesss, you need a good

idea? You also need to create a detaled business plan.

A plan for businesses are important and essential to

obtaining financing. Also, a plan will also help guide

your work after you get started. It will provide, to

your copmany's growthe, a helpful map. PlanWrite, the
```

usefullest business planing software ever sold, will organise your plan from A to Z. You will quick identify your short and long-term objectives and invents good stratejies to meet them. You will plan the firm's operations, sales and marketing, and administration very affective. And you will prepare a more perfect profit and lost projection in less hours that ever possible before. don't start that new business without this good tool for producting professional, useful planes!

APPLIED LANGUAGE

A. Write the comparative and superlative forms of each adjective given below.

Positive:	Comparative:	Superlative:
1. efficient	_____	_____
2. useful	_____	_____
3. clean	_____	_____
4. organized	_____	_____
5. early	_____	_____

B. Underline the compound adjectives in the following sentences. Then, add any necessary hyphens.

1. The application is a multiple part form.

2. The seven day meeting exhausted the participants.

3. The interview was one third complete when the power went out.

4. Post office regulations require that you include a ZIP code.

5. The business was established to help low income families.

C. In each of the sentences below, add an adjective to modify the noun. The type of adjective to add is given in parentheses before the sentence.

EXAMPLE: (Numerical) May has __30__ days.

1. (Numerical) The managers attended the convention for

_____ days.

2. (Participle) The _____ exhibit arrived in St. Louis last night.

3. (Predicate) Ms. Simms is _____ .

4. (Adjective clause) Mr. Hsu, who _____ , signed the check.

5. (Compound adjective) _____ jobs are hard to find.

D. Write a sentence that uses each of the words or phrases given below as an adjective.

EXAMPLE: (desktop) **Desktop publishing may be cost effective.**

1. (mysterious) _____

_____

2. (dynamic) _____

_____

3. (computer company) _____

_____

4. (West Virginia) _____

_____

5. (error free) _____

_____

Name _____

Date _____

ADJECTIVES

Part A. Comparison of Adjectives

Circle the adjective in parentheses that is appropriate for the sentence.

1. The increase in the price resulted in (fewer/less) dollars in sales.

2. Lucy's data-processing skills are (better/gooder) than her colleague's.

3. Which clerk is (efficientest/most efficient)?

4. July's sales are the (higher/highest) of the entire year.

5. I liked Harry's report, but I was (more impressed/most impressed) with the one Howard wrote.

6. The new production line is (faster/more faster) than the old one.

7. Carole's machine was the (less/least) expensive of the three.

8. The computer in the graphics department is the (most new/newer) of the two.

9. Helen's assembly of the equipment was (perfect/the most perfect).

10. Of the four applicants, Jesse demonstrated the (best/good) abilities.

Part B. Compound Adjectives

Rewrite each group of words to create a compound adjective containing at least one hyphen.

EXAMPLES:

a report stained with coffee

a building that was constructed well

        a coffee-stained report

        a well-constructed building

1. a machine operated by hand   _____

2. a room filled with smoke   _____

3. a team of six persons   _____

4. news that was received well   _____

5. the report that was written well   _____

6. rain driven by wind   _____

7. a release ordered by the court   _____

8. a beverage free of sugar   _____

9. an expense that was related to work   _____

10. a decision from middle management   _____

Part C. Strong Adjectives

For each of the pairs below, circle the adjective that is stronger. Remember, an adjective is stronger when it provides more information or more precise information than another adjective.

1. (great, efficient) work

2. (bad, arithmetical) error

3. (executive, large) council

4. (important, vital) issue

5. (annual, short) report

6. (15 percent, good) commission

7. (morning, 7:00 A.M.) news

8. (significant, Fulbright) scholarship

Part D. Adjective Phrases and Clauses

Rewrite each group of words below to create an adjective phrase or clause.

EXAMPLE: a single-minded person   **a person who is single minded**

1. a well-maintained engine   _____

2. the red-headed man   _____

3. an often-repeated phrase   _____

4. a quick-thinking reporter   _____

5. the poorly written report   _____

6. sugar-free dessert   _____

7. a factory-installed radio   _____

8. stain-resistant carpet   _____

Part E. Rules About Adjectives

Indicate whether each statement is true or false by writing T or F in the space provided.

_____   **1.** An adjective may describe a noun or a verb.

_____   **2.** Nouns and pronouns in the possessive case often function as adjectives.

_____   **3.** Adjectives must precede the words they modify.

_____   **4.** *This, that, these,* and *those* are always used as adjectives.

_____   **5.** The article *an,* rather than *a,* should be used whenever the following word begins with a vowel.

_____   **6.** A subject complement may be an adjective.

_____   **7.** Phrases cannot be used to modify nouns.

_____   **8.** An adjective clause may be introduced with *who, which,* or *that.*

_____   **9.** The adjective *later* is the comparative form of late.

_____   **10.** *Farther* and *further* can be used interchangeably.

CHAPTER 14

ADVERBS AND PREPOSITIONS

Don't use adverbs unless they do some work.
—William Zinsser, *On Writing Well*

LEARNING UNIT GOALS

LEARNING UNIT 14-1: RECOGNIZING ADVERBS

- Distinguish between adjectives and adverbs. (p. 284)
- Identify adverbs that modify verbs, adjectives, and other adverbs. (pp. 284–86)

LEARNING UNIT 14-2: ADVERB USAGE

- Contrast adjectives with adverbs. (pp. 287–89)
- Know how to use five troublesome pairs of adjectives and adverbs. (pp. 289–91)
- Define adverb phrases and clauses. (p. 291)

LEARNING UNIT 14-3: PREPOSITION USAGE

- Know the correct usage of prepositions and prepositional phrases. (pp. 292–94)
- Recognize clauses that function as objects of prepositions. (p. 294)
- Use correctly troublesome prepositions and word groups used as prepositions. (pp. 295–97)

Do you recall the nonsensical memory device given in Chapter 3 to help you remember the function of adverbs? If you do, you know that adverbs answer the questions *how?*, *to what extent?*, *where?*, or *when?* about a verb form, an adjective, or another adverb (or adverb substitute). This means you can find adverbs almost anywhere in a sentence.

Learning Unit 14-1 will help you recognize adverbs. In Learning Unit 14-2, we explain the various uses of adverbs. Learning Unit 14-3 discusses the first connecting word that you will study—the preposition. Chapter 15 discusses conjunctions—the second connecting word.

LEARNING UNIT 14-1
RECOGNIZING ADVERBS

If you have problems with adverbs, it probably is because they are so versatile. Adverbs modify (describe or limit) verbs, adjectives, or other adverbs (or adverb substitutes). Adverbs can also modify an entire sentence. Table 14-1 gives an overview of the various types of adverbs.

Adjectives modify nouns and pronouns and usually appear before the noun and pronoun they modify. This makes it fairly easy to spot adjectives after you have analyzed a sentence for its major elements. Once you recognize the adjectives in a sentence, you know that the modifiers remaining must be adverbs. Now you must decide if these adverbs modify verbs, adjectives, or other adverbs.

ADVERBS MODIFYING VERBS

Adverbs relate to verbs in a similar way that adjectives relate to nouns and pronouns; they describe or limit verbs. When adverbs modify verbs, they often describe *how* the verb performs the action. They can also limit verbs by stating *to what extent*, *where*, and *when* the action occurs. Adverbs modifying verbs can occur almost anywhere in the sentence.

The door of the vault closed *slowly*. (*Slowly* explains *how* the verb *closed* preformed its action.)

The new salesperson *almost* won the sales contest. (*Almost* explains the *extent* to which the verb *won* performed its action.)

The new maintenance engineer stays *here*. (*Here* explains *where* the verb *stays* performed its action.)

The lost sales report appeared *suddenly*. (*Suddenly* explains *when* the verb *appeared* performed its action.)

Note that two of the adverbs in the above sentences end in *ly*. Many words used as adverbs end in the suffix *ly*, which is usually added to an adjective.

Don't be surprised if you cannot find a particular adverb in the main listing of your dictionary. If the word *quickly* does not appear as a main entry in your abridged dictionary, you may find it at the end of the entry for the word *quick*. Adverbs that add *ly* to adjectives usually are not followed by definitions. Since the adjective is fully defined, the meaning of the adverb is clear.

Be aware that many common adverbs do not end in *ly*. The adverbs *almost* and *here* in the above sentences belong to this group. Here are some others:

Table 14-1 Overview of the Types of Adverbs

Type of Adverb	Adverbs Belonging to Group	Examples
SIMPLE ADVERBS:		
Adverbs of Manner	Tell *how* the action occurred. Usually modify verbs or adjectives; rarely adverbs. Often are formed by adding *ly* to adjectives: *completely, greatly, quietly, rapidly,* etc. Some adverbs do not end in *ly: better, even, how, ill, just, only, otherwise, quite, thus, very, well,* and *worse.* Some words ending in *ly* are adjectives: *fatherly, friendly, kindly, lonely, lovely,* and *timely.*	The company officer spoke *eloquently.* Lisa *cautiously* opened all the file drawers.
Adverbs of Degree	Answer *to what extent?* and variations—*how much?, how small?, how big?* Usually modify adjectives or other adverbs; can modify verbs. Most common: *enough, less, little, much, partly, quite, rather, so, too,* and *very.*	The new office space was *rather* small. The box was *quite* large. Jack moved *too* slowly to finish the job.
Adverbs of Place	Tell *where* the action occurred. Usually modify verbs. Most common: *here* and *there.* Others: *away, backward, downstairs, everywhere, far, forward, near, where, downtown, upstairs,* and *uptown;* also *down* and *up* when they are not followed by nouns.	James lectured *here.* The answer to your question is *there.*
Adverbs of Time	Tell *when* the action occurred. Usually modify verbs. Most common: *again, already, always, before, finally, formerly, never, now, often, seldom, soon, still, then, today, tomorrow,* and *yesterday.*	The president left *yesterday.* John *formerly* worked for the company's biggest competitor.
OTHER TYPES OF ADVERBS:		
Conjunctive Adverbs	Function as adverbs and connect clauses; may appear within the clause. Include the following: *accordingly, therefore, consequently, furthermore, hence, however,* etc.	The budget committee met; *however,* several members were absent. The new employee, *however,* is young.
Affirmative and Negative Adverbs	Common affirmative adverb is *yes.* Common negative adverbs are *no* and *not.* Other adverbs are *but, hardly, never, only,* and *scarcely.* Commas usually follow *yes* and *no. Not* is not part of verb phrase; negates verb phrase making opposite true. Do not use double negatives such as *don't* and *never* or *not* and *hardly* in the same sentence.	Elaine *never* sings. *Yes,* the stock clerk has left. Rachel did *not* give the report to me. WRONG: I *didn't* say *nothing* to him. CORRECT: I *didn't* say *anything* to him. CORRECT: I said *nothing* to him.
Relative Adverbs	Adverbs such as *when, where,* and *why* act as relative adverbs when they introduce adjective clauses and relate to word in main clause. Also function as adverbs in the adjective clause they introduce.	He left the building *when* the siren sounded. The manager chose a day *when* I could attend. (*When* introduces adjective clause modifying *day. When* modifies *could attend* in adjective clause.)
Sentence Adverbs	Modify a whole clause or sentence: *unfortunately, usually,* and *consequently.*	*Unfortunately,* no one heard the answer.
Interrogative Adverbs	Ask questions: *how, when, where,* and *why.* Also modify a word in the sentence.	*How* many news articles did you find? (*How* modifies adjective many.) *When* will you leave the city? (*When* modifies verb phrase *will leave.*)
Nouns Used as Adverbs	Nouns expressing degree, place, time, measurement, number, or value can be used as adverbs. They can also replace a phrase.	He will leave *Sunday.* (*Sunday* can be replaced by prepositional phrase *on Sunday.*)

again	hard	near	often	so
always	how	never	quite	soon
far	late	not	rather	when
fast	much	now	since	where

ADVERBS MODIFYING ADJECTIVES

When a sentence has an intransitive linking verb, look first for the predicate adjective. This adjective can be modified by an adverb that usually precedes the adjective. Adverbs modifying adjectives answer the question *to what extent?*

The old supervisor was *always* correct. (*Always* precedes the predicate adjective and describes the *extent* that the supervisor is *correct.*)

Since adverbs can modify almost any adjective in the sentence, check the sentence for other adjectives modified by adverbs. The adverb usually comes immediately before the word it modifies.

The manufacturer priced the new toy *too* high. (*Too* modifies the adjective *high* by answering the *extent* of *high.*)

The stock clerk saw a *surprisingly* large crate on the dock. (*Surprisingly* modifies the adjective *large* by answering the *extent* of *large.*)

STUDY TIP

Be aware that some words can function as adjectives and adverbs.

ADJECTIVE: We have *hard* water in our city.
ADVERB: All the employees word *hard*.

ADVERBS MODIFYING OTHER ADVERBS

Adverbs can also modify the meaning of other adverbs by answering the question *to what extent?* These adverbs usually come immediately before the other adverbs they modify. Note the following sentences:

The employees said the new machines worked *much* faster. (*Much* modifies the adverb *faster* by answering the *extent* of *faster.*)

The new employee worked *too* slowly to finish the job. (*Too* modifies the adverb *slowly* by answering the *extent* of *slowly.*)

YOU TRY IT

Write an adverb in each of the blanks in the sentences below. The question in parentheses after the sentence will give you a clue.

EXAMPLE: Susan greeted the visitor _____**cheerily**_____. (how?)

In these sentences, the adverb modifies the verb:

1. The sales manager conducted the meeting _____. (how?)

2. The senator will introduce the bill _____. (when?)

3. Carolyn Myers _____ went _____. (how? where?)

In these sentences, the adverb modifies the adjective:

4. Mrs. Garcia wrote to the customer about the _____ late payment. (how late?)

5. The copier did not have _____ toner. (how much?)

6. By Friday, the job was _____ finished. (to what extent?)

In these sentences, the adverb modifies another adverb:

7. That printer operates _____ slowly. (how slowly?)

8. He makes many errors because he does his work _____ quickly. (how quickly?)

9. She writes _____ well. (how well?)

Possible Answers: 1. efficiently 2. tomorrow 3. cautiously, upstairs. 4. very 5. enough 6. completely. 7. very 8. too 9. surprisingly

LEARNING UNIT 14-2
ADVERB USAGE

This learning unit contrasts the usage of adjectives with that of adverbs and discusses adverb phrases and clauses. Remember that errors in adverb usage usually occur because adverbs can modify three parts of speech—verbs, adjectives, and other adverbs.

ADJECTIVES AND ADVERBS

Adjectives and adverbs (1) differ in their sentence placement, (2) are similar in their comparison forms, and (3) have similar words, such as *bad* and *badly*, that can be used incorrectly in sentences.

Placement of Adjectives and Adverbs

Adverbs may appear almost anywhere in a sentence. Note the following:

1. Sentences often begin or end with adverbs.

> *Consequently*, all managers in the building were told to tighten their budgets.

> *Silently* the employees worked to finish the job.

The new president walked *proudly.*

The earthquake began *yesterday.*

2. **Adverbs can appear before or after verbs.** (Verbs are underlined twice.)

Jordan *carefully* stepped over the crack.

Max replied *quickly* to the four questions.

3. **Adverbs can appear between parts of a verb phrase.** (Verb phrases are underlined twice.)

The secretary had *quickly* closed the door.

Pauline did *not* know that the stranger came into the office.

4. **Adverbs can appear before adjectives.** (Adjectives are capitalized.)

James was *rather* SAD that the trip was over.

Corbin felt *less* ANXIOUS about her new computer.

5. **Adverbs can appear before another adverb.** (Modified adverbs are capitalized.)

The water cooler was *almost* COMPLETELY empty.

The vacation was over *too* QUICKLY.

Adjectives are usually fixed in a sentence, but adverbs are movable. Because they are movable, they can be easily misplaced. Every adverb should be placed reasonably close to the word(s) it modifies.

The *neatly* dressed applicant walked *hesitantly* into my office. (The adverb *neatly* modifies the adjective *dressed* answering how the applicant dressed. The adverb *hesitantly* modifies the verb *walked* and tells how the applicant walked.)

When words such as *almost, even, ever, just,* and *only* are misplaced, the sentence meaning can become unclear.

AVOID: By late afternoon, I had *almost* finished every item on the list.
USE: By late afternoon, I had finished *almost* every item on the list.

Consider the meaning of *only* when you use it in a sentence. Its position in the sentence can change the meaning of the sentence.

Abby *only* operates that computer. (Abby does not operate any other computer.)

Only Abby operates that computer. (No one else operates that computer.)

Comparative and Superlative Forms of Adjectives and Adverbs

Like adjectives, adverbs have three degrees of comparison: positive degree, comparative degree, and superlative degree. (If necessary, review Learning Unit 13-1 in Chapter 13, on the comparison of adjectives.) Most adverbs use *more* to form the comparative degree and *most* to form the superlative degree.

Adverbs in the comparative or superlative degree are frequently not used. Note this sentence:

INCORRECT: The engine in the truck runs *smoother* than the engine in the car. (*Smoother* is an adjective in the comparative degree and cannot modify the verb *runs*.)

CORRECT: The engine in the truck runs *more smoothly* than the engine in the car. (*Smoothly* is an adverb. Its comparative form, *more smoothly*, is correct in this sentence.)

Note the italicized adverbs in the following sentences:

Of the two men, Jim Cowan handled the problem *more easily*.

Her colleague works *more slowly* than she does. (In informal writing, the adjective forms of *slow, quick, loud,* and *cheap* may double as adverbs. In business writing, the *ly* forms should be used.)

The comparative forms of several common adverbs follow. Note that the first two listed may also function as adjectives.

Positive	**Comparative**	**Superlative**
fast	faster	fastest
far	farther	farthest
calmly	more (or *less*) calmly	most (or *least*) calmly
promptly	more (or *less*) promptly	most (or *least*) promptly
quietly	more (or *less*) quietly	most (or *least*) quietly
hurriedly	more (or *less*) hurriedly	most (or *least*) hurriedly

S T U D Y   T I P

Remember: The dictionary is not likely to show the comparative or superlative forms of adverbs that end in *ly*. If your dictionary gives run-on entries, an adverb ending in *ly* (*peacefully*) will usually be added to the entry for the shorter form (the adjective *peaceful*).

Troublesome Pairs of Adjectives and Adverbs

1. ***Bad*** **and** ***badly***. *Bad* is an adjective; *badly* is an adverb. Note the following sentence:

INCORRECT: Dr. Whelan agrees that the new soft drink tastes *badly*. (The verb *tastes* within the clause is an intransitive linking verb that must be followed by a predicate adjective—not an adverb.)

CORRECT: Dr. Whelan agrees that the new soft drink tastes *bad*.

Once you know if a word is an adjective or adverb, you can usually determine whether to use *bad* or *badly* by analyzing the sentence.

In the following sentence, *badly* is used as an adverb modifying the transitive verb *managed*:

> Gordon managed the situation *badly*.

2. ***Good* and *well*.** *Good* is always an adjective. *Well* can function as both an adjective and an adverb. As an adjective, *well* refers to good health. As an adverb, *well* refers to the way an action is performed. So when you refer to good health, use the adjective *well*.

> This is a *good* cup of tea. (*Good* is an adjective modifying the noun *cup*.)

> The new supervisor is *well*. (*Well* functions as an adjective describing the health of the new supervisor.)

> The new music teacher plays the piano *well*. (*Well* functions as an adverb describing the action of the verb *plays*.)

3. ***Most* and *almost*.** *Most* can function as an adjective and an adverb. *Almost* is an adverb. As an adverb, *most* means to the greatest or highest degree; *almost* means *nearly*. Use *most* for the superlative degree. Use *almost* when the meaning is *nearly*. When you can substitute *almost* for *most*, use *almost*. *Almost* should be next to the word it modifies.

> Kim is *almost* ready for the party. (*Almost* modifies the adjective *ready*.)

> Of all the vegetables, Pauline likes corn *most*. (The adverb *most* modifies the verb *likes*.)

4. ***Real* and *really*.** *Real* is an adjective meaning *actually* or *genuine*; *really* is an adverb. Do not use the adjective *real* for the adverb *very*. This is a grammatical error.

> INCORRECT: The color is *real* good. (*Real*, an adjective, cannot modify *good*, which functions as an adjective in this sentence.)

> CORRECT: The color is *very* good. (*Very* functions as an adverb meaning *extremely*; as an adverb, *very* can modify the adjective *good*.)

The adjectives in the following two sentences are used correctly:

> Here is a *real* diamond. (*Real*, an adjective meaning *genuine*, modifies the noun *diamond*.)

> The book was *really* excellent. (*Really*, an adverb, modifies the adjective *excellent*.)

Use the adjective *real*, meaning *genuine*, when you want to modify a noun; the adverb *really* when you mean *actually*; use the adverb *very* when you mean *extremely*.

5. ***Sure* and *surely*.** *Sure* is an adjective meaning *absolutely certain*; *surely* is an adverb meaning *certainly*.

INCORRECT: I am *sure* grateful for the gift. (*Sure*, an adjective, cannot modify *grateful*, an adjective.)

CORRECT: I am *surely* grateful for the gift. (*Surely*, an adverb, modifies *grateful*, an adjective.)

Here are two additional examples:

You know the *sure* way to success. (*Sure*, an adjective, modifies *way*, a noun.)

This is *surely* great pizza. (*Surely*, an adverb, modifies *great*, an adjective.)

ADVERB PHRASES AND CLAUSES

Adverb phrases are prepositional phrases and infinitive phrases that function as single-word adverbs. These phrases usually modify verbs and adjectives.

Adverb clauses are dependent clauses that are introduced with subordinating conjunctions and function as single-word adverbs modifying verbs, adjectives, and other adverbs. These clauses can also modify entire independent clauses.

In the sentences that follow, two phrases and two clauses have been italicized. Each functions as an adverb modifying a verb.

Mr. Staub will telephone you *in the morning*. (The prepositional phrase functions as an adverb modifying the verb phrase *will telephone*.)

A stranger walked *into the room*. (The prepositional phrase functions as an adverb modifying the verb *walked*.)

Check my figures *before you leave*. (The adverb clause modifies the verb *check*.)

We sold an expensive painting *while you were out*. The adverb clause modifies the verb *sold*.)

Adverb clauses are discussed in detail in Chapter 15.

When you analyze sentences with clauses, look for adverb clauses first. By eliminating adverb clauses from a sentence, you have only to be concerned with noun and adjective clauses, which can begin with the same cue words.

Adverb clauses usually begin with a subordinate conjunction and are always modifiers. Learn to recognize the following words used as subordinate conjunctions to introduce adverb clauses:

after	as though	provided	until
although	because	provided that	when
as	before	since	provided that
as if	even if	so that	where
as long as	even though	than	wherever
as much as	if	though	whether
as soon as	in order that	unless	while

WRITING TIP

Watch your use of adverbs. Many adverbs are unnecessary and can clutter a sentence. Try to use verbs that say what you mean without the addition of adverbs.

YOU TRY IT

A. Column A lists adjectives. Column B lists phrases that require an adverb. Change the adjective to an adverb and insert it in the blank space to complete the phrase in Column B.

COLUMN A Adjectives:	COLUMN B Adverbs:
1. calm	handle the situation _____
2. good	writes _____
3. rare	_____ drives in the city
4. prompt	arrive _____
5. angry	shouted _____

B. Some adverbs and adverb phrases in the sentences below are italicized. Underline the word that the adverbs modify. In the space provided, indicate whether the adverb is modifying a verb, a noun, or another adverb by writing the word *verb*, *noun*, or *adverb* as appropriate.

EXAMPLE: She prepared the data *quickly.* _____**verb**_____

(The adverb *quickly* modifies the verb *prepared.*)

1. The office tower is *well* built. _____

2. Lowell communicates *very* well. _____

3. Give the papers to Sylvia *in the morning.* _____

4. Please do your work *carefully.* _____

5. The lecturer spoke *too* softly. _____

LEARNING UNIT 14-3
PREPOSITION USAGE

Prepositions do not make mental pictures in your mind, but they are strong little words. They show their strength by binding words together and relating them to another word in the sentence.

Every preposition has an object (a noun or pronoun). A preposition combines with its object and any modifiers to form a prepositional phrase. This phrase functions as an adjective or adverb. From your experience in analyzing sentences, you know how to recognize prepositions and prepositional phrases. So you should have no difficulty understanding the following sentences:

He prefers a bank *with many branches.* (Prepositional phrase functions as an adjective modifying the noun *bank.*)

He would like one *near his office.* (*Near* functions as a preposition meaning close to. The word *one* functions as a pronoun. The prepositional phrase functions as an adjective modifying the pronoun *one.* Be sure to check your dictionary when you are not sure how a word functions in a sentence.)

Please insert the paper *in the printer.* (The prepositional phrase functions as an adverb modifying the verb *insert.*)

This unit gives several suggestions for the effective usage of prepositions. These suggestions are grouped under single-word prepositions, prepositional phrases, clauses as objects of prepositions, troublesome prepositions, and word groups used as prepositions.

SINGLE-WORD PREPOSITIONS

Ending Sentences with Prepositions

Avoid ending a sentence awkwardly with a preposition. Note the following sentences:

AVOID: Whom did you give the letter *to?*

USE: *To* whom did you give the letter*?*

AVOID: This is the pen I wrote the letter *with.*

USE: I wrote the letter *with* this pen. (Whenever possible, avoid beginning sentences with the words *this is* and *there is.*)

Some sentences will be made awkward if you avoid the prepositional ending. In such cases, you can end the sentence with the preposition or reword the sentence.

AVOID: He does not know *for* what we are looking.

USE: He does not know what we are looking *for.*

The beginning and the end of a sentence are positions of emphasis. A writer, therefore, will occasionally end a sentence with a preposition because he or she wants that word emphasized.

Don't tell me what the report is *for*; tell me what it is *about.*

Unnecessary Prepositions

Avoid the use of unnecessary prepositions. You can weaken a sentence by inserting a preposition that is not needed.

She does not know where the revised copy is *at.* (Omit *at.*)

The paperweight fell off *of* his desk. (Omit *of.*)

Where has he taken the booklet *to?* (Omit *to.*)

Avoid using the preposition *of* after *all* or *both* unless it is needed for clarity.

DON'T USE: All *of the* bids must be submitted by noon. (The words *of* and *the* may be omitted.)

USE: All bids *must* be submitted by noon.

USE: Both executives favor the proposed change. (Not *both of the executives.*)

USE: All *of* us were pleased with the results. (The preposition *of* is needed for clarity.)

PREPOSITIONAL PHRASES

Prepositional Phrases and Singular or Plural Verbs

Generally, ignore prepositional phrases in deciding whether to use a singular or a plural verb in a sentence. Prepositional phrases are modifiers and do not affect the number of the subject. Don't use a plural verb simply because the object of a preposition is plural. (Note exceptions in Chapter 12.)

A report of recent problems and grievances has been made. (Subject, *report*, and verb, *has been made*, are both singular.)

Pronoun as Object of Preposition

A pronoun used as the object of a preposition in a prepositional phrase must be in the objective case. (Objective case pronouns were discussed in Chapter 8.) Be especially careful when the preposition has a compound object.

The four paragraphs were written by Mr. Whittaker and *me.*

Lucy Gomez wants you to sit between Marie and *her.*

CLAUSES AS OBJECTS OF PREPOSITIONS

Clauses can serve as objects of prepositions. Make certain that the subject and the direct objects of these clauses are in the correct case. This is a good example of the importance of knowing how to analyze sentences.

You may assign that task to *whoever is available.* (A noun clause is the object of *to.* As subject of the verb *is, whoever* must be in the nominative case [Chapter 9].)

You may assign that task to *whomever you choose.* (A noun clause is the object of *to.* The verb *choose* is used as a transitive verb meaning to select freely. The subject of the verb is *you.* Since transitive verbs need direct objects and direct objects must be in the objective case, *whomever* is correct. You can reverse the word order and say *You choose whomever.*)

TROUBLESOME PREPOSITIONS

Be especially careful in your use of the following troublesome prepositions:

1. ***Accept* and *except*.** The verb *accept* means to take or to receive. The preposition *except* means to leave out or exclude.

> All employees *accept* your ideas.

> All employees *except* June agree with your ideas.

2. ***Among* and *between*.** Use *among* in referring to more than two; use *between* in referring to only *two* objects or people.

> The estate was divided *among* the five heirs.

> Ms. Clark sat *between* the two attorneys.

3. ***Beside* and *besides*.** These two prepositions have different meanings. *Beside* means close to; *besides* means in addition to.

> Walk *beside* me.

> Look for another report *besides* this one.

4. ***In* and *into*.** These words have many applications. In general, the preposition *in* is used with verbs that indicate place or position. *Into* is used with verbs that express motion or a change of condition.

> The signed documents are *in* the envelope.

> Put the signed documents *into* the envelope.

> He is *in* the room.

> Ms. Jackson's smile turned *into* a frown.

5. ***Like*.** The word *like* is often used as a preposition. Do not use *like* as a conjunction to join clauses.

> The new branch manager looks *like* Mr. Glass. (Last three words form a prepositional phrase.)

> It seems *as though* (not *like*) the building will never be completed. (Do not introduce a clause with *like*.)

> Caroline's new stationery is white, *as* (not *like*) she had hoped it would be.

6. ***Per*.** This preposition is used primarily before Latin nouns or before weights and measures.

> The cost *per* capita is small.

> The maximum rate allowed is 55 miles *per* hour.

> She wants to talk with you about (not *as per*) your request.

WORD GROUPS USED AS PREPOSITIONS

Word groups (called idioms) are often used as prepositions. Here are a few examples of such prepositions with some possible objects:

because of the circumstances   *apart from* this minor problem
according to the report   *instead of* the new edition

Be aware that many verbs or adjectives must be followed by a particular preposition to convey clearly its intended meaning. Sometimes another acceptable meaning will result if the preposition is changed. We could list hundreds of such verbs and adjectives, and the prepositions that should be used with them. The few that follow will serve as examples.

1. accompany by (a person)   The president was *accompanied by* the firm's chief counsel.

2. accompany with (an object)   The blueprints were *accompanied with* the necessary specifications.

3. acquaint with   I will be pleased to *acquaint* you *with* your new surroundings.

4. acquit of (a charge)   He was *acquitted of* the charge against him.

5. adapt from   The new model will be *adapted from* this sample.

6. adapt to   Attempt to *adapt* yourself *to* this new environment.

7. agree to (to consent or accede) They *agreed to* the terms of the contract.

8. agree with (to have same opinion) I *agree with* what you have said.

9. allude to   Please don't *allude to* my inexperience again.

10. cognizant of   The auditor did not seem *cognizant of* the error.

11. compare to (show similarity)   This office building may be *compared to* a small town.

12. compare with (examine two objects or people) Please *compare* this typewriter *with* the other.

13. compensate for   The broker must be *compensated for* her services.

14. comply with   He was willing to *comply with* my orders.

15. consistent with   That error is not *consistent with* her usual performance.

16. convenient for   Will four o'clock be *convenient for* you?

17. convenient to (near)   Our office is *convenient to* the Capitol.

18. conversant with   Few people are *conversant with* this new technique.

19. correspond to   This machine *corresponds to* our specifications.

20. correspond with (write)   We often *correspond with* our salespeople in the Midwest.

21. deal in (merchandise)   Their *firm* deals *in* earth-moving equipment.

22. deal with (subjects)	The speaker seemed reluctant to *deal with* capital punishment.
23. depend on	You can *depend on* our company to deliver the merchandise.
24. differ from (to be unlike)	A living trust *differs from* a testamentary trust.
25. differ with (disagree)	I *differ with* what she just said.
26. different from (not than)	My desk is *different from* yours in many respects.
27. encroach on	A portion of this building *encroaches on* your neighbor's property.
28. equivalent to	Her income is roughly *equivalent to* mine.
29. identical with	Their computer is *identical with* ours.
30. indicative of	That excellent sales total is *indicative of* the firm's efficient management.
31. insist on	We will *insist on* a new contract.
32. liable for (responsible)	You will be *liable for* any damage done to the vehicle.
33. liable to (susceptible)	His bout with pneumonia has made him *liable* to a heart attack.
34. parallel to	The fence runs *parallel to* the sidewalk.
35. reminiscent of	Today's meeting is *reminiscent of* one held two years ago.
36. responsible for	All workers are held *responsible for* the tools they use.
37. similar to	His bookcase is *similar to* mine.
38. specialize in	We plan to *specialize in* sporting goods.
39. talk to (one speaker)	The candidate *talked to* a large, attentive audience.
40. talk with	The judge and the attorney *talked with* each other for several minutes.

YOU TRY IT

A. In each of the items below, circle the correct pronoun from the pair provided in parentheses. Reminder: A pronoun used as the object of a preposition must be in the objective case.

1. Refer the applicants to Mr. Hubert and (I/me).

2. The check was made out to my son and (I/me).

3. Place the printer between Ms. Gwon and (I/me).

B. In each of the items below, circle the correct preposition from the pair provided in parentheses.

1. The installer positioned the computer (beside/besides) my desk.

2. Complete all of the form (accept/except) the shaded areas.

3. The task was divided (among/between) the ten members of the committee.

4. The executives agreed (with/to) the attorney's opinion.

5. The formal agreement differs (from/with) our original understanding.

Answers: **A.** 1. me 2. me 3. me. **B.** 1. beside 2. except 3. among 4. with 5. from

CHAPTER SUMMARY: A QUICK REFERENCE GUIDE

PAGE	TOPIC	KEY POINTS	EXAMPLES
284	**Recognizing adverbs**	See Table 14-1. **Adverbs modifying verbs** describe *how* a verb performs the action. Limit by stating *to what extent*, *where*, and *when* the action occurs. **Adverbs modifying adjectives** answer *to what extent*? Adverb is usually immediately before the adjective it modifies. **Adverbs modifying other adverbs** answer *to what extent*? Adverb is usually immediately before the adverb it modifies.	All workers *slowly* walked out of the building. The secretary is *nearly* ready for the lecture. Barbara talked *rather* strangely.
287	**Adjectives and adverbs**	**Placement:** 1. Sentences often begin or end with adverbs. 2. Adverbs can appear before or after verbs. 3. Adverbs can appear between parts of a verb phrase. 4. Adverbs can appear before adjectives. 5. Adverbs can appear before another adverb.	Jean turned *quickly*. Eve *actually* ran to the store. She did *not* come. Jack was *very* happy. You can walk *much* slower.
288	**Comparative and superlative forms**	Adverbs have three degrees of comparison the same as adjectives. Comparative and superlative are frequently neglected in adverbs. Use *more* for comparative and *most* for superlative.	Curt moved *more* quickly than Elaine. The president spoke *most* convincingly.
289	**Troublesome adjective and adverb pairs**	1. *Bad* and *badly*. *Bad* is an adjective; *badly* is an adverb. 2. *Good* and *well*. *Good* is an adjective. *Well* is an adjective and adverb. As adjective, *well* means good health. As adverb, *well* refers to the way an action is performed.	This is a *bad* job. James managed the team *badly*. This is a *good* apple. The employee is not *well*. The company team plays *well*.

PAGE	TOPIC	KEY POINTS	EXAMPLES
		3. *Most* and *almost*. *Most* is an adjective and adverb. *Almost* is an adverb. As adverb, *most* means greatest degree; *almost* means *nearly*. Use *most* for superlative; *almost* for *nearly*.	This is the *most* challenging job she had. I am *almost* at the door.
		4. *Real* and *really*. *Real* is an adjective meaning *genuine*; *really* is an adverb. Do not use the adjective *real* for the adverb *very*.	This desk is a *real* antique. The cake is *really* good.
		5. *Sure* and *surely*. *Sure* is an adjective meaning *absolutely certain*; *surely* is an adverb meaning *certainly*.	Look for a *sure* victory. Sukie will *surely* be happy.
291	**Adverb phrases and clauses**	**Adverb phrases** are prepositional phrases and infinitive phrases; they usually modify verbs and adjectives. **Adverb clauses** are dependent clauses introduced with subordinate conjunctions; they function as single-word adverbs and can also modify entire independent clauses.	The store owner will return *at noon*. *After the president left*, the meeting continued.
293	**Single-word prepositions**	**1.** Avoid ending a sentence awkwardly with a preposition. **2.** Avoid using unnecessary prepositions. **3.** Avoid using the preposition *of* after *all* or *both*.	This is the paper I am writing *on*. Where has Linda gone *to*? All *of the* tables must be cleaned.
294	**Prepositional phrases**	**1.** Ignore prepositional phrases when deciding to use a singular or plural verb. **2.** A pronoun used as the object of a preposition must be in the objective case.	A *group* of new locations has been chosen. The meeting place was chosen by Jeff and *me*.
294	**Clauses as objects of prepositions**	When clauses serve as objects of prepositions, use the nominative and objective cases correctly.	You may look for *whoever is qualified*.
295	**Troublesome prepositions**	**1.** *Accept* and *except*. *Accept* means to take; *except* means exclude. **2.** *Among* and *between*. *Among* is used when referring to more than two; *between*, only two objects or people. **3.** *Beside* and *besides*. The preposition *beside* means close to; *besides* means in addition to.	Please *accept* my thanks. All left *except* Kathy. The pizza was divided *among* the four employees. This argument is *between* the two managers. Stay *beside* me. Look for other answers *besides* these.

PAGE	TOPIC	KEY POINTS	EXAMPLES
		4. *In* and *into*. *In* is used with verbs indicating place or position; *into* is used with verbs expressing motion or change of condition.	I found the pen *in* the desk. Put this item *into* the box.
		5. *Like*. *Like* is often used as a preposition. Do not use *like* as a conjunction to join clauses.	It looks *like* rain.
		6. *Per*. It is used before Latin nouns or weights and measures.	The cost is $10 *per* case.
295	**Word groups as prepositions**	See the list in the chapter.	

QUALITY EDITING

Last month, your supervisor ordered a new photocopying machine from CopiesPlus. When CopiesPlus delivered the machine, your supervisor was out of the office. CopiesPlus installed the wrong machine. Your supervisor has written a letter informing CopiesPlus of the error. Edit the letter. Watch for incorrect adverb forms, prepositions, and pronouns as well as other errors.

While I was out of the ofice last week, you're compnay delivered and installed the CopyRite Photo Copier A35, which is different. than the one we ordered. We had selected copier A90 very careful because we needed an automatic document feeder We choosed the A90 from between the five copiers in the A series becuase this model has only that feature. Wood you please look in on this matter immediate and replace the A35 for the A90? Without the feeder, document coping is more time consuming for my assistants and I. Thank you.

APPLIED LANGUAGE

A. Write a sentence using each of the adverbs listed below.

1. correctly_____

_____

2. badly_____

_____

3. almost_____

_____

4. really_____

_____

5. well_____

_____

B. Your firm, Skiller and Company, recently changed travel agencies. Your new agency is Skyline Travel. Write a letter to the president of Skyline Travel to compliment George Rivas, the agent who handles travel for your executives. Use the following adverbs and adverb phrases in your letter.

efficiently

in a pleasant manner

in advance

last month

whenever I call

WORKSHEET 14

Name _____

Date _____

ADVERBS AND PREPOSITIONS

Part A. Distinguish Adverbs from Adjectives

Write an appropriate word in the blank at the end of each sentence. Then, in the space provided on the right, indicate whether it is an adjective by writing Adj. or an adverb by writing Adv.

EXAMPLE:

Sales for the past year have been _____**slow**_____.   Adj.

1. Several of our office machines are not working _____.   _____

2. His reply to my question was somewhat _____.   _____

3. He may be a brilliant scientist, but he does not express himself _____.   _____

4. The new fax machine is remarkably _____.   _____

5. The new fax machine operates _____.   _____

6. Harriet Miles spoke _____ about the new database.   _____

7. Art Garza designs _____ buildings.   _____

8. A report is sent to the state treasurer _____.   _____

9. The manual was printed _____.   _____

10. The plan worked _____.   _____

Part B. Comparative Forms of Adverbs

Circle the correct adverb from the pair in parentheses.

1. Of the two attorneys, the prosecutor presented her case (clearer/more clearly).

2. The defendant spoke (calmer/more calmly) than we expected.

3. The new lecturer spoke (longer/more long) than the previous one.

4. The sales of the recycled paper are increasing (more quickly/more quicker) than the new paper.

5. Which of the six staff members enters data (quickest, most quickly)?

6. Karen's computer operates (better/more better) than John's.

7. Martin Chang's plan worked (perfectly/more perfectly).

8. Which of the two truckers makes deliveries (more promptly/prompter)?

Part C. Troublesome Adverbs and What They Modify

Circle the correct adverb from the pair in parentheses. Then, underline the word that it modifies. In the space provided, indicate whether the word being modified is a verb, adjective, or adverb.

EXAMPLE:

He *writes* very (good/well).   _____**verb**_____

1. Mr. Fletcher's brief talk was (real/really) informative.   _____

2. The new sales representative always speaks (real/very) courteously   _____
   to customers.

3. Anita has done (well/good) in her extremely challenging position.   _____

4. Dr. Drake will (sure/surely) cooperate with us.   _____

5. I was (real/really) confused about the contract.   _____

6. He apparently did not feel (good/well) about the project.   _____

7. You will find the new offices to be (real/very) convenient.   _____

8. The new scanner functioned (most/almost) perfectly.   _____

9. Edgar prepared the graphics for the brochure (real/very) carefully.   _____

Part D. Prepositions

Complete each sentence by writing a prepositional phrase to modify the word in italics.

EXAMPLE:

The agent wrote ____**to the governor**____ .

1. We need a *warehouse* _____.

2. The auditors *will return* _____.

3. We *agreed* _____.

4. She *signed* the contract _____.

5. The vans *delivered* the freight _____.

6. The purchasing department ordered three *desks* _____.

Part E. Rules and Definitions

In the space provided write T if the statement is True or F if it is False.

_____   **1.** Adjectives can modify nouns or adverbs.

_____   **2.** Adjectives and adverbs differ in their sentence placement.

_____ **3.** Adverbs always appear before the word they modify.

_____ **4.** All adverbs end in *ly.*

_____ **5.** Adverbs modifying adjectives or other adverbs answer the question *to what extent?*

_____ **6.** An adverb is sometimes the first word of a sentence.

_____ **7.** Some prepositions have objects; some do not.

_____ **8.** Sometimes the object of a preposition is a clause.

_____ **9.** It is never proper to end a sentence with a preposition.

_____ **10.** A preposition may have only one object.

CHAPTER 15

CONJUNCTIONS

Conjunctions are a more general and versatile kind of connective than prepositions: their function is to join together or connect words or sentence elements which may have several different grammatical constructions.

—Charles H. Vivian and Bernetta M. Jackson,
English Composition

LEARNING UNIT GOALS

LEARNING UNIT 15-1: COORDINATING CONJUNCTIONS

- Explain simple and compound sentences. (pp. 308–9)
- Define coordinating conjunctions. (p. 309)
- Use coordinating conjunctions correctly in compound sentences. (p. 310)
- Know how to punctuate compound sentences without coordinating conjunctions. (pp. 310–11)

LEARNING UNIT 15-2: CONJUNCTIVE ADVERBS; CORRELATIVE CONJUNCTIONS

- Define and use conjunctive adverbs. (pp. 312–14)
- Define and use correlative conjunctions. (pp. 314–15)

LEARNING UNIT 15-3: SUBORDINATING CONJUNCTIONS

- Define complex sentences. (p. 316)
- Explain adverb, adjective, and noun clauses. (pp. 316–19)
- Define compound-complex sentences. (p. 320)

Prepositions show the relation between their objects (nouns or pronouns) and some other word(s) in the sentence. Conjunctions do not have objects. They make a connection of thought between two words or groups of words in a sentence.

Conjunctions belong to two major groups:

1. *Conjunctions joining coordinate sentence elements of equal rank.* Coordinate sentence elements of equal rank are words, phrases, and clauses that have the same function in a sentence.

   a. Coordinating conjunctions (Learning Unit 15-1).

   b. Conjunctive adverbs (Learning Unit 15-2).

   c. Correlative conjunctions (Learning Unit 15-2).

2. *Conjunctions joining subordinate sentence elements of unequal rank.* Subordinate sentence elements of unequal rank are independent and dependent clauses (Learning Unit 15-3).

LEARNING UNIT 15-1
COORDINATING CONJUNCTIONS

This learning unit begins with a definition of simple and compound sentences. Understanding the construction of a compound sentence is necessary before coordinating conjunctions and their use in compound sentences can be explained.

SIMPLE AND COMPOUND SENTENCES

The four types of sentences are simple, compound, complex, and compound-complex. This classification is based on the number and kind of clauses that a sentence contains.

Simple and compound sentences use coordinating conjunctions (introduced in Chapter 3). Complex sentences use subordinating conjunctions (also introduced in Chapter 3). Compound-complex sentences use coordinating and subordinating conjunctions.

A **simple sentence** contains only one independent clause. A clause, you will recall, is a group of related words arranged grammatically that contains both a verb and its subject. An *independent clause* can stand alone as a simple sentence. Independent clauses were explained in Chapter 4.

The following sentences are simple sentences:

The directors meet on the first Friday of each month.

The new textile plant in Jacksonville is now open.

Mr. Crawford, having hired the guilty bookkeeper, felt responsible.

The stockholders' meeting was canceled.

STUDY TIP

Remember that a simple sentence may be long and have several phrases or compound elements, but it can have *only one independent clause.*

A **compound sentence** contains two or more independent clauses that are closely related in meaning. Since compound sentences are often connected by coordinating conjunctions, let's look first at coordinating conjunctions and then see how they are used in compound sentences.

DEFINING COORDINATING CONJUNCTIONS

The word *coordinate* means *of equal importance or rank.* A **coordinating conjunction** joins two or more sentence elements (words, phrases, or clauses) that are of equal importance or rank. The following words may function as coordinating conjunctions:

and   but   or   nor   for   yet   so

The principal function of the first four conjunctions (*and, but, or, nor*) is to connect sentence elements. The other three conjunctions (*for, yet, so*) appear frequently as other parts of speech. Occasionally, however, they join independent clauses and function as coordinating conjunctions.

The conjunction *so* (meaning *so that*) is considered by some authorities to be too informal for business writing. The words *therefore* and *consequently* can sometimes be used for *so.* Often, it is best to rewrite the sentence.

> AVOID: He leased the car *so* he would have transportation to Utah.
>
> USE: He will have transportation to Utah because he leased the car.
>
> AVOID: Marta left the office early *so* she could meet with a client.
>
> USE: Marta left the office early because she had to attend a meeting with a client.

Most authorities agree that it is standard to use *so* before a clause that states a result or consequence.

> The document contained several glaring errors, *so* I returned it to your office.

Note how the following simple sentences use coordinating conjunctions to join sentence elements that are equal grammatically.

> Mr. Thorne *and* Ms. Krebs will share the responsibility. (The coordinating conjunction joins two nouns.)
>
> He *or* she will prepare the figures for the MacLean report. (The coordinating conjunction joins two pronouns.)
>
> We sang *and* talked at the office picnic. (The coordinating conjunction joins two verbs.)
>
> The day proved to be sunny *but* cool. (The coordinating conjunction joins two predicate adjectives that are somewhat opposite.)
>
> You may park the company truck on the driveway *or* in the garage. (The coordinating conjunction joins two prepositional phrases.)

STUDY TIP

You should memorize the seven coordinating conjunctions. This memory trick may help you. Remember the words *fan boys*. Take a letter from these two words and match it to the first letter of one of the coordinating conjunctions. You will have *f* for *for*, *a* for *and*, *n* for *nor*, *b* for *but*, *o* for *or*, *y* for *yet*, and *s* for *so*.

COORDINATING CONJUNCTIONS USED IN COMPOUND SENTENCES

Use the following coordinating conjunction rules for joining two independent clauses in a compound sentence:

RULE 1.   Place a comma before a coordinating conjunction that joins two independent clauses.

Our Racine plant produces these parts, *and* a local wholesaler distributes them.

Cynthia Bradshaw was enthusiastic about our plans, *but* the others seemed uninterested.

Jack may answer your question, *or* he may ignore it.

The new supervisor had no time to check old records, *nor* could she order the needed supplies.

Brad agreed to work overtime, *for* he needs the extra money.

They can't meet our deadline, *so* we canceled our order.

RULE 2.   The internal comma can be omitted in a compound sentence of less than ten words if the independent clauses are joined by *and* or *or*, providing no misunderstanding in the meaning of the sentence results.

We worked *and* they read the newspaper.

The clock simply stopped *or* someone damaged it.

The job is difficult *and* it requires patience.

BUT: Rex wants the job, but he isn't eligible.

There were no customers, *so* we left early. (If the connective is something other than *and* or *or*, you need the comma.)

PUNCTUATING COMPOUND SENTENCES WITHOUT COORDINATING CONJUNCTIONS

Some compound sentences that have a closely related independent clause are more emphatic without the coordinating conjunction.

RULE 3. Use a semicolon to separate independent clauses of a compound sentence that are not joined by a coordinating conjunction.

He gets the easy assignments; I get the difficult ones.

Johnson does the work; his partner sends out the bills.

The company delivered the message electronically; the package came by United Parcel Service.

YOU TRY IT

A. Answer the following questions about conjunctions in the space provided.

1. What type of conjunction is used in simple and compound sentences? _____

2. What type of sentence contains two or more independent clauses that are closely related in meaning? _____

3. List the seven coordinating conjunctions. (Remember the mnemonic *fan boys*.)

_____   _____   _____   _____

_____   _____   _____

4. When a coordinating conjunction joins two independent clauses, do you place a comma before the conjunction, after it, or not at all? _____

5. What does a coordinating conjunction do? _____

_____

6. When can you omit the comma before the conjunction in a compound sentence?

_____

7. What punctuation mark can sometimes replace the coordinating conjunction in a compound sentence? _____

B. Indicate whether each of the following sentences is simple or compound by writing S or C in the space provided before each sentence.

_____ **1.** The English class meets in the evening on Monday and Wednesday.

_____ **2.** The English class that Myra attends meets in the evening on Monday and Wednesday.

_____ **3.** The old restaurant in Beaumont, Texas, burned down.

_____ **4.** Mary Kirkpatrick, having purchased the used car, hoped it would keep running.

_____ **5.** Mary Kirkpatrick purchased the used car and hoped it would keep running.

C. Add any needed punctuation to the following sentences. If the sentence needs no punctuation, write *correct* at the end of the sentence.

1. Vista Books publishes the workbooks but Hopkins Marketing distributes them.

2. Loretta writes the book Elizabeth publishes it.

3. The book is long and it is boring.

4. I wanted to go to the lecture but my colleagues were not interested.

D. Write a compound sentence for each of the conjunctions listed below. Remember: Each sentence must contain two independent clauses.

1. but _____

2. and _____

3. or _____

LEARNING UNIT 15-2
CONJUNCTIVE ADVERBS; CORRELATIVE
CONJUNCTIONS

Learning Unit 15-1 discussed the first group of conjunctions that join equal grammatical elements in a sentence—coordinating conjunctions. Learning Unit 15-2 discusses the second and third groups—conjunctive adverbs and correlative conjunctions.

CONJUNCTIVE ADVERBS

This section begins with the use of conjunctive adverbs and transitional phrases in compound sentences. As you will see, conjunctive adverbs must be chosen carefully.

Use of Conjunctive Adverbs and Transitional Phrases

You may choose to join the clauses of a compound sentence with a **conjunctive adverb** (refer to Table 14-1 in Chapter 14) or a **transitional phrase**. They both serve as *modifiers* and as *connectives*. Conjunctive adverbs and transitional phrases modify the clause they introduce and relate the preceding independent clause to the independent clause in which they appear. The common conjunctive adverbs and transitional phrases follow.

Conjunctive Adverbs		**Transitional Phrases**	
accordingly	moreover	after all	in fact
also	namely	as a consequence	in other words
besides	nevertheless	as a result	in the meantime
consequently	next	as a rule	in summary
finally	otherwise	at the same time	it seems
first	still	by the way	of course
furthermore	then	for example	on the contrary
hence	thereafter	for this purpose	on the other hand
however	therefore	in addition	on the whole
incidently	thus	in brief	that is
indeed	undoubtedly	in conclusion	to be sure
likewise	whereas	in contrast	to summarize

RULE.   Use a semicolon before a conjunctive adverb or transitional phrase that separates two independent clauses. The conjunctive adverb is *usually* followed with a comma. The transitional phrase is *always* followed with a comma.

We have several problems; for example, our computer is not working. (Comma after the transitional phrase is required.)

January sales were excellent; in fact, we set a new store record. (Comma after the transitional phrase is required.)

Jessica Hadecki sold $50,000 in merchandise; thus, we managed to meet our quota. (The comma after *thus* is optional. Some writers prefer not to use a comma after a conjunctive adverb with only one syllable.)

The auction was a complete success; consequently, Mr. Wenig is in a fine mood. (Some authorities say the comma after *consequently* is optional. Usually, however, using the comma avoids misunderstanding.)

Mr. Fenske did visit this office; however, I was not here at the time. (A comma is usually used after the conjunctive adverb *however.*)

The trust agreement contains one minor typographical error; otherwise, it has been prepared perfectly.

Words used as conjunctive adverbs or transitional phrases in compound sentences also can be used as parenthetical (interrupting) words in simple and compound sentences. Note how we punctuated the italicized words in the following simple sentences.

The program, *however*, did not begin at the scheduled time.

When he spoke, *for example*, his voice sounded weak and hoarse.

The authorities, *therefore*, have stationed guards at all entrances.

The construction worker was new on the job; his performance, *however*, was excellent.

Choosing Conjunctive Adverbs

Because conjunctive adverbs both modify and connect, choose them carefully. Let's consider the following definitions:

Conjunctive Adverb	Meaning
Therefore, consequently, hence	As a result; for that reason
However, nevertheless	In spite of this (or that); still; yet; but
Moreover, furthermore, besides	In addition to that
Otherwise	In other circumstances; in other respects; or else
Thus	In this (or that) manner or way; because of this (or that)

Although some of these conjunctive adverbs are shown here as synonymous, a good dictionary may show shades of difference in meaning. The word *however*, for example, suggests that the reader consider a moderate concession or a second point; *nevertheless* emphasizes direct opposition. Note the use of conjunctive adverbs in these sentences:

The defendant was proved guilty; nevertheless [in spite of that fact], he was allowed to go free.

The new equipment is of superb quality; moreover [in addition to that fact], it will fit perfectly into the space available.

The shipment came late; however [in spite of that fact], this did not inconvenience us.

She wrote the report; therefore [for that reason], she is responsible.

WRITING TIP

Think of connectives as the glue used to hold sentences and paragraphs together. In compound sentences, they serve as bridges that help readers move smoothly from one independent clause to another. The readability and flow of your writing depends in part on your skillful use of connectives. Without connectives, your writing may appear to be choppy.

CORRELATIVE CONJUNCTIONS

Correlative conjunctions are used in pairs to join coordinate words, phrases, and clauses. Occasionally writers use an adverb and a conjunction as a pair to accomplish the same purpose. Here are a few common correlatives:

both . . . and	neither . . . nor	whether . . . or
either . . . or	not only . . . but also	

Place a pair of correlatives so that the sentence elements following them are *parallel in form*. This means that what is on one side of a single correlative conjunction must be balanced with the same grammatical structure on the other side. The items on both sides do not have to be similar, only the sentence structure must be similar. The following sentence illustrates a sentence that is *not* parallel in form:

Mr. Johnson should *either* ship the cement *or* he should give us a reason for the delay.

Note that the verb *ship* follows the first correlative, *either*; but a pronoun that introduces an independent clause follows the second correlative, *or*. We can eliminate any problem in parallelism by rewording the sentence.

Mr. Johnson should *either* ship the cement *or* give us a reason for the delay.

In this improved version, we have maintained a parallel structure because verbs follow both correlatives, *either* and *or*. Here are more examples of how to construct sentences that are parallel in form.

DON'T USE: He was asked *either* to buy supplies from the Lansing Corporation *or* the Boynton Company.

USE: He was asked to buy supplies from *either* the Lansing Corporation *or* the Boynton Company.

DON'T USE: Mr. Kelso is *not only* experienced *but also* has had five profitable years of college.

USE: Mr. Kelso is *not only* experienced *but also* well educated.

YOU TRY IT

A. Select two of the conjunctive adverbs and two of the transitional phrases from the lists below. Write a compound sentence using each of your choices.

Conjunctive Adverbs: moreover, still, besides

Transitional Phrases: for example, on the other hand, that is

1. _____

2. _____

3. _____

4. _____

B. Select two of the correlative conjunction pairs from the list below. Write a sentence using each of the pairs you select.

not only . . . but also

both . . . and

either . . . or

1. _____

2. _____

LEARNING UNIT 15-3
SUBORDINATING CONJUNCTIONS

Learning Unit 15-1 defined a *coordinating conjunction* as a conjunction that joins words, phrases, and clauses of equal rank or importance. A **subordinating conjunction** joins clauses of unequal rank, that is, an independent clause to a dependent (subordinate) clause. Dependent clauses, you will recall from Chapter 4, cannot stand alone. They depend on, and are subordinate to, the main clause.

This learning unit begins with an explanation of complex sentences—the third group of sentences. Included under this heading is a discussion about the adverb, adjective, and noun clauses that form complex sentences. You should be familiar with these clauses from your study of earlier chapters. The learning unit concludes with a brief discussion of compound-complex sentences—the fourth group of sentences.

COMPLEX SENTENCES

A **complex sentence** has one independent clause and one or more dependent clauses. Subordinating conjunctions introduce many dependent clauses. A conjunction frequently makes a clause dependent.

Dependent clauses function as adverbs (adverb clauses), adjectives (adjective clauses), or nouns (noun clauses). The first word of a clause is often a clue to the function of the clause.

Adverb Clauses

Adverb clauses are dependent clauses that usually modify verbs. Subordinating conjunctions such as those listed in Chapter 14 introduce adverb clauses. Like adverbs, these clauses answer questions such as *how?, to what extent?* (or *under what circumstances?, for what purpose?, with what result?), where?,* or *when?*

Adverb clauses appear after or before the main clause of a sentence. It is natural to write the main clause first and then the adverb clause, which ordinarily modifies the verb in the main clause. The adverb clause performs this function in the following sentences:

We listened intently *while Mr. Nichols talked.* (Remember that adverb clauses are always modifiers.)

No one was surprised *when Senator Peal proposed the legislation.*

She rejected our offer *because she could not secure the necessary financing.*

We hurried from the auditorium *as soon as the lecture had ended.*

Note that we did not use commas in the above sentences. Now let's transpose the clauses in each of the four sentences.

While Mr. Nichols talked, we listened intently.

When Senator Peal proposed the legislation, no one was surprised.

Because she could not secure the necessary financing, she rejected our offer.

As soon as the lecture had ended, we hurried from the auditorium.

The adverb clauses in the above sentences are now *introductory adverb clauses* and have been punctuated according to the following rule:

RULE 1.   Use a comma after an introductory adverb clause.

Seasoned writers sometimes do not use commas after adverb clauses if they think the sentence cannot be misread. Most writers, however, prefer to follow this rule and set off all introductory adverb clauses with commas.

Usually, you do not need a comma if the adverb clause follows the independent clause. Sometimes, however, punctuation must be used to satisfy the requirements of the following nonrestrictive (nonessential) rule:

RULE 2.   Use commas to set off nonrestrictive adverb clauses.

STUDY TIP

Nonrestrictive **(nonessential) phrases and clauses are not necessary to the meaning of the sentence and are usually set off by commas.** *Restrictive* **(essential) phrases and clauses are necessary to the meaning of the sentence and are not set off by commas.**

When a nonrestrictive adverb clause follows an independent clause, the writer probably used it as an afterthought. Because the clause is nonrestrictive, you could remove it from the sentence without materially changing the meaning of the independent clause. We have italicized the nonrestrictive adverb clauses in the following sentences:

Mr. Taft was unhappy with the results, *although we fail to understand why.*

Tell us where we should go and what we should do, *since you have all the necessary information.*

Prentice Hall's *Words into Type* states that a nonrestrictive adverb clause should be preceded by a comma; a restrictive adverb clause limiting the action of the main verb should not be preceded by a comma. Then, *Words into Type* gives the following helpful guidelines:

1. Clauses introduced by *though* or *although* are always nonrestrictive.

The students had difficulty with the test, although the answers were easy.

2. Clauses introduced by *if* are restrictive.

Publication can be hastened if the paper is made as concise as possible.

3. Clauses beginning with *because* are usually restrictive, but they may be nonrestrictive.

The accountant was pleased because the company showed a profit.

The company enforced its new restrictions, because several employees had ignored the old restrictions.

4. Clauses introduced by *unless* and *except* are usually restrictive.

The tense and voice of the verbs in a composition should not be changed unless the meaning demands it.

All may leave the meeting except the company officers.

5. Do not use a comma before a clause beginning with *before, when, while, as,* or *since* restricting the time of the action of the principal verb.

> You should make it a point to master each of these facts before you attempt to write the assignment.

6. When a clause introduced by *as, while,* or *since* does not restrict the verb but expresses cause or condition, use a comma before it.

> The president did not attend the meeting, since she had a severe cold.

> The United States was concerned with wages and prices, while Japan concentrated on productive quality and efficiency.

7. Use a comma before a clause of result (a clause that gives the result of the event discussed in the independent clause) introduced by *so that* but not before a clause of purpose (a clause that gives the purpose of the event discussed in the independent clause) introduced by *so that.*

> The owner opened all the windows, so that is why the room is cold.

> The manager posted the new positions so that all employees could apply.

Adjective Clauses

Adjective clauses appear frequently in complex sentences. These dependent clauses are introduced most often by the relative pronouns *that, which, who, whom,* and *whose* (see Chapter 13). A conjunction or an adverb such as *when, where,* and *why* may occasionally introduce adjective clauses.

If the adjective clause is needed to identify the word it modifies, the clause is *restrictive.* The clause is *nonrestrictive* if it simply supplies additional information and is not needed to identify the word to which it refers. As stated earlier, you can remove a nonrestrictive clause from a sentence without materially changing the meaning of the main clause. Note the italicized adjective clauses in the following sentences and the punctuation that has been used.

The attorney *who filed the complaint* is not available for comment. (The adjective clause is restrictive because it identifies the attorney discussed in the sentence.)

Mr. Ramsey, *who filed the complaint,* is not available for comment. (The adjective clause is nonrestrictive because it is not needed to identify Mr. Ramsey.)

The story *that the witness told* is obviously untrue. (Many writers follow the practice of using *that* to begin restrictive clauses and *which* to begin nonrestrictive clauses.)

The job requires a person *who has had sales experience.*

She returned the Apperson contract, *which must be revised.*

Our head office is in Boston, *where much history has been made.*

The auction, which will be held tomorrow in this auditorium, should attract hundreds of people.

The preceding sentences are punctuated according to Rule 2, which states nonrestrictive clauses should be set off with commas. Two commas are needed, of course, if the clause appears within a sentence; only one comma is needed if the clause appears at the end of the sentence.

Proper nouns are names, and names are used for identification; therefore, an adjective clause that follows a proper noun is usually nonrestrictive and set off with commas.

James Johnson, *who is president of our company,* can answer all budget questions.

Note that a pronoun introducing an adjective clause may be simply *implied* rather than expressed.

The plan (that) you suggested has great merit. (When the pronoun is expressed, it usually follows its antecedent [*plan* in this sentence] as shown here.)

In determining whether the word *that* is a relative pronoun or a subordinating conjunction, note that the conjunction is simply a connective, as in this sentence:

He knew that the bank was open.

The pronoun *that* usually follows its antecedent, as in this sentence:

He went to the only bank that was open.

Noun Clauses

Recall from Chapter 6 that a noun clause is not a modifier and may function as a major sentence part—subject, object, or complement. Noun clauses often begin with *that, what, which, who, whom,* and *whose.* (*Ever* can be added to most of these words, such as *whatever* and *whoever.*) As you can see in the example sentences below, noun clauses can also begin with *why.*

Usually, you cannot divide a sentence that contains a noun clause into two distinct clauses because noun clauses are part of the independent clause. Note how noun clauses are used in the following sentences:

Subject:	*What you just said* is not true.
Direct Object:	I cannot understand *why the company failed.*
Complement:	His excuse is *that he overslept.*
Object of Preposition:	She became excited about *whatever you said.*

Adverbs, pronouns, or conjunctions introduce noun clauses. They are easy to use and do not require any special punctuation.

COMPOUND-COMPLEX SENTENCES

A **compound-complex sentence** has two or more main clauses and at least one dependent clause. Following are two examples:

If you invest in this new company, you may lose your money; therefore, you should not decide hastily. (The dependent clause *if you invest in this company* is an adverb clause modifying the verb *lose*.)

He looked toward me when he entered the room, but he did not realize that I had the manuscript. (The first dependent clause, *when he entered the room*, is an adverb clause modifying the verb *looked*; the second dependent clause, *that I had the manuscript*, is a noun clause functioning as a direct object. Note that the noun clause is part of the independent clause.)

Punctuate compound-complex sentences according to the rules suggested for compound and complex sentences.

YOU TRY IT

A. Answer the following questions in the space provided.

1. What type of sentence contains one independent clause and one or more dependent clauses? _____

2. What type of conjunction is *because*? _____

3. What does a subordinating conjunction do? _____

   _____

4. What does an adjective clause modify? _____

5. What type of dependent clause answers the question *How*? _____

6. What type of sentence has two or more main clauses and at least one dependent clause?

   _____

B. Write one sentence for each of the four sentence types.

1. Simple: _____

2. Compound: _____

3. Complex: _____

4. Compound-complex: _____

C. Select two of the following subordinating conjunctions: because, unless, although, after. Write one sentence for each conjunction you select.

1. _____

2. _____

CHAPTER SUMMARY: A QUICK REFERENCE GUIDE

PAGE	TOPIC	KEY POINTS	EXAMPLES
308	**Simple sentence**	A simple sentence contains only one independent clause.	Every salesperson received a bonus.
309	**Compound sentence**	A compound sentence contains two or more independent clauses closely related in meaning.	The agent found the property, and the seller bought it.
309	**Defining coordinating conjunctions**	Coordinating conjunctions join two or more words, phrases, and clauses of equal importance or rank (*and, but, or, nor, for, yet, so*).	Both men *and* women called the officer for help.
310	**Coordinating conjunctions used in compound sentences**	**Rule 1.** Place a comma before a coordinating conjunction joining two independent clauses. **Rule 2.** The internal comma can be omitted in a compound sentence of less than ten words if the independent clauses are joined by *and* or *or*, providing no misunderstanding in the meaning of the sentence results.	Jacob found all the pens, *but* Jennie found all the pencils. The owner left and the manager stayed.
310	**Punctuating compound sentences without coordinating conjunctions**	**Rule 3.** Use a semicolon to separate independent clauses of a compound sentence that are not joined by a coordinating conjunction.	Kathy addressed the envelopes; Jim mailed them. Doris answers all messages; Eva answers only urgent messages.
312	**Use of conjunctive adverbs and transitional phrases**	Conjunctive adverbs and transitional phrases serve as modifiers and connectives. See the list in Learning Unit 15-2. **Rule.** Use a semicolon before a conjunctive adverb or transitional phrase that separates two independent clauses. A conjunctive adverb is *usually* followed with a comma. A transitional phrase is *always* followed with a comma. Set off conjunctive adverbs and transitional phrases with commas when used within sentences.	Everything went according to schedule; however, the crowd was small. The answer could be anywhere; for example, you may find it in this report. The meeting, however, was canceled.
313	**Choosing conjunctive adverbs**	Choose conjunctive adverbs carefully. They often have shades of meaning.	

PAGE	TOPIC	KEY POINTS	EXAMPLES
314	**Correlative conjunctions**	Correlative conjunctions are used in pairs to join coordinate words, phrases, and clauses. Place the pairs so the following sentence elements are *parallel in form.*	Elsie found *not only* the original *but also* several copies.
315	**Defining subordinating conjunctions**	Subordinating conjunctions join clauses of unequal rank—independent clause to dependent clause.	After the owner left, the entire staff held a meeting.
316	**Complex sentences**	Complex sentences contain one independent clause and one or more dependent clauses.	
316	**Adverb clauses**	Adverb clauses are dependent clauses that usually modify verbs. They appear after or before main sentence clause and begin with subordinating conjunctions. *Rule 1.* Use a comma after an introductory adverb clause. *Rule 2.* Use commas to set off nonrestrictive adverb clauses.	When the customer left the store, the salesperson found the keys. We are not going to buy the extra cable, since it is too expensive.
318	**Adjective clauses**	Adjective clauses are dependent clauses introduced most often by the relative pronouns *that, which, who, whom,* and *whose.* They can be restrictive and nonrestrictive.	The book *that the agent needed* was lost.
319	**Noun clauses**	Noun clauses may function as subjects, objects, complements, or appositives. They often begin with *who, whom, whose, which, what,* and *that.* Usually noun clauses and independent clauses can't be divided into two distinct clauses.	Ronald is *that person in the red jacket.*
320	**Compound-complex sentences**	Compound-complex sentences contain two or more main clauses and at least one dependent clause.	When you buy this suit, you may waste your money; therefore, think about your purchase carefully.

QUALITY EDITING

You recently bought some office supplies at the new Office Center store. Later you drafted the letter below to Mr. Myers, the manager. Edit your letter. Look for spelling errors, clause punctuation errors, and other general problems. Then, underline all the conjunctions in the letter.

Dear Mr. Myers:

Yesterday I was in the new store that you jest opened on Second Avenue and I had sevral problems. Frist the carts were not available and a clerk told me to go out side and get mine own. Second your new store does not carry the Stevens writing papper that the old store carried and I definately prefer the Stevens paper. third a clerk informed me that the Brick pens, the steel rulers, the pink erasers, the transparant tape, and the large paper clipes are on order and the warehouse will not deliver them until next week. Finally when the whele fell off the cart before I finished my shoping I left. Mr. Myers I think your new store is very atractive but you opened it to soon.

APPLIED LANGUAGE

A. Write a compound sentence using each of the coordinating conjunctions listed below. Remember: Each sentence must contain two independent clauses.

1. (for) _____

2. (yet) _____

3. (and) _____

B. Select two of the conjunctive adverbs and two of the transitional phrases from the lists below. Write a compound sentence for each of your choices.

Conjunctive Adverbs: otherwise, however, therefore

Transitional Phrases: as a result, in addition, in other words.

1. _____

2. _____

3. _____

4. _____

C. Select two of the subordinating conjunctions listed below. Write one sentence for each conjunction you select. *Subordinating conjunctions:* before, until, when, while.

1. _____

2. _____

CONJUNCTIONS

Part A. Simple and Compound Sentences

Indicate whether each of the sentences below is simple or compound by writing S or C in the space provided.

_____ **1.** Elizabeth made the suggestion; Jaime carried it out.

_____ **2.** This firm, with Ms. Yee at the helm, will do very well.

_____ **3.** The Secure Buildings representative spoke with Mr. Jamison, and then she left.

_____ **4.** Lee Grayson, Tom Fanetti, and Linda Diaz were not in their offices at three o'clock this afternoon.

_____ **5.** To reduce our overhead, we eliminated several positions in management.

_____ **6.** Both City Office Supply and Spencer's Stationery have warehouses in Bay City and Riverwood in addition to their retail stores.

_____ **7.** No one at the meeting seemed interested in the lecture on selective objectivity, but Katia and Lawrence went nonetheless.

_____ **8.** Nathan works for Mr. Brown, but Sonja supervises him.

Part B. Punctuating Compound Sentences

Add commas and semicolons to the sentences below according to the rules given in this chapter. Insert all optional commas. In the answer space, indicate the number of punctuation marks you have added. If no punctuation needs to be added, put a 0 in the space.

Number of Marks

_____ **1.** He works for Burlington but has very little seniority.

_____ **2.** JoAnn prepared the statement but Ms. Arlington did not sign it.

_____ **3.** Last week we were awarded a lucrative government contract consequently we must now hire hundreds of new employees.

_____ **4.** Our computers were made in this company our copiers were made in Japan.

_____ **5.** The alarm sounded but no one listened.

_____ **6.** Mr. Washington gave me a copy of the report and I gave him a receipt.

_____ **7.** Gasoline prices have dropped; oil prices will soon increase.

_____ **8.** Malcolm forgot the keys to the safe this morning and it was not opened until noon.

Part C. Clauses That Modify

In the space provided, write the word(s) modified by the clause in italics.

EXAMPLE:

The airplane parts *that you want* will be sold at the next auction.   **parts** _____

1. All items may be inspected *before the auction begins.*   _____

2. Smoking will not be permitted *while the auction is in progress.*   _____

3. Mr. Gee gave me a brochure *that lists the items.*   _____

4. Bidders must pay for their purchases *as soon as the auction ends.*   _____

5. The auction, *which will begin at noon,* should attract a large crowd.   _____

Part D. Using Conjunctions Correctly

In each of the following sentences, one of the choices in parentheses is correct. Circle the one that is correct.

1. I just heard (that/where) the price of silver has stabilized.

2. I plan to invest in precious metals (except/unless) you have a valid objection.

3. Please try (and/to) meet me at three o'clock.

4. The new building is just as ornate (but/and) not as functional as the old one.

5. None of us has any doubt (that/but what) Hillary will agree to act as chairperson.

6. Your sales will increase (if/whether) you advertise nationally.

Part E. Making Compound Sentences Complex

Rewrite each of the following compound sentences to make it complex.

EXAMPLE: Mr. Wilson owns five oil wells, and he has given considerable money to charity.
   Mr. Wilson, who owns five oil wells, has given considerable money to charity.

1. Ms. Grant is an excellent landscape artist, and she plans to join the Garcia Architecture firm.

   _____

   _____

2. The tractor had broken down four times within a month, so we sold it.

   _____

3. Janet Gray has a law degree, and she visited every European country on her recent trip.

   _____

4. Our computers are rather old, but they are capable of handling this job.

   _____

5. Mr. Silverman is 65 years old, and he will retire in May.

   _____

Part F. Simple, Compound, or Complex?

Indicate the type of sentence by writing *simple, compound,* or *complex* in the answer space.

1. We should write to Roger Chisholm, who works for the software developer.   _____

2. When the truck arrived, no one was here to unload it.   _____

3. The computer hasn't been programmed; therefore, I cannot start working.   _____

4. After answering my question, Ms. Ehrhart left the office.   _____

5. Mr. Pescow signed the contract, handed it to me, and breathed a sigh of relief.   _____

6. Leroy Hathaway will recommend the project, but he will not finance it.   _____

7. Ms. Chang wrote the entire speech; Mr. Plascencia proofread it.   _____

8. Three employees have been with this firm for more than 20 years.   _____

9. Monica did some filing, and the computer person repaired her computer.   _____

10. Martin asked me to meet him in the Stern Building, which is on Ninth Street.   _____

SECTION 5

SENTENCE MECHANICS; THE WRITING PROCESS

CHAPTER 16

CAPITALIZATION AND WRITING NUMBERS

> The use of capital letters is a matter of either style or convention. In situations where writers make a stylistic choice, they should be governed by the need for clarity and emphasis.
> —Louis E. Glorfeld, David A. Lauerman, and
> Norman C. Stageberg, *A Concise Guide for Writers*

LEARNING UNIT GOALS

LEARNING UNIT 16-1: CAPITALIZATION: PERSONS, PLACES, AND THINGS

- Know the basic capitalization rules. (pp. 331–36)

LEARNING UNIT 16-2: CAPITALIZATION: GOVERNMENT AND RELIGIOUS REFERENCES; OFFICIAL TITLES; FIRST WORDS

- Know capitalization rules for specific entities and word usage. (pp. 337–39)

LEARNING UNIT 16-3: WRITING NUMBERS

- Know how to write numbers as words and figures. (pp. 340–42)
- Know how to write numbers in sentences, write dates, and write ordinal numbers for days. (pp. 343–44)

The term *consistency* is frequently used when professionals discuss writing mechanics. Consistency involves making the grammatical structures and the mechanical elements of sentences agree.

The basic principle behind consistency is that inconsistency interrupts the concentration of the reader. Have you ever read a piece of writing where a word is capitalized in one sentence and in the next sentence it is not capitalized? If you noticed the inconsistency, your train of thought was interrupted.

This chapter combines two mechanical procedures. Learning Units 16-1 and 16-2 discuss capitalization. In Learning Unit 16-3, you will learn the general rules for writing numbers.

LEARNING UNIT 16-1
CAPITALIZATION: PERSONS, PLACES, AND THINGS

Recall from Chapter 6 that proper nouns are names given to *specific* persons, places, and things (including qualities, conditions, actions, and ideas) that are important enough to have a separate name. These nouns always begin with a capital letter. Common nouns, in contrast, name a *general* class of persons, places, and things. A common noun is not capitalized. Here are some examples:

Common Nouns	**Proper Nouns**
(General Designations)	*(Specific Names)*
automobile	Chevrolet
book	*Wuthering Heights*
building	Richfield Building
church	First Baptist Church
city	Milwaukee
company	General Motors
highway	U.S. Highway 101
holiday	Thanksgiving
inventor	Thomas Edison
language	French
month	January
river	Amazon River
school	Springfield High School
state	North Dakota
street	Madison Street
theater	Granada Theater
university	Fordham University
war	Civil War

You probably know many of the capitalization rules in this chapter. Concentrate on the rules that answer capitalization problems you have had in the past. We have grouped the capitalization rules in this learning unit into three general topics: names of persons, names of places, and names of things.

Some capitalization rules are easy to remember and apply. Other rules, however, involve judgment and special study. When you cannot decide whether to capitalize an expression, consult a good reference book such as *The Chicago Manual of Style*[1] or *Words into Type*.[2] In dictionary entries, words that regularly function as proper nouns begin with capital letters.

---

[1] *The Chicago Manual of Style*, 14 ed. (Chicago: The University of Chicago Press, 1993).
[2] *Words into Type*, 3rd ed. (Englewood Cliffs, NJ: Prentice-Hall, Inc., 1974).

WRITING TIP

Words beginning with capital letters attract more attention than other words in a sentence. Although advertising writers may capitalize some words to get the attention of the reader, you should not capitalize common nouns for emphasis. To call attention to a word, use italics.

NAMES OF PERSONS

We probably all automatically capitalize the names of persons. The problem that we may have is with names that have articles and prepositions (called particles) before them. The following guidelines will help you.

The second part of two-part names that begin with *Mc*, *O'*, and *St.* are capitalized. In English-speaking countries, names with particles such as *de*, *le*, *van*, *von*, and so on are usually capitalized unless the family has shown a preference for the lowercase form.

Diane Van Bakel	James O'Connor	Joy O'Toole
Henry De Vries	Jill St. John	Lynn McCahey

Family Relationships

A name that shows a family relationship is usually capitalized unless a noun or pronoun in the possessive case precedes the name.

Uncle Ralph	my *aunt* Ada	*Dad*
Mother	Morton's *brother* Paul	my *dad*
Brother David	her *cousin* Rachel	*Cousin* Mark

Descriptive Names of Persons

Capitalize descriptive names of persons that are substituted frequently for the real names such as *Honest Abe* (Abraham Lincoln). Baseball and football players frequently have descriptive names that are capitalized. For instance, the famous basketball player Michael Jordan is called Air Jordan.

Ethnic References

Capitalize the names of races and other cultures.

African American	Asian	Caucasian
Arab	Black	Hispanic

NAMES OF PLACES

Capitalize the names of countries, sections of a country, states, counties, continents, cities, streets, avenues, oceans, rivers, lakes, valleys, and other geographical terms. A good rule to follow is to capitalize places listed on maps. Here are some examples:

Avenues:	Clarmar Avenue	Rivers:	Rum River
Cities:	Chicago	Section of	
Counties:	Cook County	Country:	the South
Countries:	Canada	States:	New York
Lakes:	Lake Superior	Streets:	First Street
Oceans:	Atlantic Ocean	Valleys:	Death Valley

Also capitalize an adjective that refers to the name of a specific place.

Canadian people	Mexican baskets
Georgian antiques	Roman legions

The following sentences give some additional examples:

These microcomputers are manufactured in *New York State*.

Our ship will leave from *San Diego Harbor*.

Several textile mills in *New England* have been closed.

Property taxes are high in *Franklin County*.

BUT: His home is near the *Canadian* border. (The word *border* is not capitalized.)

Two exceptions to the capitalization of geographical terms follow.

1. **Do not capitalize a geographical term that precedes a name unless it is a part of that name.**

   Our accountant has never been in the *state* of Florida.

   He worked as an engineer for the *city* of Cleveland.

2. **Do not capitalize a plural common noun that completes the meaning of two or more proper nouns.**

   We have houseboats on Erie and Ontario *lakes*. (If the plural noun appears first, as in *Lakes Erie and Ontario*, it should be capitalized.)

Descriptive Names of Places

Capitalize descriptive names that are substituted frequently for real geographical names such as *The Windy City* (Chicago) and *The Big Apple* (New York City).

Celestial Bodies

Capitalize names of stars, planets, and constellations. Do not capitalize the words *sun, moon, star,* and *planet.*

Venus	Jupiter	Mars
Milky Way	Big Dipper	Halley's Comet

Points of the Compass

Capitalize the names of the points of the compass (*north, south, east, west,* and their derivatives) only when they name regions. Do not capitalize points of the compass when they give directions.

These silicon chips were produced in the *East*.

Buffalo is *west* of Albany.

We plan to visit *Western* Europe.

His territory is in the *northern* part of the state.

Her property is located on the *south* shore of the lake.

Our engineer was born in the *Midwest*.

NAMES OF THINGS

Under this broad group, we have included brand names; course titles and degrees; days, months, and seasons; committees, departments, and divisions; historical events or eras; noun-number designations; and miscellaneous compound proper names.

Brand Names

Capitalize brand names and trademarked names. (A manufacturer or an advertiser is likely to capitalize the common noun following the actual brand name.)

Bic pen	Decca records	Maxwell House coffee	Palmolive soap
Coca-Cola	Ford automobile	Ovaltine	Vaseline

If words derived from proper nouns have lost their specialized meanings, they should not ordinarily be capitalized. Do not capitalize words such as the following:

arabic numbers	china dishes	roman type	plaster of paris
brussels sprouts	italics	pasteurize	watt

Course Titles and Degrees

Capitalize the names of areas of study only if they are derived from proper nouns. Note these examples:

accounting	French	history	shorthand
English	German	mathematics	typing

Joy received an A in history and an A in economics.

Capitalize the names of specific courses.

She is enrolled in *Introduction to Data Processing 21A*.

Mr. Edward Stokes teaches *Business Mathematics II*.

Capitalize academic degrees when they follow a name. Do not capitalize academic degrees when used in a general sense.

John Chung, Bachelor of Science	Joseph Johnson, Ph.D.
Loren Stone, M.D.	Nora Johnson, Bachelor of Arts

All students receiving their bachelor of science degrees applied for this job.

Days, Months, and Seasons

Capitalize the names of holidays, days of the week, and months of the year.

Independence Day	Monday	August
Labor Day	Saturday	December

Capitalize the name of a season or the word *nature* only if it is strongly personified, that is, spoken of as if it were human.

Old Man Winter left a foot of snow on our parking lot.

In the *spring* we will purchase a word processor.

Business is slow during the *summer* months.

Committees, Departments, and Divisions

Capitalize committees, departments, and divisions of your organization. Committees, departments, and divisions of other organizations are usually not capitalized.

Give these receipts to our *Accounting Department.*

Send your application to the company's personnel department.

Historical Events or Eras

Capitalize the names of important historical events or eras in history and the names of well-known political policies. Here are some examples:

the Civil War	the Louisiana Purchase	New Deal
the Dark Ages	the Monroe Doctrine	Prohibition

Noun-Number Designations

In general, capitalize a noun that is followed by a number or letter that identifies a unit or division. Most writers do not capitalize the noun, however, if the unit or division is a minor one such as page, paragraph, line, or verse reference. Note these examples:

Lot 14, Tract 7243	Volume III, Chapter 8, page 133
Catalog No. 212F	Exhibit A
Policy No. 347501	Building 5, Room 12
Section 16	paragraph 3
Account No. 220	line 15
verse 21	Appendix C

Miscellaneous Compound Proper Names

All words except articles (*a, an,* and *the*), conjunctions, and short prepositions are usually capitalized in names or titles that consist of more than one word. Always capitalize the *first* word and the *last* word. Do not capitalize *the* if it precedes the name of an organization but is not a part of that organization's name. Capitalize any word with four or more letters in a compound proper noun.

The Rise and Fall of the Roman Empire (book title)

the *Eastman Kodak Company* (business organization)

Supreme Court (judicial body)

City Council (governing body)

the *Ways and Means Committee* (standing committee)

The documentary film, *The First Ship-to-Shore Broadcast*, was well received. (Capitalize the parts of a hyphenated word in a title as you would capitalize them without hyphens.)

the *Big Board* (New York Stock Exchange)

YOU TRY IT

A. Six of the nouns in the following list should be capitalized. Write them correctly in the space provided.

minneapolis	city	halloween	buick
building	college	washington	ohio river
car	stanford	river	state

1. _____   4. _____

2. _____   5. _____

3. _____   6. _____

B. Ten of the nouns in the following list should be capitalized. Write them correctly in the space provided.

robert redford	nurse	maureen o'hara	ray garza
aunt sylvia	my cousin	honest abe	hispanic
lake erie	asian	professor	orange county
river	comet	jupiter	orange

1. _____   6. _____

2. _____   7. _____

3. _____   8. _____

4. _____   9. _____

5. _____   10. _____

C. Six of the nouns in the following list should be capitalized. Write them correctly in the space provided.

colgate	italics	english	history	maria klimek, m.d.
tuesday	month	spring	volume iii	world war ii

1. _____   4. _____

2. _____   5. _____

3. _____   6. _____

LEARNING UNIT 16-2
CAPITALIZATION: GOVERNMENT AND RELIGIOUS
REFERENCES; OFFICIAL TITLES; FIRST WORDS

This learning unit gives rules for capitalizing government references, the official titles of rank and public office, religious references, and various first words. You probably already know most of the rules for capitalizing first words.

GOVERNMENT REFERENCES

Capitalize the names of governmental agencies and departments. If the word *union* or *commonwealth* refers to a specific government, it should be capitalized. Do not capitalize the words *government*, *federal*, *nation*, or *state* unless these words appear as part of an official name or are used in official government communication.

Department of the Interior   House of Representatives
Interstate Commerce Commission   New Jersey Legislature
Springfield City Council   Orange County Board of Supervisors

We should get authorization from the *Department of Housing and Urban Development.*

He has served the *Commonwealth* for 30 years. (*Commonwealth* refers to the state of Massachusetts.)

A brilliant young attorney will represent the *government.*

He is no longer a member of the *House.* (Capitalize short forms such as the *House,* the *Bureau,* and the *Department* when they designate national or international bodies.)

OFFICIAL TITLES OF RANK AND PUBLIC OFFICE

The following four rules are important for capitalizing titles of rank, position, and public office:

1. **Titles preceding names.** Unless a comma intervenes, capitalize a title that precedes a name; generally, however, do not capitalize a title that follows a name.

> I have never met *Congressman* Nelson. (BUT: I have never met our congressman, Herbert Nelson.)

> Rita is *Dr.* Grossman's dental assistant.

> Alice Matson, *treasurer* of our organization, will report on our financial status.

2. **High-ranking national, state, or international officials.** An exception to Rule 1 is to capitalize the title of a high-ranking national, state, or international official when it follows a name or serves as a substitute for that name.

> The *President* will return to Washington next week Monday.

> The dinner is in honor of Elizabeth, *Queen* of England.

> We were present when the *Governor* signed the bill.

> BUT: Ask the *lieutenant* if he knows Lois Pitts, the *councilwoman*.

3. **Titles that follow company officials or replace personal names.** Do not capitalize titles of company officials when they follow or replace personal names.

> Both the *president* and the *chairperson of the board* are out of town. (Writers of internal communications, however, are likely to capitalize such titles when they prepare minutes of meetings, bylaws, etc.)

4. **Title in address or after signature.** Capitalize a title that appears either in the address or after the signature in a letter.

> Mr. William Covert, *District Manager*
> Shannon Machine Tool Company
> Centerville, MO 63633

RELIGIOUS REFERENCES

Capitalize any reference to a supreme being, even a personal pronoun that refers to a supreme being but has no antecedent in the same sentence. Capitalize also the name of a religion, its members, and its buildings.

Jehovah	the Holy Spirit	St. Mary's Episcopal Church
the Almighty	Allah	Temple Judea
God	Catholics	Islam

> They expressed their thanks unto *Him.*

> Do you attend the *Nazarene* church on Seventh Avenue?

FIRST WORDS

Capitalize the following first words:

1. **A complete sentence.**

   The computer was programmed by Maria Sanchez.

2. **A direct quotation that could function as a complete sentence.**

   She said, "*Please* take care of the matter at once."

   BUT: He said that Mae's weakness is "*her* inability to write well." (Quoted portion could not function as a sentence.)

3. **A word or a phrase that substitutes for a complete sentence.**

   Yes, of course. When?

4. **The salutation of a letter.**

   Dear Ms. Rosberg:

5. **The complimentary close of a letter.**

   Sincerely yours,

   Yours respectfully,

6. **An independent clause following a colon that expresses a formal rule or provides a thought that deserves special emphasis.**

   He forgot this rule: Use a dash before a word that sums up a preceding series. (Capitalize the first word of a formal rule or principle.)

   He concluded with these words: "The odds are now in our favor." (Capitalize the first word of a quoted sentence.)

   We apparently have the same goal: to increase profits. (This is a phrase and should not be capitalized.)

   This is my opinion: We must accept the loan on their terms or face possible bankruptcy. (Because the important thought in this sentence follows the colon, emphasize the thought by capitalizing the word *We.*)

   Gena spent her time on two projects: she sorted the mail and proofread your manuscript. (The second clause, which simply explains the first one, does not deserve special emphasis.)

   Note: If a single word precedes a colon, capitalize the first word of a sentence that follows the colon.

YOU TRY IT

Use proofreaders' marks to indicate which words should begin with a capital letter.

EXAMPLE: do not forget to send the book to dr. wilson.

1. the wilderness society works closely with the national parks service.

2. uncle ned will visit the smithsonian institution, which is located in washington, d.c.

3. mario gomez has a new position with the government.

4. when will professor lee's course, business accounting 102, be offered?

5. the first lady made another presentation to the department of health and human services.

6. james murphy is a catholic priest assigned to st. gregory's.

7. the report was written by tim jefferson.

8. michael redfeather asked: "who will attend the conference?"

9. the letter began, "dear dr. stein."

10. the telephone operators agreed on the title for the convention: voice recognition information services.

Answers: 1. The Wilderness Society, National Parks Service 2. Uncle Ned, Smithsonian Institution, Washington, D.C. 3. Mario Gomez 4. When, Professor Lee's, Business Accounting 5. The First Lady, Department of Health and Human Services 6. James Murphy, Catholic, St. Gregory's 7. The, Tim Jefferson 8. Michael Redfeather, Who 9. The, "Dear Dr. Stein," 10. The, Voice Recognition Information Services

---

LEARNING UNIT 16-3
WRITING NUMBERS

Finding two authorities that agree completely on how to write numbers is difficult. Writers usually vary their procedures according to the nature of a communication and the emphasis desired.

Business writers often express numbers as figures and not words. They want the figures to stand out because readers of business documents are vitally interested in these figures.

Today's business writers usually favor the use of figures. Most of the ideas in this learning unit are for business correspondence. *All* numbers used in statements, invoices, vouchers, check registers, sales slips, and the like should ordinarily appear as figures.

NUMBERS WRITTEN AS WORDS

1. **Isolated numbers from *one* to *ten*.** Generally spell out isolated numbers from *one* to *ten*. When the exact time is to be emphasized, use numerals with zeros for even hours.

The discussion lasted for *ten* minutes.

We will be gone for *six* days.

The program will begin at *eight* o'clock. (The word *eight* is spelled out before o'clock; the figure *8* would be used before A.M. or P.M.)

You can listen to the televised program at 7:00 P.M.

2. **Numbers of one or two words.** Indefinite numbers of one or two words are usually spelled out. For emphasis, figures can be used.

> The auditorium will seat several *thousand* people.
>
> She will probably retire when she is in her early *seventies.*
>
> We received *hundreds* of responses to our advertisement.

3. **Number introducing a sentence.** Spell out a number that introduces a sentence. If the number is long, recast the sentence to avoid awkwardness.

> AVOID: *21* of the instruments had been damaged.
>
> USE: *Twenty-one* of the instruments had been damaged.
>
> AVOID: *471* people were in the auditorium.
>
> USE: The people in the auditorium numbered *471.*

4. **Short common fractions.** Spell out short common fractions that are used alone.

> He refused to accept his *one-fourth* share.
>
> More than *one-third* of us accepted the challenge. (A writer should, however, use figures in writing a mixed number: *We offered him a discount of 2 1/2 percent.*)

5. **Formal invitations and announcements.** Spell out all numbers in formal invitations and announcements.

> . . . on Sunday, the *twenty-seventh* of February . . .

6. **Two consecutive numbers.** If no punctuation mark occurs between two consecutive numbers, write out the shorter number to avoid confusion.

> We plan to build 24 *twelve*-room residences in this tract.
>
> He asked for *fourteen* 79-cent pens.
>
> How long will it take you to read these *32 fifteen*-page reports?

7. **Legal documents.** In legal documents, express amounts of money in both words and figures.

> . . . Four Hundred Fifty Dollars ($450)

8. **Ordinal numbers.** In general, spell out an ordinal number (first, second, third, etc.) that can be expressed as one word. *Exceptions:* Use figures for all dates and for street names above *Tenth.*

> We are starting our *fortieth* year of operations.

> He lives on *Seventh* Street in Oklahoma City.

> Our office is on the corner of 32nd Street and *First* Avenue.

> Many of these changes took place in the *nineteenth* century.

> Our head buyer will be in New York on the *2nd* (or *2d*) of June.

9. **Isolated round numbers.** In isolated round numbers, spell out million or billion to make reading easier.

> The winning candidate received a plurality of 1.2 *million* votes.

> This tax legislation would increase revenue by $7 *billion.*

NUMBERS WRITTEN AS FIGURES

1. **Generally use figures to express exact numbers above *ten*.**

> The Novikoff file has been missing for *30* days.

> She distributed *1,325* copies of the bulletin.

> This booklet contains *100* suggestions for improving public relations.

> Our firm recently purchased *13* Canon copiers.

2. **The following are usually expressed in figures.**

   a. **Amounts of money (except *one cent*).**

> You can purchase the knife for only *89* cents. (In expressing amounts less than a dollar, generally write out the word *cents.* Do not use a decimal, such as $.*49*, except to achieve uniformity in a series of numbers.)

> The irate customer asked for a refund of her *$49.50.*

> We have only *$25* in petty cash. (Do not use a decimal point and two ciphers when an isolated figure expresses an even amount in dollars.)

   b. **Market quotations.**

> The common stock in question closed yesterday at *25 7/8.*

   c. **Dimensions.**

> Our smallest boxes are *6* by *8* by *3* inches. (When several dimensions are given, as in an invoice, express them like this: *6″ × 8″ × 3″.*)

d. Degrees of temperature.

The temperature in our freezer is now *5* degrees Fahrenheit (or 5° F).

e. Decimals and percentages.

Wanda typed the sentence in *4.25* seconds.

We have already used *35* percent of the sheet metal delivered last Monday. (In scientific and statistical material, the percent sign often follows the figure.)

Please add these figures: *2.92, 3.74,* and *5.12.*

The average annual return on our investment was a meager *.09* percent.

We spent only *0.8* percent of our gross sales for advertising. (If a zero follows the decimal point [*0.08* percent], the zero before the decimal point can be eliminated.)

f. Street numbers.

Dr. Carbo lives at *126* Willow Street; Ms. Epstein lives at *8* Wickfield Court; and Professor Lund lives at *One* Shoreview Drive (*one* is an exception to the use of numbers).

g. Pages and divisions of a book or periodical.

The picture appears in Volume *8*, page *214.*

h. Time (hour of the day) when A.M. or P.M. follows (abbreviations may also be expressed in lowercase letters: a.m., p.m.).

The class was in session from 10:00 A.M. until 1:00 P.M. (When exact time is to be emphasized, use zeros with even hours. Note the following: *We talked from 9:30 A.M. until 11:00 A.M.*)

i. Weights and measures.

This oil drum, which weighs exactly *18* pounds, will hold only *10* gallons.

j. Identification numbers.

The accident on Route *5*, according to the news commentator on Channel *7*, resulted in several deaths.

SPECIAL SITUATIONS

1. **Several numbers in a sentence.** If several numbers in a sentence perform similar functions or are related, express them as you

would express the largest number. If this number is over *ten*, write all numbers in figures.

> The inventory shows *21* ranges, *9* refrigerators, *37* washers, and *10* dryers.

> Our disposers sell for *$55.00*, *$69.50*, *$79.50*, and *$95.00*. (Note the consistent use of decimal points and ciphers to achieve uniformity and clarity.)

> BUT: The *32* tables sold in five days. (The numbers do not perform similar functions.)

2. **Writing dates.** Generally use figures in writing dates.

> We plan to meet on September *7, 1995*. (If the year follows the day, do not add *st*, *nd*, *rd*, or *th* to the figure designating the day.)

> The office was closed on August *15*. (Since the year was not expressed, August *15th* would be equally acceptable.)

3. **Ordinal number designating the day.** When an ordinal number is used to designate the day, it should be expressed as a figure.

> Business has been poor since the *9th*.

> We have not seen him since the *26th* of July.

> Call me again on the *15th* of November.

YOU TRY IT

Write the following numbers in words or figures in the spaces provided. If the number is already correct, write *correct*.

1. 3 o'clock _____

2. twelve days _____

3. 6 days _____

4. several hundred _____

5. 14 24-piece sets _____

6. Seventh Street _____

7. six dollars and 45 cents _____

8. eight percent _____

9. 406 7th Street _____

10. Volume 9, page 38 _____

Answers: 1. three 2. 12 3. six 4. correct 5. fourteen, 24 6. correct 7. $6.45 8. 8 or 8.0 9. 406 Seventh 10. correct

CHAPTER SUMMARY: A QUICK REFERENCE GUIDE

PAGE	TOPIC	KEY POINTS	EXAMPLES
332	**Names of persons**	1. **Family relationships** are capitalized unless a noun or pronoun in the possessive case precedes the name. 2. **Descriptive names** are capitalized when substituted for real names. 3. **Ethnic references** to races, tribes, and other cultures are capitalized.	her *niece* Joan; *my* sister Mr. Cub (Ernie Banks) Sioux
332	**Names of places**	Capitalize names of countries, sections of a country, states, counties, and other geographical terms. Also capitalize an adjective referring to a specific name. 1. Descriptive name of places 2. Celestial bodies 3. Points of the compass	New York City, *city* of New York Alaskan pipeline The Big Apple (New York City) the Great Bear We plan to visit the *South.*
334	**Names of things**	Capitalize: 1. Brand names 2. Course titles and degrees 3. Days, months, and seasons 4. Committees, departments, etc. 5. Historical events or eras 6. Noun-number designations 7. Miscellaneous compound proper names	 Saran Wrap Managerial Accounting 101 Labor Day, Tuesday, April our Finance Committee the New Deal Lot 224 Empire State Building
337	**Government references**	Capitalize names of governmental agencies and departments. Do not capitalize *government, federal, nation,* or *state* unless part of official name.	United States Supreme Court the *state* government
337	**Official titles of rank and public office**	1. Generally capitalize titles that precede names but not those following names. 2. Capitalize the title of a high-ranking national, state, or international official when it follows a name or serves as a substitute. 3. Do not capitalize titles of company officials when they follow or replace personal names. 4. Capitalize a title in an address or after a signature in a letter.	Superintendent Smith Did you listen to what our *Vice President* said? Gale Swenson, vice president of Walker Company Sincerely, John Hutchenson, Professor of Forestry
338	**Religious references**	Capitalize reference to supreme being, religion, its members, and its buildings.	King of Kings, Seventh-day Adventists, Baptist members, Catholic Church

PAGE	TOPIC	KEY POINTS	EXAMPLES
339	**First words**	Capitalize first word of (1) complete sentence, (2) direct quotation that could function as complete sentence, (3) word or phrase that substitutes for complete sentence, (4) salutation of a letter, (5) complimentary closing of a letter, and (6) independent clause following a colon that deserves special emphasis.	The superintendent went home. Jack said, "Look, I found the report." You said? Dear Ms. Noel: Yours truly, Every person in the room turned around: The old president stood up and greeted his replacement.
340	**Numbers written as words**	1. Spell out isolated numbers from one to ten. 2. Spell out indefinite numbers of one or two words. 3. Spell out a number introducing a sentence. 4. Spell out short common fractions that are used alone. 5. Spell out all numbers in formal invitations and announcements. 6. If no punctuation mark between two consecutive numbers, write out the shorter number. 7. In legal documents, express money in words and figures. 8. Spell out an ordinal number expressed as one word except use figures for dates and streets above *tenth*. 9. Spell out isolated round numbers.	All *four* chairs were painted. She made about a *hundred* calls. *Twenty-four* checks are missing. The file cabinet is *one half* that size. . . .Monday, the *fifth* of January. . . Get me *five* 29-cent stamps. . . .Six thousand Dollars ($6,000). . . You will find the library on *Sixth* Street. The company spent over 2.5 *million* on research.
342	**Numbers written as figures**	1. Use figures to express numbers above *ten*. 2. Use figures for: **a.** Money except *one cent* **b.** Market quotations **c.** Dimensions **d.** Degrees of temperature **e.** Decimals and percentages **f.** Street numbers **g.** Pages and divisions of a book or periodical **h.** Time **i.** Weights and measures **j.** Identification numbers	The loading department bought *25* carts. He paid *$2,000* for that computer. Your stock dropped to *12½*. Use racks that fold to *5* by *12* inches. The thermometer read *101* degrees Fahrenheit in the shade. All tables were based on *29* percent capacity. The address is *1020* Lincoln Avenue. The artwork was moved to Chapter *9*. She saw the worker leave the building at *3:00 P.M.* The package weighs *15* pounds. Listen to the news on Channel *5*.

PAGE	TOPIC	KEY POINTS	EXAMPLES
343	**Special situations**	**1.** If several numbers in a sentence perform similar functions, express them as you would the largest number.	The *5* chairs, *20* glasses, and *22* plates were sold at the auction.
		2. Generally use figures for dates.	Gordon will not be here again until January *1, 1996.*
		3. Express an ordinal number used to designate the day as a figure.	Every book will be sold by the *16th* of June.

QUALITY EDITING

Your friend Lisa has applied for a job as an accounting assistant at Sun Foods Company. The personnel officer at Sun asked Lisa to write a short biography of her relevant education and experience. Lisa prepared the draft below and has asked you to edit it. Be sure to watch for errors in capitalization and spelling, and check the format of any numbers.

Lisa kwon recieved her b.a. in business Economicks from south Point univeristy in june, 1988. She graduated with a grade point average of three point two and Her department selected lisa for the "studnet of the Year Award."

After Graduating, Lisa acepted a position as an acounts payable clerk with the Grant produce Company on 9th Street in lake Bennett City. Lisa's work at this Company included cheking aproximatly 20 invoices each day to make sure the charges were valad. She logged the invoices in the computer. Each friday she wrote the checks, about one hundred and twenty checks including the payroll. Then Lisa totled the week's spenditures and gave her Supervisor a report. The average weekly amount spent was 25 thousand dollars.

APPLIED LANGUAGE

1. Write a letter to the Cripp's Catalog Office Supplies mail order house ordering some yellow tablets, black marking pens, and an 18-inch ruler. Their address is 118 South Parkway Drive, Laguna Niguel, CA 92677. Be sure to indicate how many of each item you are ordering and the cost of each item. Include the company's address, the address to which the supplies should be sent, and a closing. See the appendix if you need guidelines for formatting your letter.

2. Imagine that you have just conducted a survey in a shopping mall. You interviewed 200 shoppers. Write a short report of your findings answering these questions: (a) How many interviews were conducted? (b) What is the average number of stores that people go to when they shop at the mall? (c) What percentage of the shoppers buy fast food in the mall? (d) What percentage of the shoppers eat in a restaurant in the mall? (e) What percentage of the people come to the mall only to see a movie?

3. You work for Theodore's Travel Agency. Write a letter to a customer to describe your agency's special tour of the month. Include the following information in your letter.

the name of the city

the name of the lake the city is near

what river flows into the lake

what hotel will be used

a well-known museum in the city

the price of the tour package

WORKSHEET 16

Name _____

Date _____

CAPITALIZATION AND WRITING NUMBERS

Part A. Capitalization Rules

Indicate whether each statement is true or false by writing T or F in the space provided.

_____ **1.** The names of seasons are never capitalized.

_____ **2.** A compound proper noun generally consists of more than one word.

_____ **3.** This capitalization is correct: the Ohio and Columbia rivers.

_____ **4.** Both words should be capitalized in the expression North Africa.

_____ **5.** The course title "Business Accounting 101" should be capitalized.

_____ **6.** The second part of a two-part name that begins with Mc, O', or St. is capitalized.

_____ **7.** Only capitalize names of races, tribes, and other cultures at the beginning of a sentence.

_____ **8.** Always capitalize academic degrees.

_____ **9.** Do not capitalize supreme court.

_____ **10.** The basic rule of capitalization is to capitalize proper nouns but not common nouns.

Part B. Capitalization of Proper Nouns

Write a sentence for each item below. Use the word given in parentheses. Note that five of the words are capitalized and should, therefore, be used as proper nouns.

1. (building) _____

_____

2. (Building) _____

_____

3. (lake) _____

_____

4. (Lake) _____

_____

5. (professor) _____

_____

6. (Professor) _____

_____

7. (street) _____

_____

8. (Street) _____

_____

9. (company) _____

_____

10. (Company) _____

_____

Part C. Writing Numbers

Each of the following sentences contains one or more numbers in parentheses. For each number in parentheses, write the number in words or figures according to the rules you have learned. Then briefly state the rule that applies to the numbers. Use your own words.

EXAMPLE: At (7) this morning, Janet left for (3) days.

| **a. (7)** | 7:00 | **Use figures if a time is not followed by o'clock.** |
| **b. (3)** | three | **Spell out numbers from one to ten.** |

1. (12) of the participants spoke for about (15) minutes.

a. (12)   _____   _____

_____

b. (15)   _____   _____

_____

2. The temperature in Fairbanks, Alaska, on January (5, 1987), was (2) degrees.

a. (5, 1987) _____   _____

_____

b. (2)   _____   _____

_____

3. Nearly (1/2) of the group answered more than (90) percent of the (120) questions.

a. (1/2)   _____   _____

_____

b. (90)   _____   _____

_____

c. (120)   _____   _____

_____

4. Sandra purchased (12) (16)-page booklets.

  a. (12)   _____   _____

    _____

  b. (16)   _____   _____

    _____

5. We think the (2) computer will cost ($999).

  a. (2)   _____   _____

    _____

  b. (999)   _____   _____

    _____

Part D. More Numbers

Assume that each of the following items appears in a sentence with no other numbers. If the form used is not acceptable, write the preferred form. If it is acceptable, write correct.

1. Volume Ten, page 62   **1.** _____

2. four hundred dollars   **2.** _____

3. in her 60s   **3.** _____

4. Twentieth Street   **4.** _____

5. exactly twelve ounces   **5.** _____

6. nine percent   **6.** _____

7. the last ten days   **7.** _____

8. .05 percent   **8.** _____

9. a 1/3 share   **9.** _____

10. half the proceeds   **10.** _____

11. January second   **11.** _____

12. 3:30 P.M. or 4 P.M..   **12.** _____

13. fourteen members   **13.** _____

14. $543.21   **14.** _____

15. 1 Beach Road   **15.** _____

CHAPTER 17

COMMAS

Keep in mind at all times that the primary purpose of the comma is to prevent misreading.

—*Words into Type*

LEARNING UNIT GOALS

LEARNING UNIT 17-1: BASIC RULES FOR COMMAS THAT SEPARATE SENTENCE ELEMENTS

- Correctly use commas to separate items within sentences. (pp. 354–56)
- Correctly use commas to separate introductory words, phrases, and adverb clauses. (pp. 356–57)

LEARNING UNIT 17-2: BASIC RULES FOR COMMAS THAT SET OFF SENTENCE ELEMENTS

- Correctly use commas to set off parenthetical expressions. (pp. 357–58)
- Correctly use commas to set off nonrestrictive expressions. (pp. 358–61)
- Use the basic comma rules in your writing. (pp. 357–61)

LEARNING UNIT 17-3: MISCELLANEOUS COMMA RULES; USING COMMAS TO PREVENT MISREADING

- Understand four miscellaneous comma rules. (pp. 362–63)
- Correctly use commas to prevent misreading. (pp. 363–64)

Since punctuation can change the meaning of a sentence, every writer should know the basic punctuation rules. Even within the framework of these punctuation rules, some sentences or paragraphs may be correctly punctuated in more than one way. Often the punctuation chosen adds force to the writing and reveals the precise relationship of thoughts.

Commas are separators. They prevent readers from running together parts of sentences that should not be read together. If you are not sure about using a comma, read the sentence aloud. Will the use of the comma in your sentence help the reader understand your thought? You will find that usually you should follow the fundamental comma rules in this chapter.

The comma guidelines in this chapter are divided into three learning units. Learning Unit 17-1 explains the basic rules for commas that *separate* sentence elements. In Learning Unit 17-2, we explain the basic rules for commas that *set off* sentence elements. Learning Unit 17-3 gives some miscellaneous comma rules and some comma rules to prevent misreading.

LEARNING UNIT 17-1
BASIC RULES FOR COMMAS THAT SEPARATE
SENTENCE ELEMENTS

This learning unit begins by reviewing the use of a comma to separate compound sentences (Chapter 15). You probably already know and follow the second comma rule—the rule for separating items in a series. Rule 3 discusses how to use the comma to separate independent adjectives. Rule 4 explains the use of the comma to separate introductory elements.

COMPOUND SENTENCES

The following comma rule on compound sentences combines Rules 1 and 2 in Chapter 15.

RULE 1.   Place a comma before a coordinating conjunction that joins two independent clauses. The internal comma can be omitted in a compound sentence of less than ten words if the independent clauses are joined by *and* or *or*.

$$
\textbf{Independent Clause} \atop \text{Subject + predicate} \quad + \quad \begin{bmatrix} and \\ but \\ for \\ or \\ nor \\ so \\ yet \end{bmatrix} \quad = \quad \textbf{Independent Clause} \atop \text{Subject + predicate}
$$

The accident occurred in a busy intersection, but no one said they saw it. (The main clauses of a compound sentence could serve as separate sentences.)

The final session came to a close, and the legislators left the building.

He walked and she rode. (You do not need a comma in a short compound sentence joined by *and* or *or*.)

STUDY TIP

Sentences with a compound predicate connected by a coordinating conjunction are sometimes mistaken for a compound sentence. The following sentence only has one subject and cannot be a compound sentence: *We reported our findings to Ms. Gentry and went on to our next job.*

ITEMS IN A SERIES

RULE 2.   Use commas to separate three or more words, phrases, or clauses in a series.

She impressed us with her poise, understanding, and honesty. (Commas separate nouns in a series. Note that to avoid confusion, a comma precedes the conjunction *and*.)

Elaine did her work quietly, quickly, and satisfactorily. (Commas separate adverbs in a series.)

George swims, reads, and plays golf. (Commas separate verbs in a series.)

The young worker was enthusiastic but uninformed, energetic but untrained, willing but not able. (Commas separate pairs of words in a series.)

We cannot complete this project in an hour, in a day, or even in a week. (Commas separate phrases in a series.)

Madelaine favored Model D, Milton favored Model A, but the company purchased Model B. (Commas separate short clauses in a series.)

Three *don'ts* on the use of the series comma follow:

1. **Do not use a comma if a conjunction joins only two words or phrases.**

   Mr. Lewis *and* his partner are excellent appraisers.

   We plan to go to Concord *or* to Manchester on Tuesday.

2. **If a conjunction is used between each item in a series, commas are not necessary.**

   John asked her to wait *and* observe *and* reconsider.

3. **If an ampersand (&) is used in a company name, do not use a comma before the ampersand.**

   Haynes, Cord & Staley is a reliable company.

INDEPENDENT ADJECTIVES BEFORE A NOUN

RULE 3.   Use commas to separate independent adjectives that appear before a noun and modify the noun equally.

This rule was given in Chapter 13. Sometimes it is difficult to decide if independent adjectives modify the noun equally. The following two tests were given in Chapter 13: (1) If you can logically place the word *and* between two adjectives, consider them to be independent adjectives and use a comma. (2) If you can logically reverse the order of the adjectives, consider them to be independent adjectives and use a comma. Note the following sentences:

Police officers do *dangerous, exciting* work. (We could say *dangerous* and *exciting* work or *exciting, dangerous* work.)

These five old manual typewriters need repair. (Since it doesn't make sense to say *five* and *old* and *manual*, the adjectives are not of equal rank.)

INTRODUCTORY WORDS, PHRASES, AND ADVERB CLAUSES

RULE 4.   Place a comma after introductory words, phrases, and adverb clauses that introduce an independent clause.

Introductory Words

Commas follow introductory words such as *yes, no, well, oh,* and so on. Commas usually follow **conjunctive adverbs** that occur at the beginning of a sentence. A list of the common conjunctive adverbs was given in Chapter 15.

Yes, we foresee a bright future for the company. (Always separate the words *yes* or *no* at the beginning of a sentence.)

Consequently, we must return the order immediately. (A comma usually follows an introductory conjunctive adverb.)

Introductory Phrases

Commas follow introductory transitional phrases (Chapter 15), introductory prepositional phrases of four or more words (some writers use commas after three-word prepositional phrases), and introductory verbal phrases.

In other words, the merger should prove profitable to both companies. (The introductory transitional phrase relates this sentence to a preceding thought.)

In January we will visit our facilities in Atlanta. (A comma after *January* is not needed to prevent misreading.)

For several seconds the speaker just stared at his audience. (A comma after *seconds* is not needed to prevent misreading.)

To make matters worse, Mr. Belnick was unable to program the computer. (Introductory verbal phrase needs a comma.)

Looking to the future, he opened a savings account. (Introductory verbal phrase needs a comma.)

In his writing, Dr. Shipley did not mention the present crisis. (Some writers prefer to place a comma after a short prepositional phrase containing a verbal.)

A verbal used as a subject is, of course, an essential part of a clause and should not be set off with a comma.

To satisfy my employer is not easy. (Verbal phrase is the subject.)

Worrying will not help matters. (Verbal, *worrying,* is the subject.)

Introductory Adverb Clauses

Transposed introductory adverb clauses should be followed by a comma (Chapter 15).

Before data processing technology came into use, clerks sorted checks by hand.

If we advertise in trade magazines, our sales will increase.

As James read the fine print in the contract, his hand began to shake.

YOU TRY IT

Insert commas into these sentences following the rules given in this learning unit. If no commas are needed, write *correct* after the sentence.

1. The machinery jammed while it was running but no one knew why.

2. The accountants entered the conference room and the meeting began.

3. He wrote the report and she presented it.

4. Eleanor collected the information analyzed it and wrote the report.

5. Chan is a computer expert and he provides statistical consulting to our researchers.

6. The applicant seemed to be intelligent hard-working and reliable.

7. Ray planned to leave on Monday or Tuesday.

8. My attorney is Martin Appleby of Appleby Barucha & Chin.

9. Before leaving Anna Black Horse taught Joey some carpentry.

10. If we make some product improvements our customers will be pleased.

Answers: 1. running, but 2. room, and 3. correct 4. information, analyzed it, and 5. expert, and 6. intelligent, hard-working, and 7. correct 8. Appleby, Barucha 9. leaving, Anna 10. improvements, our

---

LEARNING UNIT 17-2
BASIC RULES FOR COMMAS THAT SET OFF
SENTENCE ELEMENTS

You may have heard the saying, "Two commas or none." This means that commas that set off internal nonrestrictive (nonessential) sentence elements come in pairs. These comma pairs set off parenthetical expressions and nonrestrictive expressions. They also set off dates and addresses.

PARENTHETICAL EXPRESSIONS

RULE 5. Use commas to set off parenthetical expressions (words, phrases, and clauses) that interrupt the flow of the sentence.

A parenthetical expression is an interrupter that occurs in the middle of a sentence and interrupts its natural flow. Conjunctive adverbs and transitional phrases (such as those listed in Chapter 15) are set off when they are used as parenthetical expressions within sentences.

Note how parenthetical expressions interrupt the flow of thought in the following sentences:

If you attempt to sell your home now, *however*, you may experience some difficulty.

He was told, *it seems*, that no more money would be available.

Ms. Handley, *by the way*, has several excellent ideas for improving our service.

The new engine, *together with a small box of tools*, will be shipped to you Wednesday.

Our company has, *I believe*, the industry's most efficient blast furnace.

Adverb clauses may be used parenthetically. If such a clause interrupts the flow of a sentence, it should be set off with commas.

Several of us, *because we had heard the report*, expressed concern about the future.

We realized, *as the vote clearly indicated*, that Julia Denton was well regarded by her colleagues.

NONRESTRICTIVE EXPRESSIONS

Restrictive (essential) expressions, you will recall from Chapter 15, are necessary for the meaning of the sentence and are not set off by commas. Commas are needed, however, to set off nonrestrictive (nonessential) words, phrases, and clauses.

RULE 6.   Use commas to set off nonrestrictive appositives, verbal phrases, adverb clauses, and adjective clauses.

Nonrestrictive Appositives

Recall that an appositive is a noun or noun equivalent that renames, explains, or identifies the noun or noun equivalent it follows. Most appositives are nonrestrictive and should be set off with commas.

Nonrestrictive: Our company's most valuable account, *H. F. Rousseau & Sons*, will soon begin producing a new sonar device.

Nonrestrictive: Charles P. Langdon, *our leading candidate for mayor*, is a brilliant business person. (The appositive may include a phrase.)

Restrictive: The word *politician* has more than one meaning. (As an appositive, *politician* renames *word*. Since it is necessary for the meaning of the sentence, it is not set off by commas.)

Restrictive: My cousin *Wayne* plans to enter the advertising field. (Writer identifies Wayne as a cousin.)

Nonrestrictive Verbal Phrases

In Learning Unit 17-1, you learned that introductory verbal phrases are usually nonrestrictive and should be followed by a comma. When verbal phrases appear elsewhere in a sentence (not at the beginning), you must decide whether the phrase is restrictive or nonrestrictive. Nonrestrictive verbal phrases should be set off with commas.

Nonrestrictive: Grace Munson, *having worked in a law office*, answered my question. (The verbal phrase modifies the noun *Grace Munson* and is not necessary for the meaning of the sentence.)

Nonrestrictive: This right front wheel, *obviously mounted improperly*, is the source of your problem. (The verbal phrase modifies the noun *wheel* and is not necessary for the meaning of the sentence.)

Restrictive: The student *helping to address envelopes* is a volunteer. (The verbal phrase is needed to identify the noun *student*.)

Restrictive: The material *to be destroyed* has absolutely no value to the company. (The verbal phrase is needed to identify the *material*.)

Restrictive: Loren purchased the transistor radio *made in Brooklyn*. (The verbal phrase is needed to identify the *radio*.)

Nonrestrictive Adverb Clauses

Usually, an adverb clause following a main clause is restrictive and should not be set off with a comma. If, however, an adverb clause is only loosely connected to the main clause, consider the clause nonrestrictive and set it off with a comma. You will find that the idea expressed in a nonrestrictive clause could be expressed in another sentence.

We did not reach our quota of 550 sales in December, *although we made a determined effort*. (Clauses introduced by *though* or *although* are always nonrestrictive.)

The company is being forced into bankruptcy, *as you have probably guessed*. (You can see from the meaning of the sentence that the clause is nonrestrictive.)

We find it advisable to vote against Mr. Lane, *however experienced he may be*. (This clause is nonrestrictive because it is an added thought.)

You will be well pleased *when you see the new advertising layouts*. (This clause is restrictive because it is needed for the meaning of the sentence.)

Claudia Hudson will be seriously considered for promotion *if her fine work continues*. (Clauses introduced by *if* are restrictive.)

Joseph Strand talked with several eyewitnesses *before he entered his plea*. (This clause is restrictive because it restricts the time of the action of the verb.)

We hesitated *because several lives were at stake.* (Clauses beginning with *because* are generally restrictive.)

You can see from the above examples that usually you will have no difficulty in determining whether an adverb clause is restrictive or nonrestrictive. Often a natural pause comes before an adverb clause that should be set off with a comma.

Nonrestrictive Adjective Clauses

Recall from Chapter 15 that an adjective clause may be restrictive or nonrestrictive. If an adjective clause is needed to identify the word(s) modified, the clause is *restrictive* and never set off with commas. However, adjective clauses that only add information are nonrestrictive and should always be set off with commas.

Experienced writers usually introduce restrictive adjective clauses with the pronoun *that*, and they use a comma and the pronoun *which* to introduce nonrestrictive adjective clauses. Note the nonrestrictive and restrictive clauses in these sentences:

Nonrestrictive: Mr. Toland, *who will head our delegation*, is a born leader. (The nonrestrictive clause simply adds information and is not needed for identification.)

Nonrestrictive: This ornate building, *which has 28 stories and 950 offices*, was built with federal funds. (The adjective *this* identifies the building. The clause gives additional information.)

Nonrestrictive: We should meet in this office Saturday, *when no work will be in progress.* (An adjective clause introduced with *when* or *where* is generally nonrestrictive. *When* is more often used in adverb clauses.)

Nonrestrictive: The three telegrams, *all of which were mailed in Columbus*, told of her good fortune. (Adjective clauses introduced with *all of which, each of whom*, and similar expressions are always nonrestrictive.)

Restrictive: The mistake *that you made* can be corrected easily. (It is not natural to pause before reading a restrictive clause.)

Restrictive: The person *who meets all your qualifications* will never be found. (Note that the restrictive clause identifies the person being discussed.)

Restrictive: The person *whose property I damaged* treated me with kindness and understanding. (If the restrictive clause were omitted, the reader would not know what person the writer had in mind.)

Adjective clauses that follow proper nouns are usually nonrestrictive. Occasionally, however, you will see a construction such as the following:

I am talking about the Albert Jones who works in your office.

The adjective clause in such a sentence is restrictive because the name (Albert Jones) does not adequately identify the person under discussion.

DATES AND ADDRESSES

RULE 7. Use commas to set off the items in a date or an address.

On May 14, 1994, we celebrated our tenth anniversary. (Note that a comma follows the year.)

The historic meeting took place at 10:00 A.M. on Tuesday, July 2, 1990, in the Blackstone Hotel.

The Springfield, Illinois, trip occurred last January. (Names of states are set off by commas.)

You can send to Jamerson Company, 1800 Ridge Road, Chicago, Illinois 60655, for a replacement part. (Do not put a comma between the state and the zip code.)

All the inquiries occurred during May 1994. (Preferred style is not to use a comma between the month and year when no day is given.)

Dates in military correspondence generally appear in this form: *14 August 1990* (no comma needed).

Y O U   T R Y   I T

Insert commas into the following sentences according to the rules given in this learning unit. If no commas are needed, write *correct* after the sentence.

1. If you read the report however you will find that the company is losing money.

2. The lectures together with the textbook information will prepare you for the test.

3. The supervisor because she observed the incident was asked to testify.

4. The price of our best-selling product the dinosaur egg will increase in June.

5. Hilda's staff member Jeremy Burns will attend the conference in her place.

6. The school's main library Talbot Library will be closed for the holiday.

7. The books to be returned were not appropriate for her children.

8. Jim Kelly will receive the award if his next painting is as powerful as his others.

9. Alicia Garcia de Lopez who is chair of our department is a skillful administrator.

10. The Englewood Colorado conference was held on May 18 1993.

Answers: 1. report, however, you 2. lectures, together; information, will 3. supervisor, because; incident, was 4. product, the; egg, will 5. member, Jeremy Burns, will 6. library, Talbot Library, will 7. correct 8. correct; 9. Lopez, who; department, is 10. Englewood, Colorado, conference; May 18, 1993.

LEARNING UNIT 17-3
MISCELLANEOUS COMMA RULES; USING
COMMAS TO PREVENT MISREADING

This learning unit begins with a discussion of four miscellaneous comma rules. Then, you will be given ten suggestions to help prevent misreading.

MISCELLANEOUS COMMA RULES

The four miscellaneous comma rules explain how to use commas (1) with words of direct address, names, and titles; (2) after the salutation of an informal letter and the complimentary close of a letter; (3) to separate quoted words from other words; and (4) with various numbers.

Direct Address, Names, and Titles

RULE 8.   Use commas to separate and set off words of direct address, names, and titles.

Direct Address.   Note how commas are used with nouns of address in the following sentences:

> *Christine*, please hand me the general ledger. (Noun of address is followed by a comma when it occurs at the beginning of the sentence.)

> Look, *Adam*, this statement is false. (Noun of address as a parenthetical expression is set off with commas.)

> Has it occurred to you, *my friend*, that some people may disapprove?

Names and Titles.   A comma is used after a surname when the given name and surname are reversed.

> Walton, David   Candelaria, Anthony

Abbreviations that follow names to indicate titles, degrees, and the like are set off in commas.

> Sidney K. Vaccaro, *Sr.*, is not a licensed broker. (Do *not* use commas, however, if Mr. Vaccaro writes his name without them.)

> Foster & Mencken, *Inc.*, will not supply the zinc. (Do *not* use commas if the official company name omits them.)

> Thomas D. Souchak, *Ph.D.*, began his brilliant writing career in 1978.

> Sal Fensky, Jr.'s resourcefulness saved the company from ruin. (Do *not* place a comma after the possessive form Jr.'s or Sr.'s.)

Letters

RULE 9.   Use commas after the salutation of an informal letter and the complimentary close of a letter.

> *Dear Susan*, (A colon is used after the salutation of a business letter.)

> *Sincerely yours*,

Best regards,

Very truly yours,

Quoted Words

RULE 10.   Use commas to separate quoted words from the other words in a sentence.

"Our fiscal year will soon end," she said.

He asked, "Where are we going to store the 500 cartons?"

"Where," he asked, "are we going to store the 500 cartons?" (Note that two commas are needed in this sentence.)

"Where is Ms. Thompson?" he asked. (Observe that a comma is not used in a sentence of this type because the quoted portion is a question.)

Numbers

RULE 11.   Use commas to separate three digits of long numbers and the digits of unrelated numbers. Do not use commas for numbers such as telephone numbers, address numbers, page numbers, and serial numbers.

Long Numbers.   Start at the right of the number and insert commas between every three digits.

Sales last year amounted to *$156,476,500.*

Unrelated Numbers.   Separate unrelated numbers in a sentence by a comma.

Of the 150, 24 employees received bonuses. (It is understood that the *150* refers to *150 employees.*)

Telephone, Address, Page, and Serial Numbers.   These numbers are not separated by commas.

Telephone: (201) 555-7539 (Area code is in parentheses.)
Address: 2416 Pascal Avenue
Page: page 2631 or p. 4233
Serial: 47-2345-1214569

COMMAS TO PREVENT MISREADING

RULE 12.   Use commas when they are needed to prevent misreading.

The following constructions clearly suggest the need for punctuation even though none of the first 11 rules apply:

a. Adjectives that follow the nouns they modify.

> Our accountant, *tense and exhausted,* finally located the error.

b. Contrasting expressions that begin with *not*, *never*, or *seldom*.

> She told me to use a hard disk, *not a diskette.*

> His lectures are always given in English, *never in Spanish.*

c. Brief questions appended to preceding statements.

> You did telephone Mr. Krause, *didn't you?*

d. Long phrases that begin with *of* and indicate a person's residence or business affiliation.

> Richard Melikoff, *of General Dynamics in Pomona, California,* questioned our ability to meet the deadline.

> Richard Melikoff *of General Dynamics* expressed his views forcefully. (Most writers today do not set off short phrases of this kind.)

e. A word repeated for emphasis.

> It seems that all she ever does is *work, work, work.*

> It may be a *long, long* time before we meet again.

f. Identical verbs that come together in a sentence.

> Most of my friends who *read, read* only for pleasure.

g. Words that can easily be misread with words that follow.

> For a *while,* longer skirts were in vogue.

h. A long introductory phrase or combination of phrases (unless the verb in the main clause immediately follows).

> At that particular moment in the life of poor Reverend Hempstead, nothing seemed real.

> In the latest edition of *The Wall Street Journal* is an article about our competitor. (BUT: In the latest edition of *The Wall Street Journal,* I found this article about our competitor.)

i. The omission of important words (if sentence could be easily misread).

> The person who comes in first will be given $500; the person who comes in second, $300; the person who comes in third, $100.

j. Clauses built on contrast.

> The longer I spoke with the professor, the more inspired I became.

YOU TRY IT

Insert commas into the following sentences according to the rules given in this learning unit.

1. Andrew has the report from the building inspector arrived?

2. Have you as the person responsible made the report to the proper authorities?

3. Ask Loretta Stevens D.D.S. about the dental insurance she accepts.

4. Dear Cousin Bea Have you heard from Tom Burroughs?

5. The manager said "Please finish the analysis by Friday."

6. "Who," asked Martin "won the employee-of-the-month award?"

7. The speaker nervous and tired spoke without energy.

8. All our accountant does is analyze analyze analyze.

9. Susan Jacobs analyst for the Department of Health and Human Services in Washington presented her report to a congressional committee.

10. Professor Shin will you be commenting on the architecture of the new bridge?

Answers 1. Andrew, has **2.** you, as; responsible, made **3.** Stevens, D.D.S., about **4.** Bea, Have **5.** said, "Please **6.** Martin, "won **7.** speaker, nervous; tired, spoke **8.** analyze, analyze, analyze **9.** Jacobs, analyst; Washington, presented **10.** Shin, will.

CHAPTER SUMMARY: A QUICK REFERENCE GUIDE

PAGE	TOPIC	KEY POINTS	EXAMPLES
354	**Compound sentences**	**Rule 1.** Place a comma before a coordinating conjunction that joins two independent clauses. The internal comma can be omitted in compound sentences of less than ten words if clauses are joined by *and* or *or*.	Our manager needed the sales figures for the meeting, but the accountant said the figures were lost.
355	**Items in a series**	**Rule 2.** Use commas to separate three or more words, phrases, or clauses in a series. Do not use a comma if **1.** The conjunction joins only two items. **2.** A conjunction is used between each item in a series. **3.** An ampersand (&) is used in a company name.	The company cafeteria serves salads, sandwiches, pizza, and hot dishes. James bought a new computer and printer. She found the knife and fork and spoon. Graham, Slater, Johnson & Newman Company will help you.
355	**Independent adjectives before a noun**	**Rule 3.** Use commas to separate independent adjectives that appear before a noun and modify the noun equally.	The new employee had worked in a closed, sheltered environment.

PAGE	TOPIC	KEY POINTS	EXAMPLES
356	**Introductory words, phrases, and adverb clauses**	**Rule 4.** Place a comma after introductory words, phrases, and adverb clauses that introduce an independent clause.	*Yes*, the conference was canceled. *On the contrary*, we must answer all questions. *In every situation*, the worker received support from others. *Before talking*, listen to me. *If we all cooperate*, the job will soon be finished.
357	**Parenthetical expressions**	**Rule 5.** Use commas to set off parenthetical expressions (words, phrases, and clauses) that interrupt the flow of the sentence.	You can understand, *of course*, why you must complete the work today.
358	**Nonrestrictive expressions**	**Rule 6.** Use commas to set off nonrestrictive appositives, verbal phrases, adverb clauses, and adjective clauses.	Our favorite customer, *Mrs. Curtis*, gave the store the painting. Loren, *having experience in trusts*, found this trust inadequate. You can use this paper, *although it may be too heavy*. Mr. Wang, *who is a favorite with customers*, will help you.
361	**Dates and addresses**	**Rule 7.** Use commas to set off the items in a date or an address.	Last *May 12, 1994*, the contract was signed. Look for Lindal Corporation, *1427 Snelling Ave., St. Paul, Minnesota 55113*, to send the samples.
362	**Direct address, names, and titles**	**Rule 8.** Use commas to separate and set off words of direct address, names, and titles.	You will find, *James*, that the answer is in this book. Jeffery Curtin, Ph.D., came to visit us.
362	**Letters**	**Rule 9.** Use commas after the salutation of an informal letter and the complimentary close of a letter.	*Dear Kimberly,* *Sincerely,*
363	**Quoted words**	**Rule 10.** Use commas to separate quoted words from the other words in a sentence.	"The room," he said, "needs painting."
363	**Numbers**	**Rule 11.** Use commas to separate three digits of long numbers and the digits of unrelated numbers. Do not use commas for numbers such as telephone numbers, address numbers, page numbers, and serial numbers.	23,406,200 The 400, 1995 delegates will all have seats. (613) 623-1092; 1428 Woodland Avenue; p. 6200; 84-23-60159

PAGE	TOPIC	KEY POINTS	EXAMPLES
363	**Commas to prevent misreading**	**Rule 12.** Use commas when they are needed to prevent misreading. **a.** Adjectives that follow the nouns they modify. **b.** Contrasting expressions beginning with *not, never,* or *seldom.* **c.** Brief questions appended to preceding statements. **d.** Long phrases beginning with *of* and indicating a person's residence or business affiliation. **e.** A word repeated for emphasis. **f.** Identical verbs that come together in a sentence. **g.** Words that can be misread with words that follow. **h.** Long introductory phrases or combination of phrases (unless verb in main clause immediately follows). **i.** Omission of important words. **j.** Clauses built on contrast.	The chairperson, *nervous and anxious,* continued the meeting. This is the correct book, *not that one.* The room is cold, *don't you think?* James Lund, *of the Apex Corporation in Grand Rapids, Michigan,* came to the office. This is a *big, big* day. In this department, those who *sing, sing* professionally. For every *worker, John* found a helper. *In every situation found by the managers and the supervisors,* the answer was difficult to comprehend. Jacob was given four tickets; *Dawn,* three tickets; and *Ralph,* nine tickets. The longer I looked at the picture, the more I wanted it.

QUALITY EDITING

Your colleague, Eva, is writing a training manual to upgrade the writing skills of entry level employees in your company. To emphasize the use of commas, she typed the paragraph below without commas. The printer will write them in, in red. Now Eva needs your help. Insert the missing commas in red for the printer to follow. Also look for faulty capitalization, end-of-sentence punctuation, and spelling. Make these corrections in a different color.

most people it seems find puncuation rules difficult to understand and as a result dificult to aply. Some rules of course are more chalenging than others? Most people can easy recognize items in a Series but the same people may have truoble with berbal frases or non-restrictive clauses. if you sincerely want to learn the rules you can. After you have studied the Comma rules you should start thinking in terms of those rules evry time you write. When you prepare letters memos reports or any other doucment keep the rules in mind and apply them. If you are unsur of yourslef look at the rools again? study the exampels of adverb clauses nonrestrictive elements parenthical expressions and compound sentences. Learn to recognize them at a glance. If you aply a rule it will become more familiar to you. Eventtualy all 12 rules will be yours to commandand writing will be less chalenging.

APPLIED LANGUAGE

Complete each of the following sentences. Be sure to include material that requires using at least two commas.

1. Fred's outstanding characteristics _____

_____

2. Yesterday I called Yoshi Tanaka _____

_____

3. Next week Ms. Bhanda will go to Chicago _____

_____

4. To be sure that the papers reach us _____

_____

5. On top of my desk was _____

_____

WORKSHEET 17

Name _____

Date _____

COMMAS

Part A. Comma Rules

Indicate whether each of the following statements is true or false by writing a T or F in the space provided.

_____ **1.** A series of clauses should never be separated by commas.

_____ **2.** There is only one way to punctuate a sentence correctly.

_____ **3.** The comma can be omitted in a compound sentence of less than ten words if the independent clauses are joined by *and* or *or*.

_____ **4.** Use commas to separate three or more words, phrases, or clauses in a series.

_____ **5.** If a conjunction is used between each item in a series, commas are not necessary.

_____ **6.** Do not use a comma after introductory words or phrases that introduce an independent clause.

_____ **7.** The following sentence is correctly punctuated: To satisfy my supervisor is not easy.

_____ **8.** Commas usually follow conjunctive adverbs, such as *consequently*, that occur at the beginning of a sentence.

Part B. Commas That Separate Sentence Elements

In the space provided, indicate which rule (or rules) justifies the punctuation used in the sentence. Use these letters (note that more than one rule can apply):

A: Items in a series

B: Two independent clauses joined by a coordinating conjunction

C: Introductory verbal or verbal phrase

_____ **1.** To facilitate the work, he purchased a desktop computer.

_____ **2.** We ordered a fax machine and a cellular telephone, but they have not arrived.

_____ **3.** This cordless telephone has a built-in speaker, a two-way intercom, and a security system.

_____ **4.** Mr. White Feather was pleased with the new clerk's performance, and he told him so.

_____ **5.** In making the presentation, he changed a few details, but Mr. Inge did not notice.

_____   **6.** Yesterday I called on Jose, apologized for the late shipment, and sold him three more cartons of computer paper.

_____   **7.** To be honest, I enjoyed the product demonstration, but it lasted too long.

_____   **8.** Encouraged by the recent increase in sales, we hired another assistant.

Part C. Commas That Set Off Sentence Elements

Insert commas as appropriate in the following sentences. Then, justify your decisions by writing the correct letter in the space provided, as follows:

A: No commas are needed

B: Commas are used to set off a nonrestrictive clause

C: Commas are used to set off an introductory adverb clause

D: Commas are used to set off a parenthetical expression

_____   **1.** This fax machine which is now on sale would serve our needs.

_____   **2.** The cellular phone that Celia lost must be replaced before Friday of next week.

_____   **3.** Unless the office remains open on Saturday we will not be able to meet their deadline.

_____   **4.** The customer is certain that Mr. Sabo is the claims processor with whom she spoke yesterday.

_____   **5.** This car stereo system which has several outstanding features comes as standard equipment.

_____   **6.** Their new contract administrator will I understand meet with our attorneys in Houston.

_____   **7.** Before our senior financial analyst recommends an investment he spends many hours studying the available options.

_____   **8.** A more efficient tax accountant it has been said could have saved the company many hours studying the available options.

_____   **9.** The materials that were lost on the subway can be replaced easily.

_____   **10.** If we had the right applications software we could complete this project before noon.

_____   **11.** Adam Smith who wrote *The Wealth of Nations* contributed greatly to the field of economics.

_____   **12.** You may discover of course that this grade of lumber will not serve your purpose.

_____   **13.** Because a bear market seemed to lie ahead we chose not to invest our funds.

_____   **14.** Our logical course of action it seems is to return the faulty merchandise and request the return of our check.

_____   **15.** Usha Abdul who was acting as temporary chairperson of our group showed excellent understanding of parliamentary procedure.

Part D. Miscellaneous Comma Rules

Indicate whether each of the following statements is true or false by writing a T or F in the space provided.

_____   **1.** Expressions used in direct address must always be set off with *two* commas.

_____   **2.** Contrasting expressions that begin with *not, never,* or *seldom* use a comma.

_____   **3.** Commas are not needed to separate adjectives when they follow the noun they modify.

_____   **4.** When an abbreviation such as *Jr.* or *Ed.D.* follows a name, a comma should never be placed after the abbreviation.

_____   **5.** Restrictive elements in a sentence should not generally be set off with commas.

Part E. Using Commas in Sentences

Insert commas as appropriate in the following sentences.

1. Let me know Luis if you decide to sell your cellular telephone.

2. The more I read about information processing the more I am inclined to accept your findings.

3. The lost items three videocassette tapes will be replaced won't they?

4. For example Renee Dubois does almost nothing but talk talk talk.

5. Lena Madorski the new technician came to us from IBM not Honeywell.

6. You will note gentlemen that all parts of the model have been painted black.

7. In fact our chief engineer Carlos Washburn assures that no corrosion will occur.

8. "The present sales campaign" Allen Hill declared "will prove to be an outstanding success."

9. In January our assets totaled $21174510.

10. Before long letters of this type will be mailed to all customers who still owe us money.

CHAPTER 18

SEMICOLONS AND COLONS, END PUNCTUATION, AND DASHES

In the writing . . . of good English, there is a silent partner: punctuation.

—John Simon

LEARNING UNIT GOALS

LEARNING UNIT 18-1: SEMICOLONS AND COLONS: BASIC RULES

- Correctly use basic rules for semicolons and colons. (pp. 374–78)

LEARNING UNIT 18-2: SEMICOLONS AND COLONS: SPECIAL USES

- Correctly use the special rules for semicolons and colons. (pp. 378–80)

LEARNING UNIT 18-3: PERIODS, QUESTION MARKS, AND EXCLAMATION POINTS

- Correctly use periods, question marks, and exclamation points. (pp. 380–83)

LEARNING UNIT 18-4: DASHES

- Correctly use dashes. (pp. 384–86)

Correct sentence structure and correct punctuation make the meaning of a sentence clear to the reader. A knowledge of sentence structure makes it possible for writers to form the habit of punctuating correctly as they write. This reduces the time spent in rewriting. Good sentences are easy to punctuate. Wilson Follett said, "The workmanlike sentence almost punctuates itself."

In Chapter 17, you studied the weakest punctuation mark, the comma. This chapter begins with a discussion of semicolons and colons in Learning Units 18-1 and 18-2. Learning Unit 18-3 discusses periods, question marks, and exclamation points. Then, in Learning Unit 18-4, you will learn how to use dashes.

LEARNING UNIT 18-1
SEMICOLONS AND COLONS: BASIC RULES

Semicolons are stronger separators than commas and weaker than periods. They usually separate equal grammatical sentence elements. **Colons** are introducers that point to something that follows. This learning unit explains the basic rules for using semicolons and colons.

SEMICOLONS

Semicolon Rules 1 and 2 are similar to the rules for punctuating compound sentences without coordinating conjunctions given in Chapter 15. Semicolon Rules 3 and 4 explain the use of semicolons for clarity.

Semicolons That Join Independent Clauses

SEMICOLON RULE 1.   Use a semicolon to separate two or more related independent clauses that are not joined by a coordinating conjunction.

> Our Akron plant supplied the raw materials; our Los Angeles branch made the finished product.

> Cambridge is near Boston; it is not a part of that city.

> Certain situations point to the individual as superior to the group; other situations point to the group as superior.

WRITING TIP

Semicolon Rule 1 is important. If you use a comma between two independent clauses instead of a semicolon, you commit the error of a *comma splice*, which means you splice two independent clauses together with only a comma. If you use no punctuation between two independent clauses, you commit the error of a *run-on sentence.*

> **David likes the green chair, I like the blue chair. (A comma splice.)**

> **David likes the green chair I like the blue chair. (A run-on sentence.)**

SEMICOLON RULE 2.   Use a semicolon to separate independent clauses followed by conjunctive adverbs or transitional phrases. Use a comma after the conjunctive adverb or transitional phrase.

The meeting has been in progress for two hours; however, the chairperson has several more important subjects to introduce.

The Federal Housing Administration has appraised the property at $300,000; therefore, we do not feel that Clark's offer is reasonable.

Production at our Louisville plant came to a standstill during the recent strike; nevertheless, we managed to earn a small profit for the fiscal year.

He tried to make amends; that is, he offered me the use of his car.

Monday is a national holiday; hence, we will have a three-day weekend. (The comma after *hence* is optional. Some writers do not use the comma after words of one syllable such as *hence*, *thus*, or *then*. The comma, of course, results in a pause. Be consistent in whatever method you choose.)

Note: Recall from Chapter 17 that conjunctive adverbs and transitional phrases sometimes appear in sentences that are not compound. Note the absence of semicolons in these sentences:

The company needs some new leadership, *that is*, a more knowledgeable president.

Roy Tortelli, *in other words*, is not well suited to his job. (The expression *in other words* has been used parenthetically.)

Semicolons Used for Clarity

SEMICOLON RULE 3.   Use a semicolon before a coordinating conjunction that separates two independent clauses when the clauses contain internal commas.

Mr. Woodson, as you may have heard, gave $10 to our cause; but Mr. Manley gave nothing.

His determination, his courage, and his sincerity could not be denied; but his methods, as you know, were often questioned by the people.

SEMICOLON RULE 4.   Use a semicolon to separate items in a series when the items themselves contain commas.

Readers are confused when they see commas *within* items in a series and also *between* the series items. For clarity, use semicolons between groups of series items separated by commas.

We will present 20-year service pins to Kay Witham, an economist; Bert Kaplan, an accountant; Sally Covello, a teller; and Calvin Haynes, a loan officer.

The four most important dates in our company's history are January 12, 1888; August 2, 1905; September 25, 1922; and December 1, 1960.

We were told that Mr. Renko, Ms. Hayden, and the others would act favorably on our application; that Gene Poston, our chief engineer, would receive full credit for his invention; and that our company would soon begin producing this revolutionary new product.

COLONS

Since the basic usage of colons is to direct a reader's attention to what follows, this learning unit begins with the rule for using colons after formal introductions. Colons are also used to introduce formal quotations (Colon Rule 2) and to explain or amplify a statement in an independent clause (Colon Rule 3).

COLON RULE 1.   Use a colon after a formal introduction to the words that follow, such as *the following* and *as follows*. These words can be stated or implied.

We hope to open a branch office in each of the following states: Nebraska, Wisconsin, Arizona, and Colorado.

Only three pieces of furniture remain in our showroom: a sofa, an end table, and a bookcase. (In this sentence *the following* is implied: "Only *the following* three pieces. . . .")

I have one serious objection to your project: it may prove to be very expensive. (This sentence could read: "I have *the following* serious objection. . . .")

Mr. Wong distributed the money collected during the campaign as follows: $2,000 for clerical help, $5,000 for postage, $7,000 for stationery and printing, and $87,500 for the underprivileged.

Do not place a colon after a verb or a preposition that is followed by a series.

Our equipment will undoubtedly include a tractor, a large crane, and a tree spade. (Do not place a colon after *include*.)

The company president sent copies of the memorandum to the production supervisor, purchasing manager, and engineering designer. (Do not place a colon after *to*.)

If the items in a series following a verb or a preposition are on separate lines, however, use a colon. Also, capitalize the first word of each item.

At tomorrow's meeting we will probably discuss:

1. The new union contract

2. Our revised office procedures

3. Our declining production figures

COLON RULE 2.   Use a colon to introduce a formal quotation.

Ms. Olson used these exact words: "Remember that we must make a cumulative profit and loss statement."

The new president expressed his philosophy in this way: "Make every customer a satisfied customer."

All workers said the "wrong tool was used" to tighten the bolts. (No punctuation is needed because the quotation is part of the sentence.)

When introductory words such as *He said* or *The agent replied* precede a short direct quotation, a comma usually follows the verb. Note the following example:

The accountant said, "You are correct."

If the quotation is long (at least two sentences) or is attributed to something inanimate, as in *The last line stated* or *The final paragraph read*, most writers use a colon.

In concluding his speech, Senator Vance said: "The young people of today will be our nation's leaders of tomorrow, and in their hands will rest the destiny of the greatest republic that this world has ever known. We are confident that they will prove equal to the monumental task that lies before them."

The report concluded: "We strongly recommend that you take the appropriate action before the end of this fiscal period." (Conclusion of the report names something inanimate or lifeless.)

COLON RULE 3. Use a colon after an independent clause when an explanation or amplification of the clause follows.

He expressed one minor complaint: his office needs more file cabinets. (Do not capitalize the first word of an independent clause that follows a colon if the clause simply explains or illustrates the idea.)

She claimed that only one kind of work would ever satisfy her: acting. (The use of a comma or a dash would be informal.)

He has forgotten this rule: Use commas to separate items in a series. (If you want to emphasize a sentence that follows a colon, use a capital letter.)

Capitalizing after a Colon

Use a capital letter if the information following a colon meets any of these criteria:

1. It is a *quoted sentence*. (See the example on page 376.)
2. It requires *special emphasis*, possibly because it states a rule or principle. (See the example on this page.)
3. It consists of *two or more complete sentences*.

The new schedule has two distinct advantages: It allows the technicians more time for research. It provides the sales team with excellent incentives. (Note that both sentences following the colon begin with capital letters.)

4. It *begins on a new line*, perhaps as a list. (See the example on page 376.)

5. It is *introduced with a single word* such as Note or Remember.

Note: The superintendent will delay the audit until Monday.

YOU TRY IT

A. Insert a semicolon or colon as appropriate in the following sentences. If neither a colon or semicolon is required, write *correct* after the sentence.

1. Wanda Roguri, our chief teller, is responsible for the error.

2. Tom Saunders accidentally deleted the report file from the computer nonetheless, we were able to reconstruct the report in time for the meeting.

3. Sheila Wyzynski and Associates made the arrangements for the conference meeting and exhibit rooms, lodging and meals, speaker contracts, agenda and schedules, and audiovisual equipment.

4. Our computers are the most recent Macs our printer is a LaserJet.

5. The atmosphere at Tea and Books, as you may know, is very relaxed but it is not open on weekends.

B. Each of the following sentences contains a semicolon or a colon. Review the rules in this learning unit to find the one that applies to each sentence. Then, in the space provided, write a brief justification for the use of the semicolon or colon. Use your own words.

1. The Hayes Co.'s offices are located in Stamford, Connecticut; Stockton, California; Austin, Minnesota; and Santa Fe, New Mexico.

_____

_____

2. A major U.S. automaker uses the following mission statement in its advertising: "Quality is Job One."

_____

_____

3. The history of the firm is linked to its founder: Sarah James.

_____

_____

Answers: **A.** 1. correct 2. computer; 3. conference; 4. Macs; 5. relaxing; **B.** 1. When items in a series contain commas, use a semicolon to separate them. 2. Use a colon before a formal quotation. 3. Use a colon after an independent clause to introduce an explanation of the clause.

LEARNING UNIT 18-2
SEMICOLONS AND COLONS: SPECIAL USES

This learning unit begins by discussing a special use of the semicolon that is similar to the use of the colon. Then, you will learn about five special uses of the colon.

SPECIAL USES OF SEMICOLONS

SEMICOLON RULE 5.   Use a semicolon before an expression that is used to introduce an enumeration, such as *for example (e.g.), that is (i.e.), for instance,* or *namely.* Place a comma after the expression.

Kristen used a few terms that the new cost estimator may not understand; for example, *duplexing, control character,* and *crossfooting.*

Our products have sold well in four New England states; namely, Massachusetts, Rhode Island, Connecticut, and New Hampshire. (Some writers prefer to use a colon in a sentence of this type.)

If the enumeration appears within a sentence, not at the end, use a pair of dashes or parentheses.

A few of the older employees—namely, Mae Scott, Ralph Pafko, and Fern Genovese—disapprove of these innovations. (The use of dashes places emphasis on the names.)

A few of the older employees (namely, Mae Scott, Ralph Pafko, and Fern Genovese) disapprove of these innovations. (Words that appear in parentheses are de-emphasized.)

The semicolon is a useful mark of punctuation if used sparingly. Don't allow your mastery of the semicolon to tempt you into writing long, unwieldy sentences. Don't use a semicolon when a period would be more logical or when the clauses that follow should be subordinated to the main idea.

SPECIAL USES OF COLONS

COLON RULE 4.   Use a colon after the salutation of a business letter.

Dear Mr. Ames: Ladies and Gentlemen:

COLON RULE 5.   Use a colon to separate the hours from the minutes when expressing time in figures.

10:45 A.M.   2:20 P.M.

COLON RULE 6.   Use a colon to separate a title from a subtitle.

American Business Enterprise: The Role of Retailers

COLON RULE 7.   Use a colon in footnotes and bibliographies if the accepted style suggests such usage.

Footnotes and bibliographies have more than one accepted style. Whatever style you choose, be consistent. Following is an example of an accepted style:

Krause, Frederick. *Investments,* 5th ed. (Englewood Cliffs, NJ: Prentice Hall, 1994).

COLON RULE 8.   Use a colon to separate the chapter from verse in a biblical reference or the parts of a mathematical ratio.

Genesis 2:8 $10:2 = 30:x$

YOU TRY IT

A. Insert colons as needed in the items below.

1. Dear Professor Kerry

2. The title of the book is *Corporations and Their Leaders A Twentieth Century Story.*

3. 1237 A.M.

4. 520 A.M.

5. Dear Mr. Stone and Ms. Bachurski

6. Who wrote *The Fifth World Music, Art, and Literature*?

7. Solve the following formula for x: $455 = 17x$.

8. T. McCarthy, Gunther, W., and Hoffman, C. *The Nursing Assistant Acute and Long-Term Care.* (Englewood Cliffs, NJ Regents/Prentice Hall, 1994).

B. Each of the following sentences contains a colon. Review the rules in this learning unit to find the one that applies to each sentence. Then, in the space provided, write a brief justification for the colon. Use your own words.

1. 12:15 A.M.

_____

_____

2. Dear Sirs:

_____

_____

3. *The Amazing Dr. Carlisle: His Medicine and His Magic?*

_____

_____

Answers: A. 1. Kerry: **2.** Leaders: **3.** 12:37 **4.** 5:20 **5.** Bachurski: **6.** World: **7.** $45:5 = 17:x$ **8.** *Assistant:* and NJ: **B. 1.** Use a colon between the hour and the minutes in indicating a time. **2.** In business correspondence, use a colon after the salutation. **3.** Place a colon between the main title and the subtitle of a work.

LEARNING UNIT 18-3
PERIODS, QUESTION MARKS, AND
EXCLAMATION POINTS

The punctuation marks discussed in this learning unit are all used to end sentences. Each punctuation mark, however, has other important applications.

PERIODS

RULE 1.   Use a period after a statement, a command, or a request.

You can use this software with your Atari computer.

Check every figure on this invoice.

Please let me know if you hear from the agency.

Would you please hand me the pen. (Although this sentence is worded as a question, it is actually a request. The person making the request expects a physical action, not a verbal response.)

Statements or commands enclosed in parentheses and written within another sentence should not be followed with a period.

When Mr. Redgrave enters the banquet room (you will recognize him by his red bow tie), tell him to come to my table. (Do not use a period after bow tie.)

RULE 2.   Use a period after words that logically substitute for a complete sentence.

Sentence fragments such as the following are acceptable when they serve as answers to questions.

No, not at all.

In July, of course.

RULE 3.   Use a period in writing some abbreviations.

Periods have been eliminated from many abbreviations in recent years. Abbreviations that indicate single words and are expressed in lowercase letters usually require periods.

etc.	Ms.	e.g.	Sept.	Inc.
i.e.	Sr.	qr.	Mon.	l.c.d.

Many abbreviations composed of initials do not need periods. Following is a random list of common abbreviations that are generally written as shown:

CPA	YMCA	FCC	TVA	OPEC	mpg
UFO	YMHA	NATO	ICC	USA	mph
AFL-CIO	YWCA	FTC	NAACP	UN	rpm
IRS	HMO	FBI	NFL	AMA	rps

When writing addresses, use the two-letter abbreviation for each state (MA, NY, NC, PA, OH, MO, AZ, NJ, etc.). Note that no periods are used in the abbreviation.

If you check more than one dictionary to determine how to write an abbreviation, you may find that they are not in agreement. For example, one dictionary may list *rpm* while another lists *r.p.m.* as the abbreviation. In making your choice, remember that the abbreviation without periods is probably the more current version. Also, whichever form you use, be consistent throughout a piece of writing.

RULE 4.   Use the period as a decimal point.

We discovered that 36.7 percent of our customers paid cash for furniture; the other 63.3 percent used credit.

Mr. Pulaski bought a memory typewriter for $549.95. (Whole amounts, such as $500, normally do not need decimals and zeros unless for consistency with other numbers.)

RULE 5.   In preparing an outline, place a period after a letter or a number that marks a division but is not enclosed in parentheses.

I. Type of securities
  A. Stock
    1. Preferred
      a. Participating
      b. Nonparticipating

Additional Notes on the Use of the Period

1. Do not place periods *after headings* (titles) or *after items in a topic outline* (see Rule 5 above).

2. Do not use *consecutive periods at the end of a sentence.*

Our mail usually arrives before 11:00 A.M. (The final period completes the abbreviation *and* ends the sentence.)

The display room is open each day for three hours (1:00 P.M. to 4:00 P.M.). (A final period is necessary because a closing parenthesis follows the period after the abbreviation.)

3. Do not use a period *after a contraction* (don't, can't).

4. Do not use periods *after items on a list* (or outline) unless they form complete sentences.

QUESTION MARKS

RULE 1.   Use a question mark at the end of a direct question.

How many barrels of oil did the well produce during January?

Which of our 50 states has the smallest population?

A sentence that is actually a question may be worded as a statement. The final mark of punctuation will determine the writer's intent.

You did remember the report?

The supervisor is late?

If several questions begin in the same way, they may be expressed in the following abbreviated form:

Where were you living in 1964? in 1974? in 1984?

A short question may be appended to a statement with a comma.

You still want the silverware, don't you?

Follow a question enclosed in parentheses or dashes with a question mark.

The law of diminishing returns (have you heard of it?) may provide the answer to your problem.

If a question ends with an abbreviation that requires a period, be certain to follow the period with the question mark.

Do you ever make use of the abbreviation *e.g.*?

If a statement includes an indirect question, follow it with a period, not a question mark.

Monique Bray asked me how to prepare an affidavit.

RULE 2. Use a question mark enclosed in parentheses to express doubt.

He was wearing a gray Botany (?) suit when he left the office.

EXCLAMATION POINTS

RULE. Use an exclamation point after a word or group of words that expresses strong feeling.

You will find that exclamation points appear frequently, perhaps too frequently, in sales literature. Used sparingly, the exclamation point adds emphasis or importance to the words it follows. Exclamation points may suggest anger, relief, fright, exasperation, excitement, surprise, or any other intense feeling.

What a day!	Watch out!	Save me!
How incredible a story!	Don't touch that!	

Exclamation points are placed before a closing quotation mark only if the exclamation point applies to the quoted matter.

"Halt!" the guard shouted.

What a stirring rendition of "America, the Beautiful"!

YOU TRY IT

A. In the sentences below insert periods, question marks, and exclamation points as needed.

1. Would you please send Mrs Sandoval an itemized report

2. What a great idea

3. What is the meaning of the abbreviation *ie*

4. The financial statement arrived from Loras Design, Inc

5. Yes, please

6. Are you a member of the YWCA

7. Nearly half, 489 percent, of the surveys were returned

8. Leave the building at once

B. Each of the following sentences contains a period, a question mark, or an exclamation point that is underlined. Review the rules in this learning unit to find the one that applies to each sentence. Then, in the space provided, write a brief justification for the use of the underlined punctuation. Use your own words.

1. Heavens, no!

_____

2. What time is the meeting?

_____

3. Please be sure to check the E-Mail each day.

_____

Answers: **A.** 1. Mrs.; report. 2. idea! 3. *i.e.?* 4. Inc. 5. please. 6. YWCA? 7. 48.9%; returned. 8. once! **B.** 1. Use an exclamation point after a strong expression of feeling. 2. Use a question mark at the end of a direct question. 3. Use a period after a command.

LEARNING UNIT 18-4
DASHES

The dash is a more emphatic mark of punctuation than any that has been discussed. The dash appears as two hyphens when the writing is done with a typewriter or word processor. In print, however, the dash appears as a solid line the width of the letter *M*. Typesetters call this dash an *em* dash.

In many of their functions, dashes act as substitutes for commas, parentheses, semicolons, or colons. Dashes may be used singly or in pairs to set off elements within sentences.

RULE 1.   Use dashes to set off any nonrestrictive element that deserves special emphasis.

My supervisor—as well as 3,000,000 other quality-minded Americans—owns this make of automobile.

This phenomenal new product—the most versatile and dependable computer ever produced—is exactly what your office needs.

Although a parenthetical expression or an appositive is usually set off with commas, a writer can place special emphasis on such an element by using dashes, as shown in the examples above. Advertising copywriters frequently use such emphasis.

If a nonrestrictive element contains commas, you can set it off with dashes or parentheses to avoid confusion. Dashes will also give you greater emphasis, as in the examples below.

Our company asked the three competing firms—Gaston Bros., Zenith Controls, and Century Motors—to submit bids. (Less emphasis would be placed on the appositive if it were enclosed in parentheses or commas.)

The company—as Robert Chen, our union representative, had predicted—refused to grant the requested increase in wages. (By using parentheses, the mention of Mr. Chen and his prediction would be de-emphasized.)

RULE 2.   Use dashes to emphasize an independent clause that abruptly interrupts another clause.

Four leading institutions—you know their names—will cooperate with us in this extensive research. (Use parentheses to set off such elements without emphasis.)

The parent company—no one can understand why—refused to grant financial aid to its subsidiary during the recent crisis.

RULE 3.   Use a dash before a word (*these, any, each, all,* etc.) that sums up a preceding series.

The broken boiler, the missing ledger, and the late shipment—all these problems demand my attention.

Joe Kaufman, Eileen O'Brien, and Don McGuire—each of them knows the combination to the safe.

RULE 4.   Use a dash to indicate an afterthought, an abrupt change in thought, or an emphatic pause.

The budgeting of personal finances is not an easy task—even for the very rich.

You should consider investing in Vulcan Materials, Dow Chemical, General Electric, Goodyear—but you have your personal ideas.

The farmers were ready to give up their struggle—and then the rains came.

The work will be finished by—let me check with the supervisor.

RULE 5.   Use a dash before the name of an author or a work that follows a direct quotation and indicates its source.

"By working faithfully 8 hours a day, you may eventually get to be boss and work 12 hours a day."—Robert Frost.

"Since dictionaries have changed much, we need not be surprised if they change more."— *Webster's New World Dictionary.*

RULE 6.   Use a dash before a word or phrase that is being repeated for greater emphasis.

Industry leaders should unite in seeking the solution—the solution that will lead to greater production in our mills and improved morale among our workers.

She was slow in taking dictation—painfully slow.

RULE 7.   Use a dash before introductory expressions such as *namely* or *for example* if the following words are to be emphasized.

Several unwelcome guests attended—for example, the reporter who had criticized our policies.

These generators have several serious deficiencies—in other words, we have no intention of buying one.

Dashes Used with Other Punctuation Marks

Usually, no other punctuation mark should immediately precede or follow a dash. However, several exceptions can occur. Observe the punctuation in the following sentences:

Only one local company—Murdoch & Agassi, Inc.—made a substantial contribution to the fund. (A period in an abbreviation may precede a dash.)

Ethan Grossweiler—have you heard?—will be our chief executive officer. (A question mark or an exclamation point may precede a dash.)

Mathew said—according to three reliable sources—"I will not take the position if it is offered to me." (A quotation mark may follow a dash.)

A dash is considered an informal mark of punctuation; therefore, it should be used sparingly in a formal business communication. Writers frequently use the dash for emphasis. However, if the dash is used too often, its effectiveness can be minimized. Use commas when you feel that the use of commas or dashes would be equally acceptable.

YOU TRY IT

A. Add dashes to each of the following sentences as appropriate.

1. The town's three top firms Alpha Co., Tree Trimmers, Inc., and Sheldon's Automotive moved to a neighboring community last year.

2. Accounts Payable, Accounts Receivable, and General Accounting all these departments are working overtime to finish the report by the end of the fiscal year.

3. The formatting of that complex document will be difficult even for our experienced production department.

4. "You don't make your character in a crisis; you exhibit it." Olney Arnold

5. Please send us Mr. Nguyen's resume in other words, we have decided to consider him for the position.

B. Each of the following sentences contains one or two dashes. Review this learning unit to find the rule that applies to each sentence. Then, in the space provided, write a brief justification for the use of the dash. Use your own words.

1. Nancy Hubbard—rumor has it—has submitted her resignation.

_____

_____

2. Martin Chuzzlewick, Oliver, David Copperfield—all these characters are from the imagination of Charles Dickens.

_____

_____

3. Dr. Fischer gave the orientation lecture—the same lecture that he has given for the last ten years.

_____

_____

Answers: **A.** 1. firms— 2. Automotive— 3. difficult— 4. it.”— 5. resume— **B.** 1. Use dashes to set off an independent clause that interrupts another clause. 2. Use a dash before the word *all* when it sums up the preceding series. 3. Use a dash before a phrase that repeats the previous phrase—but with more emphasis.

CHAPTER SUMMARY: A QUICK REFERENCE GUIDE

PAGE	TOPIC	KEY POINTS	EXAMPLES
374	**Semicolons: basic rules**	**Semicolon Rule 1.** Use a semicolon to separate two or more related independent clauses not joined by coordinating conjunctions.	Kim was excited; her department won the award.
		Semicolon Rule 2. Use a semicolon to separate independent clauses followed by conjunctive adverbs or transitional phrases; follow with a comma.	The shipping department filled every order; however, it made two substitutions.
		Semicolon Rule 3. Use a semicolon before a coordinating conjunction that separates two independent clauses containing internal commas.	Our manager, who everyone likes, will be leaving; but her replacement, John, is also well liked.
		Semicolon Rule 4. Use a semicolon to separate items in a series when the items themselves contain commas.	The new agents are Ms. Johnson, Vermont territory; Mr. Walker, New York territory; and Ms. Skok, Illinois territory.

PAGE	TOPIC	KEY POINTS	EXAMPLES
376	**Colons: basic rules**	**Colon Rule 1.** Use a colon after a formal introduction to the words that follow (*the following, as follows*). These words can be stated or implied.	Buy the following items: computer paper, index cards, and green pencils.
		Colon Rule 2. Use a colon to introduce a formal quotation.	The manager stated: "Every employee must report to me for a performance evaluation."
		Colon Rule 3. Use a colon after an independent clause when an explanation or amplification of the clause follows.	Look for any record you can find: financial record, health record, school record, and so on.
379	**Semicolons and colons: special uses**	**Semicolon Rule 5.** Use a semicolon before an expression that is used to introduce an enumeration, such as *for example* (*e.g.*), *that is* (*i.e.*), *for instance*, or *namely*; follow with a comma.	Loren sold the product only in crowded locations; for example, New York, Chicago, Detroit, and Cleveland.
		Colon Rule 4. Use a colon after the salutation of a business letter.	Dear Ms. Anderson:
		Colon Rule 5. Use a colon to separate the hours from the minutes when expressing time in figures.	11:30 A.M.
		Colon Rule 6. Use a colon to separate a title from a subtitle.	*Real Estate Ethics: The Difficult Agent*
		Colon Rule 7. Use a colon in footnotes and bibliographies if it is the accepted style.	Loren Scott, *Daily Reports* (Englewood Cliffs, NJ: Prentice Hall, 1995).
		Colon Rule 8. Use a colon to separate the chapter from verse in a biblical reference or the parts of a mathematical ratio.	John 4:12 1:23:20
381	**Periods**	**Rule 1.** Use a period after a statement, a command, or a request.	Please return my desk chair.
		Rule 2. Use a period after words that logically substitute for a complete sentence.	Yes, do it.
		Rule 3. Use periods in writing some abbreviations.	i.e., e.g., Co.
		Rule 4. Use periods as decimal points.	4.5 percent

		Rule 5. For outlines, place periods after letters or numbers that mark divisions but are not enclosed in parentheses.	1. Tables
			2. Chairs
			  a. Wood
			  b. Plastic

382	**Question marks**	**Rule 1.** Use a question mark at the end of a direct question.	Did you find the answer?
		Rule 2. Use a question mark enclosed in parentheses to express doubt.	Alexander returned the missing (?) book.

PAGE	TOPIC	KEY POINTS	EXAMPLES
383	**Exclamation points**	**Rule.** Use an exclamation point after a word or group of words that expresses strong feeling.	The lost key is found!
384	**Dashes**	**Rule 1.** Use dashes to set off any nonrestrictive element that deserves special emphasis.	This typewriter—the best available—will serve your purpose.
		Rule 2. Use dashes to emphasize an independent clause that abruptly interrupts another clause.	Green pens, red pens, black pens—any of these will do the job.
		Rule 3. Use a dash before a word (*these, any, each, all,* etc.) that sums up a preceding series.	Call either Curt, Elaine, or Doris—not Diane or Roger.
		Rule 4. Use a dash to indicate an afterthought, abrupt change in thought, or emphatic pause.	Look for the officers—they are listed in the directory—with the most experience.
		Rule 5. Use a dash before the name of an author or work that follows a direct quotation and indicates its source.	"Follow the money."—James Corben
		Rule 6. Use a dash before a word or phrase that is being repeated for greater emphasis.	Production has greatly increased—almost doubled.
		Rule 7. Use a dash before introductory expressions such as *namely* or *for example* if the following words are to be emphasized.	The memo was well received—for example, Sukie and Sully thanked me for the information.

QUALITY EDITING

Edit the following letter to Mr. Lingerhausen, a former regular customer of Gump's Department Store. Watch for spelling, punctuation, and capitalization errors.

```
May 30, 19xx

Dear mr Lingerhausen

We miss you Our staff tells me, that you have not vis-

ited our store since last June almost a full year.

     Your business is importnat, to us mr lingerhausen

therefore we have a favor to ask of you? Please tell

us, why we haven't seen you. A stamped reply card is

enclose for your conveneince.
```

An one-time offer—especialy for you is also enclosed. We hope this will encurage you to come in soon. just bring us the round blue sticker, and you may by any one item in the store at half-price.

Mr. lingenhauser, thank you for taking time to read our letter:

We hope to see you soon.

Very truly yours,

P S We are now open seven days a week from 7 A.M. to 8:30 P.M..

APPLIED LANGUAGE

1. You work in the marketing department of Speedy Express, an overnight package delivery company. Write a letter to Ms. Margaret Scott, a customer who has reported rudeness from a customer service representative. Include an indication of your dismay or shock at such an incident and a statement of apology. Use at least one period, one exclamation point, one question mark, one colon, one semicolon, and one dash (or pair of dashes) in your letter.

2. In the space provided, write a sentence that illustrates each of the rules given below. After you write your sentence, refer to the examples in the chapter to check your work. Finally, correct your sentence if necessary.

   a. *Semicolon Rule 3*: Use a semicolon before a coordinating conjunction that separates two independent clauses when the clauses contain internal commas.

   _____

   _____

   b. *Semicolon Rule 4*: Use a semicolon to separate items in a series when the items themselves contain commas.

   _____

   _____

c. *Colon Rule 2*: Use a colon to introduce a formal quotation.

_____

_____

d. *Question Mark Rule 1*: Use a question mark at the end of a direct question.

_____

_____

e. *Period Rule 1*. Use a period after a statement, command, or request.

_____

_____

f. *Exclamation Rule*. Use an exclamation point after a word or group of words that expresses strong feeling.

_____

_____

g. *Dash Rule 5*. Use a dash before the name of an author or a work that follows a direct quotation and indicates its source.

_____

_____

h. *Dash Rule 6*. Use a dash before a word or phrase that is being repeated for greater emphasis.

_____

_____

Name _____

Date _____

SEMICOLONS AND COLONS, END PUNCTUATION, AND DASHES

Part A. Semicolons and Colons
Insert semicolons and colons as appropriate in the sentences below.

1. Pacific Telesis Group is a holding company Pacific Bell is one of its subsidiaries.

2. The customer claims that he has not telephoned anyone in Boseman, Montana therefore, he refuses to pay for the call.

3. He called me at 810 A.M. our conversation ended at 905 A.M.

4. The telephone company representative used exactly these words "You should consider our experience and expertise before canceling our service."

5. The purpose of my inquiry is as follows to get the best possible long-distance service at the best possible price.

6. To learn more about services, I wrote to these companies NYNEX, New York, New York Bell Atlantic Corporation, Philadelphia, Pennsylvania USWest, Denver, Colorado and BellSouth, Atlanta, Georgia.

7. The two companies you mentioned do have something in common they both provide telephone service.

8. Pacific Telesis was once part of the AT&T system Century Telephone Enterprises has always been independent.

9. I selected MCI George selected Sprint.

10. Charges are divided into three categories day, evening, and night.

Part B. Internal and End Punctuation
Add the appropriate internal and end punctuation (colons, semicolons, periods, question marks, and exclamation points) to the sentences below.

1. When Paul smelled the smoke, he yelled " Fire "

2. Item 14 is out of stock, isn't it

3. Please hand me that disk

4. Ray Sullivan (do you know him) bought the firm last year

5. Help I can't get the machine to stop

6. Ms. Nakamura asked " Can the Computer Alliance provide us with ten work stations "

7. Does this copier make reductions make enlargements feed documents automatically

8. Frustrated when the copier repeatedly jammed, Lucy shouted " This machine hates me "

9. The small notebook sells for $1.49 the large one sells for $3.89

10. Tell me when it is noon

Part C. Dashes
Add dashes as appropriate to the sentences below.

1. Cincinnati Bell provides service to three states but Illinois is not one of them.

2. Lopez, Caminski, Brown, and Yee all four have been recommended for promotion.

3. Alan Sako made the sale the sale that earned him the award for "Salesperson of the Month."

4. The three major cities on this list Boston, New York, and St. Paul are ports.

5. The trial was held in Missoula or was it Boseman in August.

6. Small and large those are the only sizes we have in that style.

7. We hope the new manager is her name Schwartz? has experience with computers.

8. We are working on a new idea an idea that could revolutionize our industry.

9. "A penny saved is a penny earned." Benjamin Franklin.

Part D. Punctuation Rules
Indicate whether each of the following statements is true or false by inserting T or F in the space provided.

_____ 1. Only commas may be used to separate items in a series.

_____ 2. Semicolons should be used to separate the hours from the minutes in an expression of time.

_____ 3. Ordinarily, a period should follow a sentence beginning with Please.

_____ 4. Several semicolons may appear in a single sentence.

_____ 5. A colon may be followed by a quotation.

_____ 6. If a sentence ends in an abbreviation, two consecutive periods must be used.

_____ 7. A command is almost always followed by an exclamation point.

_____ 8. A period should never be used after words that do not constitute a complete sentence.

_____ 9. Dashes are always used in pairs.

_____ 10. Dashes give more emphasis to a word or a group of words than do commas.

CHAPTER 19

OTHER PUNCTUATION AND WORD DIVISION

Punctuation, to most people, is a set of arbitrary and rather silly rules you find in printers' style books and in the back pages of school grammars. Few people realize that it is the most important single device for making things easier to read.

—Rudolf Flesch

LEARNING UNIT GOALS

The rhythm of a sentence is the result of its structure and punctuation. If you want a fast-moving piece of writing, use short sentences with few commas. When you find your writing has become uneven, you can make it smoother by combining some sentences. You can also change the position of phrases to give your sentence a continued rhythm. Like a musician, you are in control of the melody your writing produces.

In Chapters 17 and 18, you learned about the main punctuation marks affecting sentence rhythm. Learning Unit 19-1 discusses parentheses and quotation marks. Learning Unit 19-2 discusses brackets, ellipses, and apostrophes. In Learning Unit 19-3, you learn how to divide words—the final topic in this section on sentence mechanics.

LEARNING UNIT 19-1
PARENTHESES AND QUOTATION MARKS

This learning unit discusses two punctuation marks that have been mentioned in earlier chapters—parentheses and quotation marks. You are probably already familiar with some of the rules given here.

PARENTHESES

To set off sentence elements, you have three choices: commas, dashes, and parentheses. Enclosing parenthetical elements in commas affects the flow of a sentence only slightly. Dashes emphasize the setoff sentence elements. Parentheses de-emphasize the added information. Note how the same sentence can be punctuated with commas, dashes, and parentheses.

With commas: He purchased a new camcorder, a Magnavox, for his professional use. (The writer has named the camcorder with the least interruption of sentence flow.)

With dashes: He purchased a new camcorder—a Magnavox—for his professional use. (The writer is emphasizing the purchase of the brand Magnavox for professional use.)

With parentheses: He purchased a new camcorder (a Magnavox) for his professional use. (The writer is de-emphasizing the brand of the new camcorder.)

Parentheses Used to De-emphasize

RULE 1.   Use parentheses to enclose nonrestrictive elements that should be de-emphasized.

Because Mr. Hingston did not attend the seminar (possibly because of illness), his notes are incomplete. (When a comma and a closing parenthesis come together in a sentence, always write the parenthesis first.)

She has decided to leave early Saturday morning (as you suggested) and to stay in Phoenix for five days. (You would give more emphasis to *as you suggested* if you set it off with commas or dashes.)

Mr. Crane's experience with the National Labor Relations Board (see page 502) may prove of interest to you. (Note that the word *see* does not begin with a capital letter.)

Our warehouse is on a street (Madison) that carries heavy traffic during work hours. (The word in parentheses is added information.)

The incident occurred on May 21 (or was it May 22?) at the Biltmore Hotel in Los Angeles. (Note the use of a question mark *inside* the closing parenthesis. If the item in parentheses was an exclamation, the exclamation mark would be used instead of the question mark.)

Parentheses Used to Enclose Numbers or Letters

RULE 2. Use parentheses to enclose numbers or letters that precede items in a series within a sentence.

We plan to (1) meet the Governor, (2) express our views on the water problem, (3) submit the signed petition to the proper authorities, and (4) attend at least one legislative session. (Note that a colon does not follow the preposition *to.*)

The agenda includes only three topics for discussion: (1) the company's official position on automatic wage increases, (2) our proposed expenditures for research, and (3) the possible acquisition of Lot 21 for parking.

Parentheses Used to Enclose Figures in Formal Writing

RULE 3. Use parentheses in formal writing, such as legal papers, to enclose figures that follow a spelled-out number.

The party of the first part agrees to assume the encumbrance of record, a trust deed note representing an indebtedness of Nine Hundred Eighty Dollars ($980).

Parentheses Used to Enclose Sentences

RULE 4. When enclosing an entire sentence in parentheses, begin the sentence with a capital letter and place the final mark of punctuation (period, question mark, or exclamation point) inside the closing parenthesis.

Their problems were comparable to those of the Wright Brothers at a later date. (The story of the first airplane appears in Chapter 14.)

QUOTATION MARKS

The correct use of quotation marks is extremely important. Editors, reporters, and legal secretaries must learn to use quotation marks skillfully. All writers, of course, should also know how to use quotation marks.

RULE 1. Use quotation marks to enclose material that is quoted directly.

"Our industry is vital to the American economy," said Mr. Paulson. (Note the placement of the comma inside the closing quotation mark.)

"Our industry," said Mr. Paulson, "is vital to the American economy." (Note that in this type of construction, you need two pairs of quotation marks. Also, only the first word is capitalized, not the word *is*.)

Ms. Schumacher used these exact words: "We have no intention, nor have we ever had any intention, of attempting to create a monopoly." (Note the placement of the period inside the closing quotation marks.)

Ms. Richardson said, "Send me five dozen of these units by April 15"; Mr. Darrow said nothing. (Note the placement of the semicolon after the closing quotation mark.)

In his closing remarks the speaker referred to "the magnificent contributions of these self-sacrificing men and women." (A quotation need not be a complete sentence.)

This was her answer: "We have been short of materials since our credit was curtailed in early June. We most urgently need plastic sheeting and copper wire." (Although two sentences are quoted, you need only one pair of quotation marks.)

"We are revising this multicolored brochure," she said. "The improved version will be ready by Friday." (Two pairs of quotation marks are needed since the quotation is interrupted by *she said*. Note that the second quoted sentence begins with a capital letter.)

A few notes follow about current writing practices that involve direct quotations.

1. **Indirectly quoted words.** Words that have been quoted indirectly should not be placed in quotation marks.

   She asked me whether the information will be stored on a magnetic card.

   Dr. Bates said that he had seen our advertisement in the Yellow Pages. (Indirectly quoted words are often introduced by *that*.)

   BUT: Dr. Bates said, "I saw your advertisement in the Yellow Pages."

2. **Long quotations.** When typing or using a word processor, one of the following methods can be used for long quotations:
   a. You may single space and use a shorter typing line than for the remainder of your copy. No quotation marks will be necessary.
   b. If you maintain a uniform line length, you should place a quotation mark at the beginning of each paragraph and at the end of the final paragraph only. Any quotation marks used *within* these paragraphs should be single quotation marks (').
   c. When different type sizes are available, long quotations can be indented and set in smaller type than the text surrounding the quotation (this is called an extract).

3. **Reporting a conversation.** When reporting a conversation, begin a new paragraph each time the speaker changes.

   "If the lumber arrives by noon tomorrow, how soon will the cabinets be ready?" asked Ms. Garrison.

"We have several other jobs to finish first," said the supervisor.

"Can you guarantee shipment by November 1?" persisted Ms. Garrison.

4. **Quotation within a quotation.** To punctuate a quotation within a quotation, use double quotation marks for the first quotation and single quotation marks (apostrophes if you are typing) for the quotation it contains. Avoid a third quotation within the second quotation.

> The final paragraph read: "You would do well to heed Mr. Purdin's advice: 'Give the public what the public wants, and you will be in business for a long time.'"

5. **Question at the end of a sentence.** Do not set off a question at the end of a sentence in quotation marks unless it is someone else's words.

> The question is, Does our work require a letter-quality printer? (In this sentence the writer is asking the question.)

6. **Words placed in quotation marks.** Place only the exact word or words taken from another source in quotation marks.

> Director Shantz referred to the "devastating effect" of this new directive.

7. **Quotation marks with *yes* and *no*.** Do not put the words *yes* and *no* in quotation marks unless you are quoting another person.

> We are hoping that he says yes to our offer.

> Dr. Fuentes is not likely to respond with a simple yes or no.

> BUT: When we asked her if she could program the computer, she said, "No."

RULE 2.   Use quotation marks to enclose the titles of minor works (articles from magazines, songs, essays, short stories, short poems, one-act plays, lectures, sermons, chapters, motion pictures, and television shows).

Note that the rule does not include the names of books, magazines, newspapers, long musical works, or book-length poems. These items are written in italic type (or underscored if using a typewriter).

This week's issue of *U.S. News & World Report* contains an article entitled "The Changing Role of World Powers."

If you read Graham and Dodd's *Security Analysis*, you will find the chapter on "Investment and Speculation" to be particularly interesting.

Mr. Lloyd asked, "Was the poem 'If' written by Rudyard Kipling?" (The title of the poem is in single quotation marks because it is within a quoted sentence.)

RULE 3.   Use quotation marks to enclose words used in an unconventional manner.

If we continue in our present spending habits, this firm will soon go "down the tubes." (When using a cliché for effect, enclose it in quotation marks.)

You could induce your son to go to college if you tried a little "consumer motivation." (You may enclose technical and trade terms in quotation marks when using them in unconventional contexts.)

Although you may enclose words referred to as words in quotation marks, most writers prefer to put them in italic type (or to underscore them if using a typewriter).

Does the word *run* really have 140 different meanings?

The expression each and every should be avoided in business writing.

Quotation Marks and Other Punctuation

When a closing quotation mark occurs with another mark of punctuation, use the following accepted practices:

1. Always place a period or a comma *inside* (before) a closing quotation mark.

2. Always place a colon or a semicolon *outside* a closing quotation mark.

3. If a question mark, an exclamation point, or a dash applies to the quoted material, place it *inside* a closing quotation mark; otherwise, place it *outside*.

   "What is the circulation of your magazine?" he asked. (The quoted portion is a question.)

   Did he say, "Our circulation will soon exceed one million"? (The entire sentence is a question, but the quote is not.)

   Were you present when he asked, "Why don't we discontinue production?" (Use only one question mark even though the quoted portion and the entire sentence are both questions.)

   "I can't believe it!" shouted the happy winner.

   "If the patent is granted—" are the only words I heard.

YOU TRY IT

A. Add any parentheses or quotation marks needed in the following sentences.

1. Because Allen was late for work he may have overslept, he missed the announcement.

2. Purchaser agrees to pay Seller the Amount of Four Hundred Fifty Dollars $450.

3. Who wrote The Star Spangled Banner ?

4. The accounting supervisor asked, Will the analysis be ready by Tuesday?

5. The order included three items: 1 floppy disks, 2 a mouse pad, and 3 a toner cartridge.

B. Each of the following sentences contains either parentheses or quotation marks. Review the rules in this learning unit to find the one that applies to each sentence. Then, in the space provided, write a brief justification for the use of the parentheses or the quotation marks.

1. "On Attitude and Behavior" is a short essay written by Anna Haver in 1934.

_____

_____

2. The medical technician (Kevin, maybe) provided extra training for the new staff.

_____

_____

3. What general is widely quoted as saying, "I shall return!"?

_____

_____

LEARNING UNIT 19-2
BRACKETS, ELLIPSES, AND APOSTROPHES

Often writers stumble over the usage of brackets, ellipses, and apostrophes. This learning unit should solve any problems you have.

BRACKETS

RULE 1.   Use brackets to enclose any words that are added to a direct quotation.

In preparing the works of other writers for publication, editors of newspapers, magazines, and books use brackets to enclose their own explanatory comments. Note the following two sentences.

> These were the treasurer's exact words: "Our current deficit of $925,000 [that figure has been challenged] will not affect future operations."

> An officer of the company said, "Slim [President Edward S. Brady] is the person most responsible for this firm's outstanding success."

In quoted matter the Latin word _sic_, which means thus or so stated, may be in brackets after a word or phrase that appears to be an error. A misspelling, for example, may appear in the original material.

> His last sentence read: "I will soon return to France, the county [_sic_] of my birth."

RULE 2.   Use brackets as a substitute for parentheses that would logically be placed within other parentheses.

> While he was still very young, Mr. Avery came to this country from England. (Some historians [see page 424] credit Ireland as his birthplace.)

ELLIPSES

RULE 1.   Use an ellipsis mark (three periods) to indicate that words are being omitted from a quoted passage.

If the omission of words follows a complete statement, it will be necessary to use four periods—one to end the sentence and three to form the ellipsis. When typing, space before and after each ellipsis mark.

> These were his words: "We could not have achieved final victory without the full cooperation of companies like . . . the Ballard Engineering Corporation."

> She gave you these instructions: "Check the balance of our account with Ramsey Motors. . . . Cancel my appointment with Mr. Solano." (Note the period and three ellipses after *Motors* to indicate the end of a sentence.)

Rule 2.   An ellipsis mark may be used at the end of an unfinished thought.

> If you succeed, we will reward you handsomely, but if you fail . . . (No period is used. The dash would be less formal.)

APOSTROPHES

Recall that Chapter 7 discussed the use of possessive nouns. If necessary, review the rules given for the use of possessive nouns. These rules are not repeated here. The following rule should serve as a reminder.

RULE 1.   Use the apostrophe correctly in forming the possessive case of nouns and certain pronouns.

> We could not understand Ms. Drew's attitude toward her work.

> The plumbers' union may have to increase its monthly dues.

> I did not understand Mr. Davis's reference to Socrates' death. (Why were these singular nouns made possessive in different ways?)

> Stanley and Albert's new real estate office has attracted considerable attention.

> In this office, one's work is never done.

RULE 2.   Use an apostrophe to indicate the omission of letters or numbers in a contracted form.

We *haven't* time to fill these orders today.

He graduated with the class of *'88.*

Don't send the cement pipe until October 15.

In writing contractions, make certain that you place the apostrophe where the omission has been made. In a word such as *didn't*, the apostrophe replaces the *o* in *did not*. Contractions are used most frequently in informal writing.

RULE 3.   If needed to prevent misreading, use an apostrophe in forming the plural of a letter or a word referred to simply as a word.

He does not always pronounce his *r*'s. (Plurals of lowercase letters always require apostrophes.)

She received *A*'s in her three most difficult courses. (Use apostrophes in the plural forms of capitalized *A, I, M,* and *U.*)

His last name has three *T*s and two *L*s. (Although the modern trend is away from the use of apostrophes in such readable plural forms, some authorities may still prefer to insert them, as in *T* 's and *L*'s. Be consistent.)

You would have a better paragraph if you dropped a few of the *so*'s. (You need the apostrophe to prevent misreading.)

His one short paragraph on microcomputers contained five *ands.* (The word *ands* is readable without an apostrophe.)

Lisa Molinski is not yet familiar with the *do*'s and *don't*s of data processing.

Trends of the Times

Until recent years it was common to add an apostrophe and *s* when forming the plural of a number expressed as a figure or of an abbreviation made up of individual letters. Today, however, we usually add only an *s.*

You will find *YMCAs* in most parts of this country.

We have several *M.D.s* working in our research laboratories.

The depression of the early *1930s* must never be repeated.

The column is difficult to add because his *7s* and *9s* look alike.

He quoted the *CEOs* of three major corporations.

Although the apostrophe is used less frequently today than in former years, it continues to have the following three primary functions:

1. To build possessive-case forms of nouns
2. To indicate omitted letters in contractions
3. To form plurals that would otherwise be easily misread

Typists, of course, use the apostrophe as a *single* quotation mark, and anyone preparing an invoice, order form, or similar communication may use an apostrophe as a symbol for *feet* (4′× 8′ plywood).

YOU TRY IT

A. Add brackets as appropriate in the following sentences.

1. "The report indicated that 46 percent of the respondents *sic* preferred the old program features to the new."

2. The promotion will be effective February 1. (It is retroactive from July 1 six months.)

3. Janice Morgan wrote the following comment: "The outcome of the meeting was an agreement to establishing *sic* a partnership with the supplier."

B. Add ellipses as appropriate in the following items.

1. If you do not understand the concept after all this work

2. According to Hendrix, the firm bought a new computer system, leased new furniture, installed a new phone system, painted inside and out

3. Mr. Allen's report states: "First, we recruited a president; second, we hired a new sales manager; fifth and last, we recruited five new customer service representatives." Please note this in your memo.

C. Insert any needed apostrophes in the following items.

1. Dont send the letter until I have checked it.

2. Is this machine Howards?

3. The package should arrive by four oclock.

Answers: **A.** 1. [*sic*] 2. [six months] 3. [*sic*] **B.** 1. work . . . 2. out . . . 3. manager; . . . **C.** 1. Don't 2. Howard's 3. o'clock

LEARNING UNIT 19-3
WORD DIVISION

Most word processors and computers allow you to justify the right-hand margin to make a piece of writing even. Should you not want the right-hand margin justified, you can usually set the computer to hyphenate the words. You should know the word division rules, however, since the computer may ask you questions about the hyphenation of some words. Also, you will probably want to divide some words if you use a typewriter.

The first rule of word division is a general rule.

RULE 1.   Avoid excessive word division.

Excessive word division can make writing appear choppy. Also, if a writer breaks a word illogically, the reader will have difficulty mentally joining the two parts of the word.

Common sense suggests that we should never divide abbreviations and figures, unless they are exceptionally long. It tells us also that to facilitate reading, we should type as much of a word as possible before using the hyphen to divide a word.

When it is necessary to divide a word to keep an acceptable margin, the following rules will be helpful.

HOW TO DIVIDE A WORD

RULE 2. Divide a word only between syllables.

A writer can sometimes determine the proper syllabication of a word by pronouncing it slowly. When in doubt about the proper syllabication, check your dictionary. Note how the dictionary divides each of these words.

ac-knowl-edg-ment	de-vel-op	lat-er-al
ad-mo-ni-tion	e-ro-sion	pu-ri-fi-ca-tion
be-hav-ior	har-mo-nize	sea-son-al-ly
be-liev-er	ig-ni-tion	tri-en-ni-um

RULE 3. If a consonant is doubled where syllables join, generally divide the word between the consonants.

admis-sion	rebel-lion	repel-lent
confer-ring	rebut-tal	submit-ted
plan-ning	recol-lect	suppres-sion

RULE 4. If a suffix has been added to a root word that ends in a doubled letter, ordinarily divide the word before the suffix.

This rule is an exception to Rule 3. Since a word such as *tell-ing* is made up of the root *tell* and the suffix *ing*, it would be logical to divide the word after the second *l*. Here are several other words that would be governed by this rule.

cross-ing	fall-ing	small-est
drill-ing	install-ers	spell-ers

RULE 5. Divide a hyphenated word only at the hyphen.

self-explanatory	quasi-judicial
pre-emptive	court-martial

HOW *NOT* TO DIVIDE A WORD

RULE 6. Do not divide a word that contains only one syllable.

Before attempting to divide a word, pronounce it to yourself slowly and carefully. If the word contains only one sound, or one syllable, write it in solid form. Do not divide any one-syllable words such as the following:

brought	sighed	strength	though
freight	straight	thirst	through

RULE 7. Do not divide a word unless it has at least six letters.

Even words of six letters should not be divided unless each syllable contains three letters, as in the following words:

ask-ing	lis-ten	pen-cil
fil-ter	man-tel	tab-let

RULE 8. Do not carry two letters of a word over to the next line. At the end of a line, two-letter divisions are permitted, although many writers avoid this break.

In applying this rule, words such as *a-gainst* and *a-gend-a* should be written in solid form. Words like the following can be broken only after the first two letters:

be-troth-al	ex-port-er
de-cant-er	in-sert-ed

RULE 9. Do not divide a word immediately before a one-letter syllable.

Most writers divide a word such as continuance *(con-tin-u-ance)* after, rather than before, the *u.* Here are a few other words to which this rule would apply.

co-ag-u-late	o-bit-u-ar-y	re-tal-i-ate
mu-nic-i-pal	re-cip-i-ent	spec-u-late

If the one-letter syllable is part of a common suffix, such as *able* or *ible,* carry the suffix over to the next line (*reli-able, convert-ible,* etc.). In some instances the syllabication of a word will not permit you to keep the suffix intact (*ca-pa-ble, pos-si-ble,* etc.).

RULE 10. Do not end more than two consecutive lines with hyphens.

Because words should be divided only when really necessary, you should have little difficulty in following this rule.

RULE 11. Do not divide the last word of a paragraph or the last word on a page.

RULE 12. Avoid the division of names.

Divide proper nouns only when absolutely necessary. It is generally better to leave a line slightly short than to break a word such as Prendergast or Massachusetts.

Divide proper-noun groups so readers can read them with ease. The name *Donald S. Crenshaw* might be typed with *Donald S.* at the end of one line and *Crenshaw* at the start of the next. *The Honorable Charles P. Lucas* might be divided after *Honorable* or after the *P.* You will have no difficulty breaking names if you remember that clarity to your reader is the goal.

A Reminder Concerning Compound Adjectives

You will recall from Chapter 13 that a compound adjective should usually be hyphenated when it precedes the noun it modifies. Note the use of hyphens in the following sentences:

We hope to dispose of this *hand-operated* duplicator.

He does not have change for a *ten-dollar* bill.

In writing a word such as *hand-operated* or *ten-dollar* at the end of a line, remember to divide it only at the hyphen.

YOU TRY IT

Rewrite each of the following words adding hyphens to show where the words can be divided. *Use a dictionary whenever you are unsure of the divisions.* If a word should not be divided, write *Do not divide* in the space provided.

1. administrative   _____

2. trough   _____

3. communication   _____

4. disclosure   _____

5. emphasis   _____

6. following   _____

7. transmitted   _____

8. Jefferson   _____

9. convicted   _____

10. retaliate   _____

11. yellow   _____

12. Elizabeth   _____

Answers: 1. ad-min-is-tra-tive 2. Do not divide 3. com-mu-ni-ca-tion 4. dis-clo-sure 5. em-pha-sis 6. fol-low-ing 7. trans-mit-ted 8. Do not divide 9. con-vict-ed 10. re-tal-i-ate 11. yel-low 12. Do not divide

---

CHAPTER SUMMARY: A QUICK REFERENCE GUIDE

PAGE	TOPIC	KEY POINTS	EXAMPLES
396	**Parentheses**	**Rule 1.** Use parentheses to enclose nonrestrictive elements that should be de-emphasized. **Rule 2.** Use parentheses to enclose numbers or letters that precede items in a series within a sentence.	Every person that spoke (including Jane) said the same thing. Look for (1) new stationery, (2) larger envelopes, and (3) smaller note paper.

PAGE	TOPIC	KEY POINTS	EXAMPLES
		Rule 3. Use parentheses in formal writing, such as legal papers, to enclose figures that follow a spelled-out number.	The broker hereby agrees to list the property for Eighty-Four Thousand ($84,000).
		Rule 4. When enclosing an entire sentence in parentheses, begin the sentence with a capital letter and place the final mark of punctuation (period, question mark, or exclamation point) inside the closing parenthesis.	Use the first figure for your analysis. (The second figure could result in an error.)
397	**Quotation marks**	**Rule 1.** Use quotation marks to enclose material quoted directly. **Rule 2.** Use quotation marks to enclose titles of minor works. **Rule 3.** Use quotation marks to enclose words used in an unconventional manner.	Garth said, "I can take care of everything." Did you read Elaine's last poem titled "Let's Take a Walk?" You are breaking every rule "in the book."
401	**Brackets**	**Rule 1.** Use brackets to enclose any words added to a direct quotation. **Rule 2.** Use brackets as substitute for parentheses that would logically be placed within other parentheses.	Ellery said, "A company vice president [James Anderson] will speak." (See the first book written by the author [Loren Bell].)
402	**Ellipses**	**Rule 1.** Use an ellipsis mark to indicate that words are omitted from a quoted passage. **Rule 2.** An ellipsis mark may be used at the end of an unfinished thought.	Every worker said: "I will do what you say . . . work." This may be true, but if it is not true . . .
402	**Apostrophes**	**Rule 1.** Use the apostrophe correctly in forming possessive case of nouns and certain pronouns. **Rule 2.** Use an apostrophe to indicate the omission of letters or numbers in a contracted form. **Rule 3.** If needed to prevent misreading, use an apostrophe in forming the plural of a letter or a word referred to simply as a word.	All the workers' time slips had to be rewritten. Let's go to lunch. All her grades were *E*s for excellent.
404	**Word division**	**Rule 1.** Avoid excessive word division. **Rule 2.** Divide a word only between syllables. **Rule 3.** If a consonant is doubled where syllables join, generally divide the word between the consonants.	sub-sti-tute; sub-or-di-nate progres-sive; recom-mend; transmit-tal

PAGE	TOPIC	KEY POINTS	EXAMPLES
		Rule 4. If a suffix has been added to a root word that ends in a doubled letter, ordinarily divide the word before the suffix.	dwell-ing; miss-ing; stress-ing
		Rule 5. Divide a hyphenated word only at the hyphen.	degree-day; self-taught
		Rule 6. Do not divide a word that contains only one syllable.	friend; strange; through
		Rule 7. Do not divide a word unless it has at least six letters.	out-age; pre-fer; pre-tax
		Rule 8. Do not carry two letters of a word over to the next line. At the end of a line, two-letter divisions are permitted, although many writers avoid this break.	be-tween; ex-changed; re-signed
		Rule 9. Do not divide a word immediately before a one-letter syllable.	man-u-fac-ture/manu-facture; lec-i-thin/leci-thin
		Rule 10. Do not end more than two consecutive lines with hyphens.	
		Rule 11. Do not divide the last word of a paragraph or the last word on a page.	
		Rule 12. Avoid the division of names.	

QUALITY EDITING

You work for Deena Kwon, the Vice President of Finance. Deena dictated a memo to a temporary clerk. The memo below is the one that the clerk transcribed and printed. The clerk left out most of the necessary punctuation. Also, since Deena was dictating, her sentences are unnecessarily wordy. Edit the memo by inserting appropriate punctuation; removing unnecessary words, phrases, or clauses; breaking long sentences; and correcting spelling errors. Note: You may think of several ways to edit this memo since there are many correct possibilities.

MEMORANDUM

July 21, 19xx

To: Jeff Watson, Perchasing Manager

From: Deena Kwon, Vice President, Finance

Order Number A156TN

We recieved a shipment Computer and Office Supply Center COSC yesterday July 20 that was ordered July 14 that was supposed to contian one color moniotr but it did not contain the monitor but contained instead 2000 boxes of paper clips I Know what to do with a monitor but with 2,000 boxes of paper clips

When I checked the catolog I saw that the item number for the monitor G4368 is a transposition of the number for the pa-perclips G3468 The order you sent COSC had the transposed number a common error COSCs advertising always says Customer Service is our specialty Woiuld you please call COSC and aranged for an immedeate delivery delivery of the correct item and pickup of the paper clips

The monitor for Carol Abramss' computer have been broken now for two weeks therefore Carol's work is falling behind and she needs the monitor asap

Jeff three purchasing errrors has been mae in yuor department in the past month errors which have been espensive and wasteful You must establich purchasing procedures that well help you avoid these probelms Please analize the work of the Perchasing Department and 1 provide a report to me by next monday august 3 2 include your prespectives on the sitaution and 3 be prepared to present suggestions for improvment.

APPLIED LANGUAGE

Write a sentence that illustrates each of the rules given below. Refer to the examples in the chapter to check your work, and then correct your sentence if necessary.

a. Use quotation marks to enclose material that is a direct quote.

_____

_____

CHAPTER 19   OTHER PUNCTUATION AND WORD DIVISION   **411**

b. Use parentheses to enclose nonrestrictive elements that should be de-emphasized.

_____

_____

c. Use parentheses to enclose numbers or letters that precede items in a series within a sentence.

_____

_____

d. An ellipsis mark may be used at the end of an unfinished thought.

_____

_____

e. Use the apostrophe in forming the possessive case of nouns.

_____

_____

f. Use an apostrophe to indicate the omission of letters or numbers in a contracted form.

_____

_____

g. The modern way to form the plural of a number expressed as a figure, as in the 1990s, is to add only an _s_ to the figure.

_____

_____

h. Divide most words between syllables.

_____

_____

WORKSHEET 19

Name _____

Date _____

OTHER PUNCTUATION AND WORD DIVISION

Part A. Parentheses and Quotation Marks

Indicate which punctuation marks—parentheses or quotation marks—should be used for each of the following situations. Write your answer in the space provided.

1. To set off a parenthetical expression that is to be de-emphasized. _____

2. To enclose a direct quote. _____

3. To enclose figures that follow a spelled-out number. _____

4. To enclose numbers or letters that precede items in a series. _____

Part B. Brackets and Ellipses

Each sentence below needs a set of brackets or an ellipsis mark. In the space provided, write the word or phrase with the required punctuation.

1. If you can fix it by Tuesday, do it; if not _____

2. The manager quoted the old saying "When life gives you lemons, " _____

3. "In the future, no smoking will be allowed anywheres *sic* in this building." _____

Part C. Apostrophes

One word in each sentence below needs an apostrophe. Write the word *with* the apostrophe in the space provided.

1. Mr. Mathews office was left unlocked. _____

2. The workers petition was signed by 4,500 people. _____

3. The ten customers reaction to the new store layout was very positive. _____

4. Only Ms. Morenos argument made sense. _____

Part D. Word Division

Three choices are given to break each word in this exercise. In the space provided, write the letter that indicates the best way to divide the word if it appeared at the end of a line.

1. **a.** medita-tion, **b.** me-ditation, **c.** meditat-ion   _____

2. **a.** rec-onsider, **b.** recons-ider, **c.** recon-sider   _____

3. **a.** occup-ational, **b.** occupa-tional, **c.** o-ccupational   _____

4. **a.** mani-fest, **b.** man-ifest, **c.** ma-nifest   _____

5. **a.** judi-cial, **b.** ju-dicial, **c.** judic-ial   _____

6. **a.** cont-radict, **b.** co-ntradict, **c.** contra-dict   _____

7. **a.** he-sitate, **b.** hesi-tate, **c.** hesit-ate   _____

8. **a.** begin-ning, **b.** beg-inning, **c.** beginn-ing   _____

9. **a.** friendli-ness, **b.** fri-endliness, **c.** friendlin-ess   _____

10. **a.** dep-en-dable, **b.** dependa-ble, **c.** depend-able   _____

Part E. Rules Review

Indicate whether each of the following statements is true or false by writing T or F in the space provided.

_____   **1.** Words that deserve special emphasis should be placed in parentheses.

_____   **2.** Parentheses and brackets serve the same function and can be used interchangeably.

_____   **3.** Any clause enclosed in parentheses should begin with a capital letter and end with a period.

_____   **4.** Quotation marks are used to enclose direct quotes.

_____   **5.** The title of a novel should be enclosed in quotation marks.

_____   **6.** An ellipsis sometimes consists of four periods.

_____   **7.** Ellipses indicate that words have been omitted.

_____   **8.** Apostrophes indicate possession only for singular nouns.

_____   **9.** A word such as *beginning*, which contains a doubled consonant, would be divided after the second *n*.

_____   **10.** One-syllable words may be divided when necessary.

CHAPTER 20

OVERVIEW OF THE WRITING PROCESS

The greatest possible merit of style is, of course, to make the words absolutely disappear into the thought.

—Nathaniel Hawthorne

LEARNING UNIT GOALS

LEARNING UNIT 20-1: BUILDING SENTENCES

- Know how to make the correct choice of words. (pp. 416–17)
- Understand variety, unity, and clarity in sentences. (pp. 417–19)

LEARNING UNIT 20-2: WRITING UNIFIED PARAGRAPHS

- Develop unified paragraphs. (pp. 421–22)
- Write transitions between sentences and paragraphs. (p. 422)

LEARNING UNIT 20-3: WRITING A REPORT

- Recognize the importance of planning written documents. (pp. 423–26)
- Know how to write and revise the first draft and write the final draft. (pp. 426–30)

Is there a particular author whose writing style you admire? If you like the style of a particular author, can you define this author's style? What is a strong writing style?

This text has given you the basis for a strong writing style. As you continue to write, the grammar, punctuation, and word choice you have learned will become integrated into your writing style.

Should you do more to develop your writing style? Professional writing experts give the following advice on how to develop a strong writing style.

415

You will note that much of this advice has been given in earlier chapters.

1. **Concentrate on the thought you want to convey.** This will keep your message focused.

2. **Your writing style must be appropriate to your subject and audience.** A letter written to a company president would probably be more formal than a letter written to a prospective customer.

3. **Your writing style must have simplicity, precision, clarity, directness, and be interesting to a reader.** To achieve these style characteristics, you must continually concentrate on them as you write and rewrite.

4. **Your writing style should express your individuality.** In *On Writing Well* (Harper & Row), William Zinsser says that since style is who you are, you need only relax, say what you want to say, and be true to yourself. Gradually your style will emerge from the accumulated clutter and debris, growing more distinctive every day.

You can conclude that to be a good writer with a strong style, *you should follow general writing principles, write in an interesting manner appropriate for your subject and audience, and be yourself.*

Learning Unit 20-1 discusses building sentences; Learning Unit 20-2 discusses unified paragraphs. In Learning Unit 20-3, we use a business report to explain the fundamentals of report writing.

LEARNING UNIT 20-1
BUILDING SENTENCES

Sentences move thoughts from writers to readers. Effective sentences begin with the correct choice of words. In addition, sentences should have variety, unity, and clarity.

CHOICE OF WORDS

Much of this information on word choice has been discussed in previous chapters.

1. **Increase your vocabulary.** Pay attention to the words you hear and read. Use the dictionary to learn the meanings of new words. Keep a record of these new words and their meanings. Know the meaning of every word you use.

2. **Use the word choice principles given in Chapter 1.** Prefer the active voice, avoid unbusinesslike expressions, omit unnecessary words, and use positive words and specific words.

3. **Use concrete nouns and action verbs.** Chapter 6 discussed concrete nouns. Chapter 10 discussed action verbs. When possible, replace linking verbs with action verbs. To do this, look for some action buried in the sentence and use it as the verb.

    AVOID: All employees agree that Curt has an understanding of computers.

    USE: All employees agree that Curt understands computers.

4. **Use precise and colorful adjectives.** Chapter 3 discussed adjective usage.

5. **Avoid a string of nouns used as adjectives.** This was explained in the writing tip given in Chapter 6.

6. **Avoid adding a suffix to verbs.** Adding a suffix such as *tion, ion, ity, ment, ence,* and *ness* to a verb often results in a weak abstract word. When you have such a word in a sentence, try to change it to an active verb.

> AVOID: Authorization for the absence was given by the leader. (Often, as in this sentence, writers use passive constructions when they change a verb to a noun.)
>
> USE: The leader authorized the absence.

7. **Avoid unnecessary adverbs.** Many adverbs are unnecessary and clutter a sentence. Try to use verbs that say what you mean without the addition of adverbs.

8. **Use prepositions effectively.** Chapter 14 discussed preposition usage.

9. **Avoid expletives.** Expletives are beginning words not needed for the sense of a sentence, such as *this is, it is,* and *there is (are, were)*. Instead of *There were several people in the office,* say *Several people were in the office.*

10. **Avoid slang, colloquial language, and clichés. Slang** phrases, such as *blew her away* and *far out,* are confusing because they change and vary according to locations. **Colloquial language,** such as *really neat* and *flunked,* are characteristic of casual conversation and should be avoided in writing. **Clichés,** such as *out of sight* and *last but not least,* have become commonplace and weaken writing.

11. **Avoid trite words and phrases.** Chapter 1 gave some words and phrases to avoid. Following are some additional phrases to avoid:

Avoid	Substitute
at the present time	now
at your earliest convenience	soon
due to the fact that	since, because
for the purpose of	for
for the reason that	because
in order to	to
in the amount of	for
in the final analysis	finally
in the near future	soon
in view of the fact that	because
prior to	before

VARIETY, UNITY, AND CLARITY IN SENTENCES

Using simple, compound, and complex sentences of varying lengths helps hold the interest of readers. Unity and clarity are required to build sentences that people can understand. Frequently, the components of unity also affect clarity, and vice versa. Too many words in a sentence, for example, can be detrimental to unity and clarity.

Sentence Variety

A variety of sentences avoids the choppy writing that can result from a series of short, simple sentences. Since the simple sentence is often preferred, use good judgment in varying your sentences.

Consider the following five ways that you can combine these two closely related ideas: (1) *Ward graduated from Columbia University in June.* (2) *He now works for a New York brokerage firm.*

1. Compound sentence with a coordinating conjunction:

> Ward graduated from Columbia University in June, and he now works for a New York brokerage firm.

2. Simple sentence with a participial phrase.

> Having graduated from Columbia University in June, Ward now works for a New York brokerage firm.

3. Complex sentence with an adjective clause.

> Ward, who graduated from Columbia University in June, now works for a New York brokerage firm. OR Ward, who now works for a New York brokerage firm, graduated from Columbia University in June.

4. Simple sentence with an appositive:

> Ward, a June graduate of Columbia University, now works for a New York brokerage firm.

5. Complex sentence with adverb clause:

> Since Ward graduated from Columbia University last June, he has been working for a New York brokerage firm.

Sentence Unity

Writers achieve sentence unity when they build sentences that are units in substance and form. Both types of unity are necessary for effective sentences.

Unity of Substance.   A sentence lacks substance unity if it includes ideas that are not *closely* related. A lack of substance unity confuses readers. Note how the two ideas in the first sentence are not closely related.

> AVOID: We could complete this entire mailing in less than three hours, and we may purchase a software program.
>
> USE: We may purchase a software program that would complete a mailing of this size in less than three hours.

You cannot achieve unity of substance if you have too much or too little information in a sentence. Avoid excessive detail *and* the omission of needed words.

Unity of Form. A sentence may be a unit in substance and yet be so arranged as not to be a unit in form. Unity of form involves the correct placement of modifiers and the correct use of parallel sentence structure.

1. **Correctly placed modifiers.** Place every modifier close to the word or words it modifies (see Chapters 13 and 14). A misplaced descriptive word, phrase, or clause clouds the meaning of a sentence. Concentrate on recognizing the modifier and then determine which word or words it modifies.

2. **Correct use of parallel sentence structure.** Use a parallel sentence structure for parallel ideas (see Chapter 15). If, for example, the first of three elements in a sentence begins with a pronoun, the other two elements should begin with a pronoun.

Sentence Clarity

Your sentences should mean the same to your readers as they mean to you. No reader should find it necessary to read a sentence twice. In addition to the correct choice of words, clarity involves the following:

1. **Use correct pronoun references.** Chapter 9 discussed the agreement of personal pronouns with antecedents.

2. **Avoid illogical shifts in tense.** Although you may have to use verbs of dissimilar tenses in a single sentence, do not make a change in tense that results in an awkward construction or that obscures the intended meaning.

> AVOID: She accepted the merchandise and pays the seller with a personal check. (The shift from past tense to present tense is illogical.)
>
> USE: She accepted the merchandise and paid the seller with a personal check.

3. **Use subordination correctly.** Subordinate an idea of secondary importance by expressing it in the dependent clause of a complex sentence. Ideas expressed in the main clauses of a compound sentence are given equal emphasis with coordinating conjunctions.

> AVOID: The government allows a substantial exclusion, and Lloyd will not have to pay a federal tax on his father's estate.
>
> USE: Because the government allows a substantial exclusion, Lloyd will not have to pay a federal tax on his father's estate.

4. **Use emphasis correctly.** To gain emphasis, place important words at the beginning of a sentence or, even better, at the end. Transitional words and other words and phrases of minor importance can often be placed within a sentence.

YOU TRY IT

A. Each of the following sentences lacks unity. Rewrite each sentence two ways to show a relationship between the ideas. Add information if necessary—and use your imagination!

EXAMPLE: Mario Castellani is an excellent architect, and the Draper Building has ten stories.

a. **Mario Castellani, an excellent architect, designed the ten-story Draper Building.**

b. **Mario Castellani, who designed the ten-story Draper Building, is an excellent architect.**

1. The shipment arrived at 10:00 A.M. on Tuesday, and most railroad workers were on strike.

   a. _____

   _____

   b. _____

   _____

2. Our annual earnings increased by 24.5 percent, and Melissa Loo is an outstanding chief executive officer.

   a. _____

   _____

   b. _____

   _____

3. I completed the manuscript in only ten days, and the new word processor is labeled WP7700.

   a. _____

   _____

   b. _____

   _____

B. The following sentence is too long. Break it into two or more shorter sentences. Some changes in wording may be necessary.

One of the outstanding advantages of a sole proprietorship, which allows the owner to manage his or her own business, keep all profits, and assume all risks, is its simplicity, since the only legal requirements involve securing a minimum number of licenses and the payment of certain fees, which are not large and are required by governing authorities.

Possible Answers: **A.** 1. a. The shipment did not arrive until 10:00 A.M. on Tuesday because most railroad workers were on strike. b. Most railroad workers were on strike when the shipment arrived at 10:00 A.M. on Tuesday. 2. a. Melissa Loo, an outstanding chief executive officer, increased our annual earnings by 24.5 percent. b. Melissa Loo, who increased our annual earnings by 24.5 percent, is an outstanding executive officer. 3. a. With the help of the new word processor, a WP7700, I completed the manuscript in only ten days. b. I completed the manuscript in only ten days with the help of the new WP7700 word processor. **B.** A sole proprietorship allows an owner to manage his or her own business, keep the profits, and assume all risks. One of its outstanding advantages is its simplicity; the only legal requirements are to secure licenses and to pay certain nominal government-required fees.

LEARNING UNIT 20-2
WRITING UNIFIED PARAGRAPHS

Each paragraph should have one main idea. After you have written a paragraph, reread it and ask yourself, "Do I have *only* one main idea?" If you have more than one main idea, you need more paragraphs.

Usually, the main idea of a paragraph is expressed in the first sentence—the topic sentence. This is the preferred position because readers expect the main idea to be in the first sentence. You may find, however, that the topic sentence is the second sentence or even the last sentence in the paragraph.

All sentences in a paragraph must support its main idea in a way that makes the paragraph flow. Paragraphs can be developed with examples, descriptive information, or facts. The methods used to support a main idea should make readers feel they are one with the writer as they follow the writer's thoughts.

As you write a sentence in a paragraph, mentally ask yourself, "How am I going to get to the next sentence and connect it with this sentence? The next paragraph with this paragraph?" This mental exercise will soon become automatic and make your sentences and paragraphs flow, saving you rewriting time.

The ending sentence in a paragraph should make readers feel a sense of completeness. After you have written a paragraph, read it aloud. It should make readers feel that you have completed one main idea and now are ready to give them the next main idea.

The following paragraph was taken from a first draft of a university bulletin instructing readers how to grow trees from seed:

Extracting Seed from Cones

The method for extracting seed from cones is the same for all conifer species grown in Minnesota, except eastern redcedar and some jack pine cones. The object is to get the cones to open and cones open when the moisture content of cones is reduced via warm temperatures and adequate aeration. The best method to reduce the moisture content of cones is to spread them out on a wood-framed screen. This will allow air to circulate around each cone. Support the screen off the ground and place it in an open area where the sun can heat the cones and the wind can carry off the moisture transpiring from the cones or place the screens inside a well-ventilated building and use electric fans to speed the drying. If the screen mesh is too large and the released seed drops through the screen, place a tarp, cloth, plastic, or paper underneath the screen to catch the seed.

Does this paragraph have more than one idea? Yes, and the ideas are strung together, making the paragraph difficult to follow. Let's rewrite this paragraph into two paragraphs.

Extracting Seed from Cones

The method for extracting seed from cones is the same for all conifer species grown in Minnesota, except eastern redcedar and some jack pine cones. Remember that the object is to get the cones to open.

Cones open when warm temperatures and adequate aeration reduce the moisture content of the cones to the point that they release their seeds. To reach this point, you must spread the cones out to dry on a wood-framed screen. Then, you can use one of the following procedures:

1. Support the screen off the ground in an open area where the sun can heat the cones and the wind can carry off the moisture transpiring from the cones.
2. Place the screen inside a well-ventilated building and use electric fans to speed the drying process.

If the screen mesh is too large and the released seed drops through the screen, place a tarp, cloth, plastic, or paper underneath the screen to catch the seed.

The following suggestions will help you make a smooth transition from sentence to sentence and paragraph to paragraph:

1. **Use a logical flow of ideas between sentences and paragraphs.** To keep a flow between two paragraphs, you may find it necessary to use a transitional sentence as a topic sentence of the second paragraph. You also might need a transitional paragraph (one or two sentences) between topics.
2. **Use conjunctive adverbs and transitional phrases that let the reader know how ideas are connected.** Chapter 15 gave examples of conjunctive adverbs and transitional phrases used to connect ideas.
3. **Readers should know the antecedents of pronouns.** Chapter 9 discussed the agreement of antecedents and personal pronouns.
4. **Repeat key terms (or synonyms of the terms).** This method is good for both sentence and paragraph flow.
5. **Use parallel structure when ideas are parallel.** Chapter 15 discussed parallel structure.

In *Rewriting Writing*, McCuen and Winkler say, "The principle of writing unified paragraphs is a simple one. Plainly put, it says, stick to the point. Don't introduce irrelevant issues or details that have no bearing on your topic sentence."

Unit 20-3 uses a business report to explain the procedure of writing a report. The two other common business documents—the letter and the memo—are discussed in the appendix.

YOU TRY IT

A. Read paragraphs A and B below. Then answer the questions that follow.

Paragraph A: Product brochures are prepared by manufacturers for the purpose of selling their goods. The products should be of the highest quality. Brochures must have positive comments to make about the products.

Paragraph B: Product brochures are prepared by manufacturers for the purpose of selling their goods. The content of the brochures always presents the most positive information about the products. Responsible manufacturers take care that the information is accurate and does not exaggerate the good qualities of the products.

1. Which paragraph opens with a main idea and then sticks to the point? _____

2. Which paragraph adds an unrelated idea? _____

B. Topic sentences may be supported by examples, descriptive information, or facts. For each of the topic sentences below, write a supporting sentence in the space provided.

1. This new word processing software has several outstanding features.

_____

_____

2. The Diamond line of office furniture is very finely crafted.

_____

_____

3. The population of Los Angeles has grown quickly.

_____

_____

Answers: 1. Paragraph B 2. Paragraph A

LEARNING UNIT 20-3
WRITING A REPORT

Writing a report begins with preliminary planning. Then, you gather information and use various prewriting techniques to organize your topics in preparation for writing the first draft of your report. After revising your first draft, you may be ready for your final draft. Sometimes you will find you need a second or third draft before the final draft.

PRELIMINARY THINKING

In the preliminary thinking phase of writing a report, you determine your audience and the purpose of the report. Audience and purpose affect the content, form, and tone of a report. This learning unit uses a business report to illustrate report writing.

Let's create an example of a business report by imagining that you manage the raw materials storeroom of the Carley Furniture Corporation. Carley makes special-order quality furniture and plans to build a new factory and enlarge its operations.

The cost of maintaining Carley's raw materials inventory has been steadily increasing. Maintaining excessive stocks of raw materials to meet special orders involves large inventory costs. Reducing these costs would give the company more available cash.

The company president, Jan Carley, was fascinated by a new inventory system discussed in a manufacturing journal. This system originated in Japan and is called the just-in-time (JIT) inventory system.[1]

Jan asked you to research the JIT inventory system and write a report for the officers of the company. She wants you to explain the key elements of the JIT system and its benefits.

You begin to think about your assignment. You know your audience. Your purpose for writing the report was decided by the company president.

A friend who took a managerial accounting course tells you what she knows about the JIT system. She suggests that you make an appointment with her managerial accounting professor.

GATHERING INFORMATION

Following your friend's advice, you make an appointment with her college professor. He gives you a list of books and articles on the JIT system and a list of topics that you should research. You are told by the professor to organize the information you collect by topic and write this information on 4 x 6 index cards, giving the sources of all your information.

The professor arranges for you to visit a manufacturing company that uses the JIT system. You visit the company and are impressed with its smooth operation and its low raw materials inventory costs.

Now you have a collection of topics, supporting information, and first-hand knowledge on how a company uses JIT. You are ready to enter the prewriting phase of your report.

PREWRITING TECHNIQUES

As you sit at your desk and look at the various piles of index cards you have grouped according to topics, you recall what the president asked you to do. Your visit to the company using JIT and the library research have given you many ideas about JIT.

If you needed more ideas, you could use one or more of the following techniques:

1. **Freewriting**. In freewriting, you write nonstop for a fixed period of time to generate ideas without concentrating on any formal structure. Often you will get ideas in freewriting that you were not aware of consciously.

2. **Answering questions that journalists ask.** By answering who, what, when, where, why, and how, you can usually formulate the content of your piece of writing.

3. **Brainstorming (or listing).** Brainstorming is intense thinking. You write ideas about your subject in the order that they come to you. In brainstorming, you can add more ideas or eliminate some ideas.

---

[1]The information on the JIT inventory system was adapted with permission from Ray H. Garrison and Eric W. Noreen, *Managerial Accounting*, 7th Ed. (Burr Ridge, IL: Richard D. Irwin, Inc., 1994).

The preceding prewriting techniques show no particular relationship between ideas. To organize ideas, use the following techniques:

1. **Clustering (mind mapping).** In clustering, you write an idea on the middle of a page and circle it. Then, like the spokes of a wheel, you draw lines from this main idea and circle the related ideas. A related idea may become another cluster with lines and circled ideas drawn from it.

2. **Branching.** Branching begins with writing your main idea at the top of a page. Then, you list the major supporting ideas and branch out to specific ideas. When minor ideas lead to other ideas, you continue the branching technique.

3. **Select a tentative focus and write a tentative outline.** This is the traditional method of organizing ideas. Many writing authorities still favor this method.

SELECT A TENTATIVE FOCUS AND WRITE A TENTATIVE OUTLINE

The president of Carley Furniture Company has given you the tentative focus for your JIT system report by asking for the key elements of the JIT system and its benefits. Usually, business reports are requested for a particular purpose.

Business reports can be (1) proposals, (2) progress reports, or (3) evaluation reports. **Proposals** are the result of a business problem and suggest a plan of action. **Progress reports** give the progress of a project. **Evaluation reports** examine the usefulness of proposed ideas. The JIT system report is an evaluation report. It is also a research report because you were asked to research the key elements and benefits of JIT.

Report writers usually state their focus in a **thesis statement.** This is often given in the first paragraph of the report. The thesis statement introduces and summarizes the entire report and prepares the readers for the facts and details. The following thesis statement gives the subject of your report and how you will expand it.

The just-in-time (JIT) inventory system helps managers reduce costs, increase efficiency, and expand output.

From your note cards organized by topics, you make a list of the ideas you want to include in your report. You know about the importance of organization in writing and that you must keep related ideas together. Now you are ready to group your ideas into a parallel structure that becomes your *tentative* outline.

You also realize that an outline is personal and you may use whatever style or combination of style—topic outline or sentence outline—that will help you develop a unified report. Following are the major topics of your tentative outline:

I. Key elements of JIT
   A. Number of suppliers are limited.
   B. Plant layout is improved.

  C. Setup time is reduced.
  D. Company develops total quality control over parts and materials.
  E. Work force must be flexible.
II. Benefits of JIT system
  A. Worker productivity is increased.
  B. Setup time is decreased.
  C. Total production time is decreased.
  D. Waste is reduced.
  E. Inventories of all types are reduced.
  F. Working capital is increased.
  G. Usable plant space is increased.

Now, if time permits, you should walk away from your project for a few days and give yourself some "percolation" time. The conscious part of your mind needs a break. The subconscious part of your mind will keep working.

PROFESSIONAL WRITING TIPS

Before you begin to write the first draft of your JIT system report, you should know some of the writing methods used by professional writers. These tips are practical and easily adapted to all types of writing.

1. Establish a Personal Writing Place.   With the advent of computer communication systems, many companies encourage employees to do their creative work, especially report writing, at home. Computer communication systems make possible conferences, electronic mail, and other unique options. Some employers provide employees with laptop computers so they can make a smooth transition from their office at work to their home office.

If you do some of your writing at home, set aside a special place where you write. Your writing place should center around the space allowed for your computer. Make a ritual of going to your writing place. If you like to write with background music, choose a tape that relaxes you. Make it part of your writing equipment. If you like to have a cup of coffee on your desk, get a cup before you turn on the computer.

Do you know why this ritual is helpful? The reason is that an established routine prepares you mentally for writing. As you experience success in writing in your special place, you begin to associate success with your location. Then, you become relaxed in your special place. This relaxed attitude provides the ability to concentrate.

When you write at a desk in an office, your writing place has been decided for you. You can, however, use a routine that mentally prepares you for writing. Clear your desk of everything except your reference materials and your notes. If possible, put your telephone calls on the company answering service. Each person's writing ritual is different. Learn what works for you.

In *A Practical Guide for Writers* (Winthrop Publishers), Diana Hacker and Betty Renshaw say, "If there is any magic connected with writing, it may come from your writing place. The more at home you are in the place where you write, the more free and easy you will feel."

2. Overcoming Writer's Block. At some point in their writing, most writers experience writer's block. **Writer's block** occurs when writers find themselves looking at a piece of paper or a computer screen and they have no thoughts to put into words. The following suggestions will help you avoid writer's block:

a. Begin by seeing yourself writing. Many professional writers say that they write more easily if they picture themselves doing it first.

b. Know your best time to write. If you are a morning person, do your creative writing in the morning. If you work more efficiently in the afternoon or evening, do your creative writing in the afternoon or evening.

c. Do not become discouraged by a large assignment. Break your assignment down to smaller tasks. Always remember that "inch by inch, it's a cinch; by the yard, it's hard."

d. Often writer's block is a result of poor planning and organization. Have a tentative writing plan or outline and follow it. Your plan will probably change as you write, but you will have something to get you started.

e. Do something physical. Take a brief walk or climb a couple of flights of stairs. If you have difficulty finishing a sentence, stop in the middle of the sentence. Get up and walk around repeating the first half of the sentence to yourself. Often you will get an idea on how to finish the sentence and can successfully return to your writing.

f. Do some freewriting. Relax and write whatever comes into your mind about your subject.

g. Separate writing from editing. Write first; edit later. Trying to edit while you are writing interrupts your concentration.

h. Promise yourself a reward when you finish part of a task. The reward may be small—a telephone call or a special treat.

i. Do not feel you have to start at the beginning of your writing piece and move straight to the end. If you have difficulty with the beginning, write about something in the middle that comes easy for you.

3. Working with Deadlines. Most writers work with deadlines. Often these deadlines are set by someone else—a superior or committee. When given a writing task, be sure you understand when the task is to be completed. Most important, do not procrastinate!

4. Expect to Discover New Ideas While Writing. Always begin writing with a tentative plan or outline. Outlines, however, are only guides. Few writers complete a piece of writing with the same outline they used to begin their writing because they usually get new ideas while writing.

Runners get a "runner's high"; writers get a "writer's high." The excitement occurs when you are continually surprised by your new ideas and wonder where they originated. Sometimes you become frightened that the flow of ideas will cease. Don't be. Relax, and new ideas will continue to flow.

5. Reread What You Have Previously Written.   Experts say you should not write and edit at the same time. Instead, reread and revise what you have written after you have completed a topic or at the beginning of a new day. This will gradually perfect your writing, and your final draft will require less work.

WRITING THE FIRST DRAFT

Think of your first draft as having a beginning, middle, and end. You were probably thinking about the beginning and end of your report while you did your research and have some ideas that you can use for these difficult sections.

The Beginning

Most professional writers will tell you that the first paragraph, especially the first sentence, of a piece of writing is the most difficult to write. This paragraph should create interest in the reader. If possible, it should relate to something with which the reader is familiar. The thesis or purpose of the writing is usually in the opening paragraph.

Introductory paragraphs forecast the subject about which you are writing. Not all subjects, however, have a definite thesis. To be successful, introductory paragraphs should catch the attention of the reader in an interesting and straightforward way. The following paragraph begins the sixth edition of *The Guide to Real Estate* (Regents/Prentice Hall) by William B. French, Stephen J. Martin, and Thomas E. Battle, III:

> Real estate represents a great share of the total economic wealth in this country. Like the elephant in the living room, real estate cannot be ignored or avoided. We make decisions about real estate every day of our lives. We live *on* real estate and *in* real estate. We use real estate as we travel to and from work, while we work, and for recreation.

The second sentence of this introduction is an excellent example of what is called a reader's *hook.* As you read that sentence, you probably imagined the elephant in *your* living room.

In *The Elements of Business Writing* (Macmillan), Gary Blake and Robert W. Bly say: "Your first paragraph should engage the reader by arousing curiosity or presenting important news in a clear, compelling fashion." Use this thought as a guide for the opening paragraph of your JIT report.

The first paragraph of your JIT report should give your thesis, purpose, or objectives. This could include a bird's eye view of the main topics of the report. Readers often like to know where you are going before you get there.

The Middle

Your JIT report is a formal report. Formal reports usually occur in response to a specific request. They aim at major company changes and can involve library and research work, collecting and analyzing data, interviews, and so on.

Now you have completed a tentative outline, you have written your introduction, and you are at your special writing place ready to begin the middle of your report. Since this is the first draft of your report, relax and write quickly. Do not be concerned with your choice of words or your sentence structure. Remember, however, that the usual order for nonfiction is from the general to the specific.

Be sure to add headings and subheadings as you go along. They will serve as signposts for your readers. When necessary, document your information with footnote references. Prepare a bibliography as you write.

If your first draft takes several days, begin each day by rereading and improving what you have written the day before. This will keep you focused and make the revision process easier. Do not be discouraged if you find it takes time to reread and improve your writing. Remember the words of the professional writer Richard Dowis, ". . . there is no good writing, only good *rewriting.*"

The End

Your concluding paragraph(s) should give the reader a feeling of closure. You may do this by restating in different words the main points of your report. You may also express an opinion. Your JIT research, for example, has made you an expert in this new inventory system. In your conclusion, do not hesitate to carefully present some of your ideas on the effectiveness of JIT.

In *A Manual of Writer's Tricks* (Paragon House), David L. Carroll says: "When you're stuck for an ending, go back to your beginning. . . . Since opposites tend to meet in some mysterious way, you will often discover that the ending is somehow logically implied in the beginning and that your very first ideas somehow also contain a logical conclusion."

Your first draft is completed. Once again, assuming your schedule permits, it is time to put some distance between your report and yourself. Close up your writing place. Take a few days away from writing. Visit a friend. Go to a ball game or a movie. This break will result in a fresh attitude when you return to revise what you have written.

REVISING THE FIRST DRAFT

Since you have been rereading what you have written as you have moved through your report, now you should look for definite elements that you want to improve. Let's list some of them.

1. **Check your sentence and paragraph idea flow.** Carefully read each paragraph to be sure it meets the standard of effective paragraphs discussed in Learning Unit 20-2.

2. **Check for wordiness.** Look for adjectives and adverbs that you can eliminate. Study your prepositional phrases and see if you can eliminate words by changing them to adjectives and adverbs. Eliminate sentence openers such as *There is* and *It is.* Eliminate wordy phrases and redundancies.

3. **Look for ways to improve your word choice.** Strengthen your nouns and verbs. When possible, use action verbs instead of linking verbs. Look for passive voice constructions that you can change to the active voice. Check to see that you have correctly placed your modifiers.

4. **Check your spelling and punctuation.** Do not rely completely on a speller to catch all spelling errors. Spellers do not recognize incorrect word usage. If necessary, go over the punctuation chapters again to be sure you catch all your punctuation errors.

5. **Check your headings and general organization.** You may decide you should rearrange some topics and subtopics.

Do not try to do more than one or two of the above items at one reading. You will catch more errors when you concentrate on looking for only one type of error. This means you will be reading your report several times during the revision phase.

WRITING THE FINAL DRAFT

Some writers find it necessary to make a second draft and another revision before the final draft. Do not be discouraged if this happens to you. The writer James A. Michener said: "I have never thought of myself as a good writer. Anyone who wants reassurance of that should read one of my first drafts. But I'm one of the world's great rewriters."

In the final draft, make all your corrections from the revision process. Then read the report again, aloud if possible. Make whatever changes are necessary to polish your report.

Now is the time to prepare the final pages needed for the completed report. These pages may include a cover, title page, copyright page, letter of transmittal, preface or foreword, acknowledgements, abstract or summary, table of contents, list of illustrations, bibliography, and appendix. Each company usually has its requirements for the contents of formal reports. Also, the nature of the formal report dictates the special pages needed.

Probably the most important lesson you learned while going through the writing stages of your report is the importance of rewriting. The final product is always greatly improved and usually much different from the first draft. In *Rewriting Writing* (Harcourt Brace Jovanovich), Jo Ray McCuen and Anthony C. Winkler say:

> Writing and rewriting are inseparably part of the same process as inhaling and exhaling are to breathing. When you write you invent. When you rewrite you better your invention. The process is circular and indivisible and there is a great deal of back and forth movement between writing and rewriting as a writer struggles to express an idea.

YOU TRY IT

1. Write a review of your favorite movie. A review is an evaluation report. Write the name of the movie at the top of a sheet of paper. Use two minutes to write whatever comes into your mind about this movie.

2. Write a thesis statement for your movie evaluation report. Write only one sentence. Make sure it includes the main point of your review.

3. Write an opening paragraph for your movie evaluation report by expanding on your thesis statement. Add three or four sentences to your thesis statement.

4. The day after you write the opening paragraph, read it again. Check for the flow of your ideas from one sentence to the next sentence. Check to make sure that the paragraph is unified—that it sticks to the point. Also, check for wordiness, correct spelling, punctuation, and word choice. Make any changes that you think will improve your paragraph.

CHAPTER SUMMARY: A QUICK REFERENCE GUIDE

PAGE	TOPIC	KEY POINTS	EXAMPLES
416	**Choice of words**	1. Increase your vocabulary. 2. Use word choice principles (Chapter 1). 3. Use concrete nouns and action verbs. 4. Use precise and colorful adjectives. 5. Avoid a string of nouns used as adjectives. 6. Avoid adding a suffix to verbs. 7. Avoid unnecessary adverbs. 8. Use prepositions effectively. 9. Avoid expletives. 10. Avoid slang, colloquial language, and clichés. 11. Avoid trite words and phrases.	See examples in text.
417	**Variety, unity, and clarity in sentences**	**Variety**: Use simple, compound, and complex sentences of varying lengths. **Unity**: Sentences lack unity of substance if they include ideas not closely related. To have unity of form, modifiers must be placed correctly and parallel sentence structure used correctly. **Clarity** involves (1) correct pronoun references, (2) avoiding illogical shifts in tense, (3) using subordination correctly, and (4) using emphasis correctly.	See examples in text.
421	**Writing unified paragraphs**	Unified paragraphs have one main idea—usually in a topic sentence. Sentences in the paragraph must support the main idea. Ending sentences should give a sense of completeness. **Smooth transitions**: 1. Use a logical flow of ideas. 2. Use conjunctive adverbs and transitional phrases.	Study the revised paragraph on extracting seed from cones given in the text.

PAGE	TOPIC	KEY POINTS	EXAMPLES
		3. Readers should know the antecedents of pronouns. 4. Repeat key terms. 5. Use a parallel sentence structure when ideas are parallel.	
423	**Preliminary thinking**	1. Think about your assignment. 2. Determine your audience and purpose.	
424	**Gathering information**	1. Interview an authority on your subject. 2. Study research books and articles. 3. Write topics and information on index cards. 4. Get firsthand knowledge about your subject.	
424	**Prewriting techniques**	**To get ideas**: 1. Freewriting 2. Asking questions 3. Brainstorming (or listing) **To organize ideas**: 1. Clustering 2. Branching 3. Select a tentative focus and write a tentative outline.	
425	**Select a tentative focus and write a tentative outline**	**Thesis statement**: Introduces and summarizes entire report. **Tentative outline**: Uses the conventional outline system to separate ideas into a parallel structure.	
426	**Professional writing tips**	1. Establish a personal writing place. 2. Know how to overcome writer's block. 3. Know how to work with deadlines. 4. Expect to discover new ideas. 5. Reread what you have previously written.	
428	**Writing the first draft**	**The beginning**: Should create interest, relate to something familiar to the reader, forecast the subject, and give purpose and objectives. **The middle**: Write quickly using your outline, add headings and subheadings, use footnotes, and begin a bibliography. **The end**: Give a feeling of closure; can restate the main points and express an opinion.	

PAGE	TOPIC	KEY POINTS	EXAMPLES
	Revising the first draft	1. Check the sentence and paragraph idea flow. 2. Check for wordiness. 3. Look for ways to improve word choice. 4. Check spelling and punctuation. 5. Check headings and general organization.	
430	**Writing the final draft**	1. Make corrections from revision process. 2. Read report again and polish where necessary. 3. Prepare the necessary final pages.	

QUALITY EDITING

The following memo is poorly written. It is wordy, rambles, includes unnecessary information, and is very informal. Mark up the example as much as necessary and then rewrite the memo on a computer or typewriter or by hand. Make sure to correct sentences that (1) are too long or unclear, (2) contain ideas that seem unrelated, (3) are short and choppy, (4) use colloquialisms or slang, or (5) need to be stronger and more active. Add or delete any information you wish, but make sure to convey the important main points and the request. Also check to see that all the spelling, capitalization, punctuation, subject-verb agreement, and verb tenses are correct.

```
Date:    July 18, 19xx

To:      All People working in this building

From:    Building Management

Subject: Parking Lot Messes


First, I need to be asking your coperation about the

parking lot in that it is a big mess. Please do not

through trash from your car their and also you walk

through the plants that are in the dividers between the

rows of cars and brake their leaves and branch's and

make more mess and damage them. This is a real terrible

look for visitors to see.
```

And also furthermore, only visitors park in the first row. It should always be only visitors who park there so they do'nt have to walk very far to do business with you they can go to somebody else.

Your hellp is apreciated very much.

APPLIED LANGUAGE

Write a two-page report on one of the following topics:

1. You work in a dentist's office where music is played throughout the day. The tape player has broken, and you would like to have Dr. Baldwin purchase a CD player. Obtain information on CD players under $400. Then, write a short report indicating why a CD player would be preferred to a tape player and recommend one of the players you researched. Be sure to give the reasons for your choice.

2. Why we should protect the American wilderness.

3. Why your abilities, experience, and characteristics qualify you for a job you would like to have. (Be sure to identify the abilities, experience, and characteristics and to describe the job.)

4. Why learning a foreign language is beneficial to American students.

OVERVIEW OF THE WRITING PROCESS

Part A. Sentence Variety

Combine the information from the sentence pairs below to create a new sentence. Vary the structure of the new sentence. If necessary, refer to the five examples given in Learning Unit 20-2.

EXAMPLE:

(a) The old computer on Mr. Wells' desk crashed.

(b) Mr. Wells leased a new computer.

Mr. Wells leased a new computer because his old one crashed.

When Mr. Wells' old computer crashed, he leased a new one.

1. **(a)** Harriet's report was poorly written, and her supervisor criticized it.
   (b) Harriet's supervisor told Harriet to rewrite her report.

   _____

   _____

2. **(a)** Our Dallas warehouse is considerably larger than the one in Portland.
   (b) Our Dallas warehouse was built in 1985.

   _____

   _____

3. **(a)** Mr. Franchetti is our most capable sales representative.
   (b) He did not finish high school.

   _____

   _____

4. **(a)** This book discusses the impact of electronics on our economy.
   (b) The book was written by Hanley Stokeley.

   _____

   _____

5. **(a)** George Ito is the new president of this corporation.
   (b) He is a graduate of Yale University.

   _____

   _____

6. (a) These components will not withstand extreme heat.
(b) We cannot use them in the manufacture of our appliances.

_____

_____

Part B. Sentence Clarity

Each of the following sentences contains an unnecessary or illogical shift in tense. Edit each sentence using proofreaders' marks.

1. Ms. Plascencia made 100 copies but leaves the machine uncovered.

2. Mr. Taylor looked at me and tells me to sign the contract.

3. I shall write the memorandum, and I am going to tell him about the management communication problem.

4. We have met the agent, and we gave her our documents.

5. Ms. Nehruda received the crates on Monday, but she doesn't open them until yesterday.

Part C. Writing Unified Paragraphs

In the space provided, write a sentence that might follow the one given.

1. Many sales representatives do not appreciate the importance of courtesy.

_____

_____

2. To build your next fence, use TimberLake lumber.

_____

_____

3. Miss Evans is a dedicated, hard-working employee.

_____

_____

4. The solution to our delays in mailing out reports may lie in replacing our old copier.

_____

_____

5. During the past year, our Graphic Arts Department has produced several exciting, colorful posters, but they are too small.

_____

_____

6. According to the employee survey, the offices need much more light.

_____

_____

7. The accounting department should be evaluated for efficiency and accuracy.

_____

_____

8. We firmly believe that the features on our new FAX-ALOT machine will improve your firm's communications with customers.

_____

_____

Part D. Planning a Report

You would like to promote a ride-sharing program to help reduce the severe air pollution in your city. Furthermore, you would like your firm to establish flexible work hours so that you could avoid the hours with the heaviest commuter traffic.

Write six questions for each of these two concepts that will help you organize your proposals. Begin the questions with the interrogative pronouns provided.

Promoting a Ride-Sharing Program

1. Why _____

2. When _____

3. Where _____

4. What _____

5. Who _____

6. How _____

Establishing Staggered Work Hours

1. Why _____

2. When _____

3. Where _____

4. What _____

5. Who _____

6. How _____

Part E. Writing a Report

Write four _paragraph enders._ These are sentences that sound like the end of a description, memo, progress report, proposal, or evaluation. These sentences may refer to any subject you wish.

EXAMPLES:

Therefore, I ask all of you to re-double your efforts to make Shiny-Brite the most desirable tree ornaments available.

Finally, establishing staggered work hours would leave too few customer service representatives on duty during critical hours.

1. _____

_____

2. _____

_____

3. _____

_____

4. _____

_____

APPENDIX A

BUSINESS LETTERS AND MEMOS

Letters and memos are the two most versatile business tools. Unlike the telephone, letters and memos are permanent records of events. You can take them with you to meetings and conferences. You can refer to them years later to refresh your memory. You can study them and make decisions based on their contents. Sometimes they are the basis of legal action. The effectiveness of letters and memos depends, of course, on the writing ability of the writer. To write effective letters and memos, you must use the basic writing skills taught in this text.

BUSINESS LETTERS

Business letters are communications between business people of an organization with those outside the organization. This makes the tone of a business letter so important. Use a personal, courteous, and positive tone. At all times be professional.

The seven major types of business letters include the following:

1. *Inquiry*: A letter to request information from a company
2. *Reply*: A letter in response to an inquiry
3. *Instructions*: Instructions for a company product or service
4. *Order*: An order for company goods or services
5. *Sales*: A letter offering a product or service for sale
6. *Claim*: A letter explaining the circumstances of a claim
7. *Adjustment*: Response to a claim letter that explains how the claim will or will not be satisfied

Format and Parts of a Business Letter

The two most widely used formats for business letters are the **full block** (Figure A-1) and the **modified block** (Figure A-2). The top margin of a letter is a minimum of 1 inch. If the letter is short, the top margin may be deeper. The side margins can be between 1 and 1½ inches. Figures A-1 and A-2 show the parts of a letter.

Figure A-1 Full Block Letter Style

Centerhead	**Noel Furniture Company** **2349 West State Street** **Minneapolis, MN 55411**
Dateline *File (if used)*	June 28, 19xx File: 3400/1/8
Inside address	Ms. Nancy Cole Purchasing Manager Snelling Tool Company 2415 Cornell Avenue Chicago, IL 60655
Salutation	Dear Ms. Cole:
Subject line	*Subject: Repairs of Damaged Office Furniture*
Body	We regret that four of the eight posture chairs you ordered May 2, 19xx, were damaged in shipment. These chairs were shipped from Minneapolis on June 1, 19xx. The shipping company has been contacted about the damaged chairs. It seems that an error was made in directing the shipment. This explains the long delivery time. Noel Furniture Company will, of course, replace the damaged chairs. The posture chairs shipped were Style 20, 21, and 34. Please indicate on the enclosed form which chairs were damaged. We will make arrangements to have the damaged chairs picked up. We thank you for your past business and hope to continue to serve you in the future.
Closing	Sincerely yours, Kimberly Noel President
Reference *Notations*	KN:ls enc

Figure A-2 Modified Block Letter Style

Centerhead

Noel Furniture Company
2349 West State Street
Minneapolis, MN 55411

Dateline
File (if used)

File: 3400/1/8   June 28, 19xx

Inside address

Ms. Nancy Cole
Purchasing Manager
Snelling Tool Company
2415 Cornell Avenue
Chicago, IL 60655

Salutation

Dear Ms. Cole:

Subject line

    Subject: Repairs of Damaged Office Furniture

    We regret that four of the eight posture chairs
you ordered May 2, 19xx, were damaged in shipment.
These chairs were shipped from Minneapolis on June 1,
19xx.

    The shipping company has been contacted about the
damaged chairs. It seems that an error was made in di-
recting the shipment. This explains the long delivery
time.

Body

    Noel Furniture Company will, of course, replace
the damaged chairs. The posture chairs shipped were
Style 20, 21, and 34. Please indicate on the enclosed
form which chairs were damaged. We will make arrangements to have the damaged chairs picked up.

    We thank you for your past business and hope to
continue to serve you in the future.

Closing

                                    Sincerely yours,

                                    Kimberly Noel
                                    President

Reference
Notations

KN:ls
enc

MEMOS

Memos are communications between business people within the organization. The three kinds of memos are (1) the *informative memo*, (2) the *persuasive memo*, and (3) the memo used to *seek information*. Informative memos give announcements and verify data; persuasive memos try to persuade readers to a point of view; and memos used to seek information ask questions and can ask for general comments.

Letters and memos should be organized and have a beginning, middle, and end. The tone of a memo is more informal than that of a letter.

The headings of memos may be double or single spaced and may be formatted in two ways:

DATE:	Month (spelled out), Date, Year	Month, Date, Year
TO:	Name, Title (optional)	To: Name, Title (optional)
FROM:	Name, Title (optional)	From: Name, Title (optional)
SUBJECT:	Topic of Memo	Topic of Memo

The paragraphs of memos should be in block form. The sender may initial the FROM line or sign the memo at the bottom. The sender's name should not be typed at the end of the memo.

Reference initials and other reference notations should be placed two lines below the last line of the memo at the left margin. Figure A-3 shows an example of a memo.

Figure A-3 Memo

Heading

```
DATE:     September 12, 19xx
TO:       Garth Schultz
FROM:     Kimberly Noel, President
SUBJECT:  Officer's Meeting
```

Paragraphs

You are requested to meet with the company's officers to explain the progress of your staff in developing a computer network system for Noel Furniture.

Please present the bids you have received for new company computers and printers. We would like your input on which machines you would recommend from the vendor list enclosed.

Be aware that many of the officers are not familiar with the technical language of computers. As much as possible, please use examples and terms that the officers can understand.

If you have any questions, please contact my office.

Reference Notation

enc.

INDEX

DATE DUE

PE   Barry, Robert E.
1128
A66   Applied English
1995

Overdue fines are $ 0.25 per day